Quantitative Analysis in Financial Markets

Collected papers of the New York University
Mathematical Finance Seminar

Quantitative Analysis in Financial Markets

Collected papers of the New York University
Mathematical Finance Seminar

Editor

Marco Avellaneda

Professor of Mathematics
Director, Division of Quantitative Finance
Courant Institute
New York University

World Scientific
Singapore • New Jersey • London • Hong Kong

332
N53q
1998
VOL. 1

Published by

World Scientific Publishing Co. Pte. Ltd.

P O Box 128, Farrer Road, Singapore 912805

USA office: Suite 1B, 1060 Main Street, River Edge, NJ 07661

UK office: 57 Shelton Street, Covent Garden, London WC2H 9HE

Library of Congress Cataloging-in-Publication Data
New York University Mathematical Finance Seminar (1995–1998)
    Quantitative analysis in financial markets : collected papers of
the New York University Mathematical Finance Seminar / editor, Marco
Avellaneda.
      p.   cm.
    A collection of papers presented at the weekly Mathematical
Finance Seminar at New York University's Washington Square campus
from 1995–1998.
    Includes bibliographical reference (p.).
    ISBN 981023788X (hc). -- ISBN 9810237898 (sc)
    1. Finance -- Mathematical models -- Congresses.   I. Avellaneda,
Marco, 1955–   .   II. Title.
HG106.N48   1998
332--dc21   99-44136
                                        CIP

British Library Cataloguing-in-Publication Data
A catalogue record for this book is available from the British Library.

Printed in Singapore.

INTRODUCTION

It is a pleasure to introduce this first volume of Courant Institute Lectures in Financial Modeling. The book is a collection of papers presented at the Mathematical Finance Seminar, which meets weekly at New York University's Washington Square campus.

The Mathematical Finance Seminar started in September 1995, as an extension of activities in Quantitative Finance at the Courant Institute. Raphael Douady and myself coordinated the seminar initially. Raphael was "fresh off the boat", on leave from the Ecole Normale Superieure, Paris, where he developed the well-known and well-attended Seminaire Bachelier on financial modeling. Raphael played a major role in attracting excellent speakers from both academia and industry to New York University.

The seminar is unusual in many respects. For example, it is open to the public. The majority of the audience is from financial institutions, investment banks, insurance companies and university research centers in New York. Mathematicians and physicists also attend regularly. This audience is sophisticated in terms of financial products and in the securities trading business. Presentations and ideas are therefore analyzed from the market viewpoint as much as from their intrinsic mathematical elegance. The same can be said with respect to finance theory.

A group of experts in financial modeling deliver the papers in the present volume. They cover a broad range of subjects, which I have divided into the following categories: (a) Models and Model Selection, (b) Option Pricing and Exotics, (c) Time-Series Estimation, (d) Empirical Studies Involving Options and (e) Financial Economics and Portfolio Theory.

The section on "Models and Model Selection" treats different aspects of design and implementation from a practical point of view. Subhramanyam, Ho and Stapleton describe the construction of a lattice model for pricing derivative securities which depend on two correlated indices or prices. Levin analyzes fixed-income models and their calibration. He derives closed-form solutions for pricing loans with uncertain amortization rate, deposits, swaptions, caps and floors and other fixed-income derivatives. Kortanek and Medvedev propose a new approach to interest-rate modeling that uses dynamical systems and optimization to estimate forward rates. The theory is applied to estimate the volatility of Treasury-bill yields and to forecast rates. Avellaneda, Friedman, Holmes and Samperi discuss the calibration, or fitting to market prices, of models for pricing and hedging over-the-counter options. Their method uses the Kullback–Leibler information distance to determine volatility surfaces.

The papers on "Option Prices and Exotics" discuss technical aspects of pricing and hedging derivatives. Carr, Ellis and Gupta present a methodology for hedging exotic options, such as barrier options, using combinations of standard (and hence more liquid) option contracts. Douady deals with double-barrier options and the variants known as "corridors", "boost options", etc. He discusses the distribution of the first hitting time for a barrier, the pricing of options with rebates and the construction of hedges based on the expected time for hitting the barrier. Posner and Milevsky's elegant paper deals with average-rate options, in which the payoff depends on the average value of an index over a period of time. They show that the value of this option can be obtained by a modification of the Black–Scholes formula, in which the normal distribution is replaced by the reciprocal-gamma distribution. This paper provides a solution to the heretofore-open problem of finding "closed-form solutions" for average-rate options. The computation of the early-exercise premium and the exercise boundary of American-style options is discussed by Subhramanyam, Huang and Yu, who present a new computational scheme and compare it with existing methods.

The papers on applied estimation of time series by Riedel and by Riedel and Sidorenko treat the problem of estimation and filtering in the presence of noisy data. This is an important area of research that will allow us to better understand the structure of financial data.

Empirical studies are an important testing ground for any theory. Zhu and Avellaneda analyze the term structure of implied volatilities of foreign-exchange options traded in the interbank market. They show that the relative variations of the implied volatilities of 13 major currency pairs can be described with a few principal components following E-ARCH statistics. In another study, Guo compares different stochastic volatility models and simulates the delta-hedging strategies for currency options under different statistical assumptions. Guo and Esipov study the effects of hedging long-term stock-index puts using a GARCH specification for the underlying index.

The final section contains two theoretical presentations. Milne and Jing show the existence of a general equilibrium in a model for a financial market with transaction costs, in which the agents include consumers, production forms, brokers or dealers. Finally, Fernholz proposes a new method for portfolio selection using "mathematical generating functions". He establishes, under certain assumptions, a dominance relationship between the portfolio generated using his method and the market portfolio.

In my opinion, this volume reflects the diversity of quantitative research in finance which exists in the late 1990's. It spans a broad spectrum of financial applications and has strong theoretical content and applicability to trading and risk management. I hope that you find the book as enjoyable as the original talks.

Marco Avellaneda
New York, September 1998

ACKNOWLEDGEMENTS

The seminar benefited from the generous support of the Board of Trustees of New York University, and especially of Mr. Alan C. Greenberg and Mr. William H. Heyman.

I would like to thank co-organizers Raphael Douady and Peter Carr, and especially Ms. Dawn Duffy, from New York University, who assisted in the production of this volume.

Finally, I would like to thank Ms. Yubing Zhai and Ms. Siew-Hooi Gan and others at World Scientific Publishing Co. for editing this book.

THE CONTRIBUTORS

Marco M. Avellaneda is Professor of Mathematics and Director of the Division of Financial Mathematics at the Courant Institute of New York University. He earned his Ph.D. in 1985 from the University of Minnesota. His interests center around mathematical modeling in the sciences. He has published in the fields of partial differential equations, the design of composite materials, turbulence modeling, and mathematical finance. He was consultant for Banque Indosuez, New York, where he established a quantitative modeling group in FX options in 1996. Subsequently, he moved to Morgan Stanley & Co., as Vice-President in the Fixed-Income Division's Derivatives Products Group, where he remained until 1998. He is the managing editor of the *International Journal of Theoretical and Applied Finance*, and an associate editor of *Applied Mathematical Finance* and *Communications in Pure and Applied Mathematics*.

Peter Carr is Vice-President in the Quantitative Equity Department of Morgan Stanley. Prior to joining Morgan Stanley he was a Professor of Finance at Cornell University, having received his Ph.D. in Finance from UCLA in 1989. He has published articles in the *Journal of Finance*, the *Review of Financial Studies*, and various other finance journals. He is currently an associate editor for seven academic journals and a director of the Bachelier Society and the Financial Management Association. His research interests are primarily in the field of derivative securities, especially American style and exotic derivatives.

Raphael Douady is Professor of Mathematical Finance at the Ecole Normale Superieure of Cachan, France, and senior consultant for the Canadian Imperial Bank of Commerce (CIBC) in New York. A former student of the Ecole Normale Superieure of Paris, he holds a Ph.D. in Mathematics and specialized in chaos theory and dynamical systems. Douady's professional experience spans 10 years, during which he began as a consultant in the space industry, and then for Societe Generale, Caisse des Depots et Consignations (Paris), NatWest (London), Murex (Paris) and KPMG (Paris).

Sergei Esipov holds positions at Centre Solutions NY and Centre Risk Advisors, which are both members of the Zurich Group. He earned his Ph.D. in 1986 from the Landau Institute for Theoretical Physics and the Solid State Physics Institute in Chernogolovka, Russia. His interests include field-theoretical modeling in the sciences and markets. He is the author of over 50 publications in the theory of condensed matter, phase transitions, turbulence, dark matter detection, sedimentation, granular dynamics, bacteriology, and theoretical finance. He also works as a consultant for the aviation industry. Since 1996 he has been responsible for modeling insurance, financial derivatives and investments at Centre Risk Advisors.

E. Robert Fernholz is Chief Investment Officer of INTECH, a subsidiary of The Prudential Insurance Company of America, which uses mathematical strategies to manage institutional equity portfolios. Formerly he was the Director of Research at Arbitrage Management Company, where he developed option pricing and risk control systems. Earlier in his career he was on the faculty of Princeton University and the Universidad Nacional de Buenos Aires, and he has published articles on mathematics, statistics, and finance in academic and professional journals. He received his A.B. in Mathematics from Princeton University and his Ph.D. in Mathematics from Columbia University. His recent research on equity market disequilibrium led to the development of functionally generated portfolios, for which he was issued a patent in 1998.

Craig Friedman is Vice-President at Morgan Stanley Dean Witter, where he develops relative value trading strategies for the Global High Yield Group (Emerging Market and High Yield Corporate Debt). He has extensive Wall Street experience in derivatives modeling and fixed income. He expects to complete his Ph.D. in Mathematics at the Courant Institute in Fall 1998.

Dajiang Guo is a quantitative analyst at Centre Risk Advisors and Centre Solutions NY, which are both members of the Zurich Financial Service Group. He earned his Ph.D. in 1995 from the University of Toronto. His interests center around asset/derivatives pricing, hedging and arbitraging, risk management, investment and portfolio management, credit/insurance derivatives and securitization, and financial econometrics. He has published in the *Journal of Business and Economics Statistics*, the *Journal of Fixed Income*, *Financial Engineering and Japanese Markets*, the *Journal of Futures Markets*, and the *Canadian Journal of Economics*. Before joining Centre Solutions he was Assistant Professor at the University of Guelph, Canada, and in the Treasury Analytics Group of the Bank of Montreal.

Xing Jin is Associate Professor of Mathematics at the Chinese Academy of Sciences. He earned his Ph.D. in 1991 from the Chinese Academy of Sciences. His interests center around mathematical finance and queueing theory, where he has published extensively. He was a visiting scholar at the Department of Economics of Queen's University, the Department of Finance at the University of Maryland and the Department of Finance at the Hong Kong University of Science & Technology.

Kenneth O. Kortanek received a Ph.D. in Engineering Science in 1964 from Northwestern University. He is currently John F. Murray Professor in the Department of Management Sciences of the University of Iowa. He was Assistant Professor at the Graduate School of Business of the University of Chicago, where he taught Corporate Finance in the MBA program, and then Associate Professor of Operations Research at Cornell University. He joined Carnegie Mellon University in 1969, where he held professorships in the School of Urban and Public Affairs and the Graduate School of Industrial Administration, before becoming Professor of Mathematics. Kortanek's research focuses on the fields of operations research, mathematical programming, mathematical finance, and semi-infinite optimization. Since 1962 he has published about 130 papers, a book, and edited several volumes on optimization theory.

Alexander Levin is senior quantitative developer of The Dime Bancorp, Inc. He holds Soviet equivalents of an M.S. in Applied Mathematics from the University of Naval Engineering, and a Ph.D. in Control and Dynamic Systems from the Leningrad State University (St. Petersburg). His career began in the field of control systems engineering. He taught at The City College in New York City and worked as a quantitative system developer at Ryan Labs, Inc., a fixed income research and money management company, before joining The Dime Bancorp. His current interests include numerical and analytical tools for pricing complex term-structure-dependent contingent claims, risk-measurement and management and modeling mortgages and deposits. He has recently published a number of papers in this field and is the author of Mortgage Solutions, Deposit Solutions, and Option Solutions, which are proprietary pricing systems at The Dime.

Vladimir G. Medvedev is Associate Professor in the Optimal Control Methods Department in the Department of Applied Mathematics and Informatics at the Byelorussian State University. He holds a Ph.D. in Physical-Mathematical Science from the Byelorussian Academy of Sciences. His Ph.D. thesis was devoted to developing numerical methods for solving linear semi-infinite programming problems. He was Engineer and Research Associate at the Department of Optimal Control Methods and Assistant Professor of Applied Mathematics and Informatics at the Byelorussian State University. He was a Postdoctoral Associate in the Department of Management Sciences at the University of Iowa. Medvedev's research focus includes the fields of semi-infinite optimization, optimal control, observation and identification of dynamical systems and mathematical finance. Since 1987 he has published about 26 papers.

Moshe Arye Milevsky has a B.A. in Mathematics and Physics, an M.A. in Mathematics and Statistics, and a Ph.D. in Finance. He is currently an Assistant Professor of Finance at the Schulich School of Business, York University, Toronto, Canada. The focus of his research, teaching, and consulting work is on the subjects of risk management, insurance, and derivative securities. Dr. Milevsky has published in the *Journal of Financial and Quantitative Analysis*, the *Journal of Derivatives*, the *Journal of Financial Engineering*, the *Journal of Risk and Insurance*, as well as *RISK*.

Frank Milne earned his Ph.D. in Economics and Finance Theory at the Australian National University in 1975. Since then, he has held positions at the University of Rochester, Australian National University, and is currently Professor of Economics and Finance at Queen's University, Canada. He has published widely in economics and finance theory. He is an associate editor of *Mathematical Finance*. Currently he is working on problems concerning financial markets with frictions and the theory of the firm.

Steven E. Posner is Vice-President in Marsh & McLennan Securities Corp's capital markets group, involved in pricing and structuring insurance-risk securities. He received his Ph.D. from Princeton University and was Assistant Professor of Statistics at the University of Toronto, where he taught and conducted research in financial derivatives pricing and investment theory. He was the recipient of the Best Paper Award in Investments from the Southern Finance Association. Until June 1998, he was a quantitative analyst on the emerging market fixed-income derivatives trading desk at ING Barings Securities in New York, responsible for pricing and structuring financial products. Dr. Posner has also published papers in information theory and control theory.

Kurt S. Riedel is a senior research scientist at SAC, a leading hedge fund since June 1996. From 1991 to 1997, he was a Research Associate Professor at the Courant Institute. He is the author of over 40 refereed publications in statistical analysis, signal processing, and plasma physics. His present address is: SAC Capital Management, LLC 540 Madison Ave., New York, NY 10022, USA.

Dominick Samperi is an independent consultant providing mathematics and software support to investment firms. He worked for Citibank for six years, where he managed a front-office team providing pricing and risk management in equity derivatives. He designed an interest-rate swap pricing system, and worked on inverse problems in finance, like the construction of implied forward rate curves and implied diffusion models (or "implied trees"). He introduced the idea that the latter problems should be viewed as ill-posed inverse problems (requiring regularization for stability) in 1994, which led to the completion of his Ph.D. in Mathematics from the Courant Institute in 1998. Prior to joining Citibank, he was Assistant Professor of Mathematics and Computer Science at Manhattan College.

Marti G. Subrahmanyam is Charles E. Merrill Professor of Finance and Economics in the Stern School of Business at New York University. He holds a degree in Mechanical Engineering from the Indian Institute of Technology, Madras, a post-graduate diploma in Business Administration from the Indian Institute of Management, Ahmedabad, and a doctorate in Finance and Economics from the Massachusetts Institute of Technology. Professor Subrahmanyam has published numerous articles and books in the area of corporate finance, capital markets, and international finance. He has been a visiting professor at leading academic institutions in England, France, Germany, and India. Most recently, he held a Visiting Chair at INSEAD, France, and was an Overseas Fellow at Churchill College, Cambridge University. He has served as a consultant to several financial institutions in the U.S. and abroad and sits on many boards of directors, such as Nomura Capital Management Inc., Sanwa International plc., Infosys Technologies Ltd., and Deutsche Software India Ltd., a subsidiary of Deutsche Bank A.G. He has also served as an advisor to international and government organizations. He has taught extensively on executive programs in over a dozen countries around the world. Professor Subrahmanyam currently serves or has served as an Associate Editor of the *Journal of Banking and Finance*, the *Journal of Finance, Management Science*, the *Journal of Derivatives*, the *Journal of International Finance and Accounting*, and *Japan and the World Economy*. He is the Editor of a new academic journal specializing in derivative securities and markets, entitled *Review of Derivatives Research*. His current research interests include valuation of corporate securities, options and futures markets, and equilibrium models of asset pricing, and market micro-structure. His recent books include *Recent Advances in Corporate Finance* (Irwin, 1985) and *Financial Options: From Theory to Practice* (Dow Jones-Irwin, 1992). He is currently working on a new book, "Interest Rate Derivative Products."

Yingzi Zhu is Vice-President of Risk Analytics at Citibank. She earned her Ph.D. in 1997 from the Courant Institute, and holds an M.S. in Physics from New York University. Her interest centers around mathematical modeling of financial derivatives, as well as identifying and quantifying financial risk. She has published several papers in the field of mathematical finance. Her present responsibilities at Citibank include model validation, credit risk exposure calculation, and quantitative risk advisory.

CONTENTS

Reprinted from Rev. Financial Studies 8(4) (Winter 1995) 1125–1152

MULTIVARIATE BINOMIAL APPROXIMATIONS FOR ASSET PRICES WITH NONSTATIONARY VARIANCE AND COVARIANCE CHARACTERISTICS*

TENG-SUAN HO and RICHARD C. STAPLETON

Department of Accounting and Finance, The Management School, Lancaster University, Lancaster LA1 4YX, England, UK

MARTI G. SUBRAHMANYAM

New York University, Stern School of Business, 44 West 4th Street, New York City, NY 10012, USA

In this article, we suggest an efficient method of approximating a general, multivariate lognormal distribution by a multivariate binomial process. There are two important features of such multivariate distributions. First, the state variables may have volatilities that change over time. Second, the two or more relevant state variables involved may covary with each other in a specified manner, with a time-varying covariance structure. We discuss the asymptotic properties of the resulting processes and show how the methodology can be used to value a complex, multiple exerciseable option whose payoff depends on the prices of two assets.

In practice, many problems in the valuation of the derivative assets are solved by using binomial approximations to continuous distributions. In this article, we suggest an efficient method of approximating a general, multivariate lognormal distribution by a multivariate binomial process. There are two features of such multivariate lognormal distributions that are of interest. First, the state variables may have volatilities that change over different time intervals (i.e., exhibit a term structure of volatility), either because of changing volatility or mean reversion. Second, there may be two or more relevant state variables involved (e.g., a commodity price and a foreign exchange rate) which may covary with each other in a specified manner.

The binomial approach to the valuation and hedging of options has become increasingly important with the creation of new exotic derivative products. Many options with path-dependent payoffs (e.g., American options and Asian or average-rate options) may be valued using the binomial methodology.[a] Also, in the analysis of

---

*Earlier versions of this article have been presented at the European Finance Association Conference, Lisbon, Portugal; Aarhus University, Denmark; Melbourne University, Australia; University of Michigan, and Virginia Polytechnic Institute, USA. We are grateful to participants at these presentations and to John Chang, Chi-fu Huang (the editor), Apoorva Koticha, Ser-Huang Poon, Steven Tsay, and an anonymous referee for comments on previous drafts.

[a] In the case of many path-dependent options, the problem, however is that the the number of paths explodes with the number of stages in the binomial tree. Hence, such options can only be valued for quite small *n*-size binomial trees.

1

interest rate derivative products, it is often useful to model the construction and evolution of the term structure of interest rates using a binomial process. The present paper provides a general methodology for the construction of binomial approximations of multivariate lognormal distributions, both across state variables and across time.

In order to illustrate the type of problems that can be solved using the methods presented in this paper, consider the following example of an option on oil.[b] The contract is an American-style option which allows the holder to buy oil at a specified price denominated in Japanese yen. The payoff on the option in yen depends on two variables: the U.S. dollar-denominated oil price and the yen/dollar foreign exchange rate. At time 0, suppose that the volatility of the oil price is high, but is expected to decline over the life of the option. Also, suppose that there is significant mean reversion in the price of oil. In contrast, in this example, the yen/dollar exchange rate is assumed to have constant volatility and no mean reversion. Further, suppose that the correlation between the price of oil and the yen/dollar exchange rate is high, but is expected to fall.

We wish to value this oil option by approximating the true joint distribution of the variables with a bivariate binomial process. The method developed in this article is designed to solve this general problem of changing volatilities, mean reversion, and multiple state variables, whereas existing models in the literature only deal with one or two aspects of the problem at a time.[c] For example, Nelson and Ramaswamy (1990) show how to approximate various univariate distributions with binomial process. Amin (1991) derives an alternative approach to the problem of a changing volatility function. However, neither article deals with changing volatility in a multivariate context. Amin (1991) and Boyle (1988) provide models of the multivariate case, but they deal with either one asset in the context of changing volatility or multiple assets with a constant correlation structure. Amin, for example, assumes a constant covariance matrix when dealing with the multivariate problem explicitly. The incremental contribution of our article is that it deals with the general multivariate problem, where the assets may have volatilities changing over time but differently across assets, a changing covariance matrix, and differential mean reversion.

In the literature, one method suggested to solve both the univariate and multivariate problem involves a change in the conditional probability computed on a node-by-node basis. In this article, we explicitly recognize that the two problems are really two aspects of the same general issue. In other words, we need, in general, to choose the probabilities so that both the time series (term structure of

[b]Other examples of multivariate options include the "delivery option" that is embedded in many bond futures contracts, a "cocktail" option to receive the principal payment on a bond in one of many currencies at specified exchange rates, and a long-term American-style currency option under stochastic interest rates. For other examples, see Boyle, Evnine, and Gibbs (1989) and Stulz (1982).
[c]Note that a term structure of volatility can result either from mean reversion, changing conditional volatility, or both.

volatility) and cross-sectional (correlation structure) characteristics of the underlying asset prices are satisfied. We show that an application of the linear multiple regression property of joint normally distributed variables leads to a formula for the appropriate conditional probabilities. Application of this linear relationship yields a multivariate binomial distribution which is simple to compute and apply.

In Sec. 1, we introduce the problem in a formal manner and define our notation. In Sec. 2, we review the literature on the topic of binomial approximation of lognormally distributed variables and relate our work to the previous research. Section 3 describes the method in the context of a single variable that has both volatility changes and mean reversion. Section 4 derives and demonstrates a similar methodology for the general multivariate case described above. Our conclusions are summarized in Sec. 5.

1. Statement of the Problem and Notation

We assume that the prices of each of the underlying assets, $X_1, X_2, \ldots, X_j, \ldots,$ X_J follows a lognormal diffusion process:

$$d \ln X_j = \mu_j(X_j, t)dt + \sigma_j(t)dZ_j, \quad j = 1, 2, \ldots, J, \tag{1}$$

where μ_j and σ_j are the instantaneous drift and volatility of $\ln X_j$, and dZ_j is a standard Brownian motion. The instantaneous correlation between the Brownian motions dZ_j and dZ_k is $\rho_{j,k}(t)$. The instantaneous drift in Eq. (1) is a function of X and t, Which allows for mean reversion that may change over time. We assume that $\mu(X_j, t)$ is linear in X and the instantaneous variances and covariances are non-stochastic functions of time. Hence, the asset prices are lognormally distributed at any time t. There is a finite number (m) of future dates in the time interval $[0, T]$ at which we are interested in the asset prices. The dates are numbered $t_1, t_2, \ldots, t_i, \ldots, t_m$, where $t_m = T$, and these are determined by the requirements of the option valuation problem that is being solved.[d] We are interested in the joint distribution of the prices of the assets on these dates.

We denote the unconditional mean (at time 0) of the logarithmic asset return at time i as $\mu_{i,j}$. The conditional volatility over the period $i - 1$ to i is denoted $\sigma_{i-1,i,j}$, and the unconditional volatility is $\sigma_{0,i,j}$. Also, the covariances across assets at time t_i are, respectively,

$$\text{cov}_0(\ln X_{i,j}, \ln X_{i,k}) = \rho_{0,i,j,k}\sigma_{0,i,k} \equiv \sigma_{0,i,j,k},$$

$$\text{cov}_{i-1}(\ln X_{i,j}, \ln X_{i,k}) = \rho_{i-1,i,j,k}\sigma_{i-1,i,j}\sigma_{i-1,i,k} \equiv \sigma_{i-1,i,j,k},$$

where the correlation coefficients $\rho_{0,i,j,k}$ and $\rho_{i-1,i,j,k}$ are specified exogenously.[e]

---

[d]For example, if, as in Breen (1991), we are approximating an American option using the Geske and Johnson (1984) technique, using a trivariate distribution, we would be interested in three dates: t_1, t_2, T.
[e]Note that, given our assumptions, the volatilities and covariances are state independent.

The conditional and unconditional volatilities depend, in general, on the functional form of $\mu_j(X_j,t)$ and $\sigma_j(t)$. For example, if μ_j is a constant, $\ln X_j$ follows a random walk with

$$\sigma_{0,i,j} = \sqrt{\frac{1}{t_i}\int_0^{t_i}\sigma_j^2(t)dt}$$

and

$$\sigma_{i-1,i,j} = \sqrt{\frac{1}{t_1-t_{i-1}}\int_{t_{i-1}}^{t_i}\sigma_j^2(t)dt}\,.$$

In general, however, the volatilities $\sigma_{0,i,j}$ and $\sigma_{i-1,i,j}$ are complex functions of the instantaneous volatilities $\sigma_j(t)$. In this article, instead of starting with a particular specified process for X_j, we assume, more generally that the relevant means, volatilities, and conditional volatilities are given exogenously.[f]

We wish to approximate the process in Eq. (1) with a sequence of binomial distributions such that the mean, variance, and covariance characteristics converge to these given values. The sequence of binomial distributions yields vectors of outcomes of $X_{i,j}$, where $X_{i,j}$ is the price of asset j at time t_i, $i = 1,2,\ldots,m$. Conditional probabilities have to be so chosen that the volatility specifications and correlations are satisfied for each time period.

Since we are concerned only with a finite set of dates (m), the data input required consists of m vectors of exogenously given means $(\bar{\mu}_1,\bar{\mu}_2,\ldots,\bar{\mu}_m)$, where $\bar{\mu}_i = (\bar{\mu}_{i,1},\ldots,\bar{\mu}_{i,J})$, and m unconditional covariance matrices $(\Sigma_{0,1},\Sigma_{0,2},\ldots,\Sigma_{0,m})$, where[g]

$$\sum_{0,i} = \begin{bmatrix} \sigma_{0,i,1}^2 & \cdots & \sigma_{0,i,1,k} & \cdots & \sigma_{0,i,1,J} \\ & \cdots & \cdots & \cdots & \cdots \\ & & \sigma_{0,i,k}^2 & \cdots & \sigma_{0,i,k,J} \\ & & & \cdots & \cdots \\ & & & & \sigma_{0,i,J}^2 \end{bmatrix}, \qquad (2)$$

and $m-1$ conditional covariance matrices $(\Sigma_{1,2},\Sigma_{2,3},\ldots,\Sigma_{m-1,m})$, where

$$\sum_{i-1,i} = \begin{bmatrix} \sigma_{i-1,i,1}^2 & \cdots & \sigma_{i-1,i,1,k} & \cdots & \sigma_{i-1,i,1,J} \\ & \cdots & \cdots & \cdots & \cdots \\ & & \sigma_{i-1,i,k}^2 & \cdots & \sigma_{i-1,i,k,J} \\ & & & \cdots & \cdots \\ & & & & \sigma_{i-1,i,J}^2 \end{bmatrix}, \qquad (3)$$

The exogenously given drift and volatility terms are satisfied asymptotically by choosing the number of binomial stages between 0 and t_1, t_1 and t_2, etc. There are

[f]There are, however, natural restrictions that have to be imposed on the variance–covariance matrices. In particular, the relevant variance–covariance matrices have to be positive semidefinite.
[g]Note that it is not restrictive in any sense to confine our attention to the series of dates $t_1,t_2,\ldots,t_m$ because the time intervals between the dates can be made as small as desired.

n_1 binomial stages between 0 and t_1, n_2 stages between t_1 and t_2, and so on. As we discuss below, the ability to choose the "denseness" of the binomial tree between any two dates represents a significant difference between our model and the previous binomial approximations available in the literature.

In our method, we first construct a binomial approximations for each of the state variables on which the option payoff depends. We then model the given covariance stricture by adjusting the conditional probabilities of $X_{i,k}$ given the outcomes of $X_{i,j}$.[h] Since our methodology is closely related to, but significantly different from, the previous work cited above, we now discuss the contribution of these papers.

2. Relationship to the Literature on Binomial Approximations and Option Pricing

The method we use to approximate the multivariate process is closely related to previous contributions by Amin (1990, 1991), Amin and Bodurtha (1995), Boyle (1988), He (1990), and Nelson and Ramaswamy (1990) [henceforth, NR].[i] The basic idea in NR is to approximate a given univariate process for the price for the underlying asset by a "simple" binomial process. In the NR context, "simple" means that the number of nodes of the binomial process increases linearly with time. In order to ensure that the process has the desired variance characteristics, while remaining simple, NR suggest an adjustment in the conditional probabilities of the binomial process over time.

In this article, we also construct simple binomial processes. However, in contrast to NR, we allow the number of binomial stages n_i between any two points t_{i-1} and t_i to be greater than 1. This means that, in the univariate case, the NR method can be regarded as a special case of ours, where $t_i - t_{i-1}$ is the same for all i and also $n_i = 1$ for all i. In our approach, even in the multivariate case, we can, if we wish, accommodate a change in volatility by allowing n_i to vary. This follows from a suggestion of Amin (1991), but was an important difference. We do not need to change the time intervals (t_{i-1}, t_i) in our more general setting. This means that our method is readily extendible to the multivariate case.

It is well known that a multivariate process across two or more assets can also be modeled by changing the conditional probabilities associated with the nodes. Boyle (1988) shows how to construct a trinomial distribution for each variable and then combine them in a lattice so that the variables have the required variance and covariance characteristics. Amin (1991) models the covariance characteristics of the assets assuming a constant variance–covariance matrix (i.e., one where the asset prices have constant volatilities over time). He then compares the efficiency of Boyle's (1988) method against a range of alternatives, including two where the

---

[h] An alternative method when the correlation between the variables is high is to orthogonalize the variables and then construct the binomial tree.

[i] A method similar to that in Nelson and Ramaswamy (1990) was suggested and implemented in Stapleton and Subrahmanyam (1988).

variables are joint binomially distributed. He finds that, from the standpoint of computational efficiency, the latter approaches are marginally inferior. However, if the factors generating the asset prices ("shocks" in his Example 2) are chosen so as to be independent, the binomial model has the advantage that it does not suffer from the problem of "negative probabilities" that arises in the Boyle approach. Amin (1990) and He (1990) use a trinomial distribution to model a bivariate distribution of two factors. They do it in such a way that the factors are uncorrelated, but not independent. He (1990) shows that the distribution converges to the multivariate Brownian motion assumed and that option prices also converge. These models do not incorporate changing conditional variances or mean reversion.

In our model, we choose the multivariate binomial method as suggested by Amin's work. The major difference between our model and the work by Amin and Boyle is that unlike their approaches, the n_i in our case is not constrained to be one through time. We are then able to model simultaneously the effects of both changing volatility and mean reversion in the case of two or more underlying assets.

In summary, we extend the existing literature in two directions. First, in the univariate case, we are able to extend the NR and Amin models by allowing the number of binomial stages to exceed one and to vary within the time intervals (t_{i-1}, t_i). This means that we can control the "denseness" of the binomial tree between the multiple exercise dates of the option. Our technique can then be used, for instance, to extend Breen's (1991) accelerated binomial option pricing model to the case where volatility change and/or the asset price mean reverts. Secondly, we are able to extend Amin's work by modeling a general multivariate process, where the individual assets have different rates of change of variance and mean reversion.

3. A Method for Constructing a Univariate Binomial Process with Specified Variances

The problem, in the univariate case, is to approximate with a binomial process the true process for X_i, given the means μ_i, conditional volatilities $\sigma_{i-1,i}$, and the unconditional volatilities $\sigma_{0,i}$.[j] The conditional volatilities of the approximated binomial process will be denoted $\hat{\sigma}_{i-1,i}(n_i)$ since they will be a functions of n_i, the number of binomial stages. We require that

$$\lim_{n_i \to \infty} \hat{\sigma}_{i-1,i}(n_i) = \sigma_{i-1,i}, \quad \forall\, i. \tag{4}$$

The unconditional volatility of the approximated process over the period $(0, t_i)$ is similarly denoted $\hat{\sigma}_{0,i}(n_i, n_2, \ldots, n_i)$, since it is, in general, a function of the number of binomial stages over each of the subperiods $t_1, t_2, \ldots, t_i$. Here, we require

$$\lim_{n_l \to \infty} \hat{\sigma}_{0,i}(n_i, n_2, \ldots, n_i) = \sigma_{0,i}, \quad \forall\, i, l, \quad l = 1, \ldots, i. \tag{5}$$

[j]We drop the subscript j in this section. For instance, instead of $\sigma_{0,i,j}$ and $\sigma_{i-1,i,j}$ we simply write $\sigma_{0,i}$ and $\sigma_{i-1,i}$, respectively. Also, instead of $X_{i,j}$ we simply write X_i.

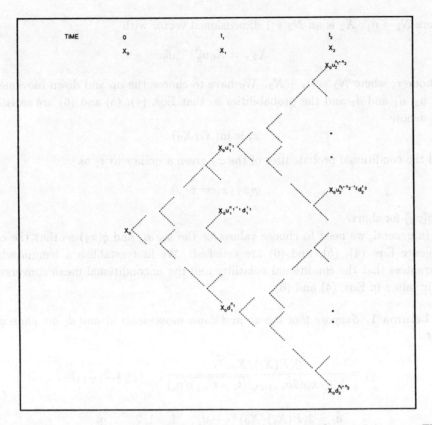

Fig. 1. A discrete process for X_1, X_2. There are $n_1 + 1$ nodes at t_1 number $r = 0, 1, \ldots, n_1$. There are $n_1 + n_2 + 1$ nodes at t_2 numbered $r = 0, 1, \ldots, n_1 + n_2$. X_0 is the starting price, X_1 is the price at time t_1, X_2 is the price at time t_2. u_1, d_1, u_2, and d_2 are the proportionate up and down movements. Although X_1 and X_2 are generated by binomial processes, this method gives two vectors ($n_1 + 1$ dimensional for X_1 and $n_2 + n_1 + 1$ dimensional for X_2). Intermediate values at time $t \in (0, t_1)$, $t \in (t_1, t_2)$ are not defined.

In addition, of course, we constrain the mean of the approximated process to be equal to μ_i for each i. In other words,

$$\lim_{n_i \to \infty} \hat{\mu}_i = \mu_i, \quad \forall\, i. \qquad (6)$$

Our method involves the construction of m separate binomial distributions, where the time periods are denoted $t_1, t_2, \ldots, t_i, \ldots, t_m$. The set of these distributions form a discrete stochastic process for X_i:

$$(X_1, X_2, \ldots, X_m),$$

where X_i is only defined at the times t_i. An example, where $m = 2$ is shown in Fig. 1. X_1 is an $N_1 + 1$ vector where at node r

$$X_{1,r} = X_0 u_1^{N_1 - r} d_1^r,$$

where $N_1 = n_1$. X_2 is an $N_2 + 1$ dimensioned vector with

$$X_{2,r} = X_0 u_2^{N_2-r} d_2^r$$

at node r, where $N_2 = n_1 + N_2$. We have to choose the up and down movements u_1, u_2, d_1 and d_2 and the probabilities so that Eqs. (4), (5) and (6) are satisfied. We denote

$$x_i = \ln(X_i/X_0)$$

and the conditional probabilities of the x_2 given a node r at t_1 as

$$q(x_2 \mid x_1 = x_{1,r})$$

or $q(x_2)$ for short.

In general, we need to choose values for the u_i, d_i, and $q(x_2)$ so that the convergence Eqs. (4), (5), and (6) are satisfied. We first establish a lemma which guarantees that the conditional volatility and the unconditional mean converge to their values in Eqs. (4) and (6).

Lemma 1. *Suppose that the up and down movements u_i and d_i are chosen so that*

$$d_i = \frac{2(E(X_i)/X_0)^{\frac{1}{N_i}}}{1 + \exp(2\sigma_{i-1,i}\sqrt{(l_i - t_{i-1})/n_i})}, \quad i = 1, 2, \ldots, m, \tag{7}$$

$$u_i = 2(E(X_i)/X_0)^{\frac{1}{N_i}} - d_i, \quad i = 1, 2, \ldots, m, \tag{8}$$

where $N_i = \sum_{l=1}^{i} n_l$, then if, for all i, the conditional probability $q(x_l) \to 0.5$ as $n_l \to \infty$, for $l = 1, 2, \ldots, i$, then the unconditional mean and the conditional volatility of the approximated process approach respectively their true values:

$$\lim_{\substack{n_l \to \infty \\ l=1,2,\ldots,i}} \frac{\hat{E}(X_i)}{X_0} \to \frac{E(X_i)}{X_0}, \quad \lim_{n_l \to \infty} \hat{\sigma}_{i-1,i} \to \sigma_{i-1,i}.$$

Proof. See Appendix A.   □

The up and down movements u_i and d_i are analogous to those in Cox, Ross, and Rubinstein (1979), in that they are chosen to match the true mean and conditional volatility. Of course, in our case, since the conditional volatilities are allowed to change over time, the u_is and d_is also change correspondingly.

Given Lemma 1, we can guarantee that the mean and conditional volatility of the approximated process converge to their required values by choosing u_i and d_i using the given conditional volatilities $\sigma_{i-1,i}$.[k] The remaining problem is to choose

---

[k]Note that the convergence of the approximated mean to the true mean implies that the approximated logarithmic mean $\hat{\mu}_i$ also converges to the true logarithmic μ_i.

the conditional probabilities $q(x_i)$ in such a manner that the *unconditional* volatility converges to the true value as in Eq. (5). First note that, by Lemma 1, we are free to choose $q(x_i)$ without affecting the convergence of the conditional volatility, as long as $q(x_i) \to 0.5$ as $n_i \to \infty$. Thus we can choose the probabilities so that both the unconditional and conditional volatilities converge.

Since $x_i = \ln(X_i/X_0)$ is normally distributed, it follows that the regression

$$x_i = a_i + b_i x_{i-1} + \varepsilon_i, \quad E_{i-1}(\varepsilon_i) = 0,$$

is linear with

$$b_i = \sqrt{[t_i\sigma_{0,i}^2 - (t_i - t_{i-1})\sigma_{i-1,i}^2]/t_{i-1}\sigma_{0,i-1}^2},$$

and[1]

$$a_i = E(X_i) - b_i E(X_{i-1}).$$

We now choose the conditional probabilities $q(x_i)$ so that

$$E_{i-1}(x_i) = a_i + b_i x_{i-1} \tag{9}$$

also holds for the approximated variables x_1 and x_{i-1}[m] In Appendix B we show how to choose the conditional probability $q(x_1)$ so as to guarantee that Eq. (9) holds. We can now establish:

Theorem 1. *Suppose that the X_i are joint lognormally. If the X_i are approximated with binomial distributions with $N_i = N_{i-1} + n_i$ stages and u_i and d_i given by Eqs. (7) and (8), and if the conditional probability of an up movement at node r at time $i - 1$ is*

$$q(x_i \mid x_{i-1} = x_{i-1,r}) = \frac{a_i + b_i x_{i-1,r} - (N_{i-1} - r)\ln u_i - r\ln d_i}{n_i(\ln u_i - \ln d_i)}$$

$$- \frac{\ln d_i}{\ln u_i - \ln d_i}, \quad \forall\, i, r \tag{10}$$

then $\hat{\mu}_i \to \mu_i$ and $\hat{\sigma}_{0,i} \to \sigma_{0,i}$ and $\hat{\sigma}_{i-1,i} \to \sigma_{i-1,i}$ as $n_i \to \infty$, $\forall\, i$.

Proof. First note that, as $n_i \to \infty$

$$q(x_i) \to \frac{-\ln d_i}{\ln u_i - \ln d_i} \to 0.5.$$

Hence, the conditions of Lemma 1 are fulfilled and $\hat{\sigma}_{i-1,i}$ converges to $\sigma_{i-1,i}$.

Now we prove the convergence of $\hat{\sigma}_{0,i} \to \sigma_{0,i}$ by induction. First, we choose for $i = 1$, $q(x_1) = 0.5$. Hence from Lemma 1, the theorem is true for $i = 1$. We need

[1]Also, since X_i is lognormally distributed, $E(x_i) = \ln\left(\frac{E(X_i)}{X_0}\right) - \frac{1}{2}\sigma_{0,i}^2$.
[m]Note that Eq. (9) has to hold for the approximated process, while it is true by assumption for the original variables.

now to show that $\hat{\sigma}_{0,2} \to \sigma_{0,2}$ as n_1 and $n_2 \to \infty$. Then, it follows by induction that $\hat{\sigma}_{0,i} \to \sigma_{0,i}$ for all i. Lemma 1 guarantees that $\hat{\sigma}_{1,2} \to \sigma_{1,2}$ if $q(x_2)$ is chosen by Eq. 10. In Appendix B we show that, with this choice of $q(x_2)$, then[n]

$$\hat{E}_{t_1}(x_2) = a_2 + b_2 x_1 \,.$$

The volatility $\hat{\sigma}_{0,2}$ is therefore

$$\hat{\sigma}_{0,2} = \sqrt{[b_2^2 \hat{\sigma}_{0,1}^2 t_1 + (t_2 - t_1)\hat{\sigma}_{1,2}^2]/t_2} \,.$$

Now, since $\hat{\sigma}_{0,1} = \sigma_{0,1}$ and $\hat{\sigma}_{1,2} \to \sigma_{1,2}$ as $n_2 \to \infty$, then

$$\hat{\sigma}_{0,2} \to \sqrt{[b_2^2 \sigma_{0,1}^2 t_1 + (t_2 - t_1)\sigma_{1,2}^2]/t_2} \,,$$

and substituting

$$b_2^2 = [t_2 \sigma_{0,2}^2 - (t_2 - t_1)\sigma_{1,2}^2]/t_1 \sigma_{0,1}^2 \,,$$

we find that $\hat{\sigma}_{0,2} \to \sigma_{0,2}$. It follows that if $n_i \to \infty$, $\hat{\sigma}_{0,i} \to \sigma_{0,i}$, $\forall\, i$.   □

3.1. *Properties of the approximation for small n*

In practical applications we are forced to approximate the stochastic process of X_i using small values of n_i. The smaller the n, the more rapid will be the computation of option prices. When n_i is small, the conditional probability at a node suggested by the formula in Eq. (10) could be outside the natural bounds for probability (i.e., q could exceed 1 or be less than 0).[o]

In Table 1, we show the effect of the restriction $0 \le q(x_i \,|\, x_{i-1} = x_{i-1,r} \le 1$, $\forall\, i, r$. We simulate the model for three cases. In the first case, the process for X_t has the *mean reversion* property. In the second case, the process for X_t exhibits *declining conditional volatility*. Finally, the third case combines the mean reversion property of Case 1 and the changing volatility of Case 2.

The simulations all show, as we would expect, that the accuracy of the approximation improves as n increases. The first panel indicates that, for a constant variance process, mean reversion is easily captured by the method, even for relatively small n. This is also true for cases of declining volatility in Panel 2. The method is accurate for quite small n size and converges rapidly for reasonable changes in volatility. When the volatility declines more rapidly from 10 percent in the first period to 7 percent in the second, then the accuracy begins to decline somewhat. This is due to the reliance of the method on the conditional volatility for the

---

[n]Note that the proof in Appendix B refers to the trivariate case with coefficients α, β, and γ.
[o]Note that this is always a "small n" problem, since $q(x_2)$ in Eq. (10) limits to 0.5 as $n_i \to \infty$. This can be seen from Eq. (10) where the second term goes rapidly to 0.5 as n_i increases and the first term can be kept within the range -0.5 to 0.5 by choosing a large enough size of n_i.

Table 1. Convergence of binomial approximation in a two-period example with (1) mean reversion, (2) changing volatilities, and (3) mean reversion and changing volatilities.

			Volatility of approximated process $\hat{\sigma}_{0,2}$[1]		
$\sigma_{0,1}$	$\sigma_{1,2}$	$\sigma_{0,2}$	$n=10$	$n=30$	$n=50$[2]
(1) Mean reversion[3]					
0.10	0.10	0.10	0.10000	0.10000	0.10000
0.10	0.10	0.09	0.08987(13)[4]	0.08996(4)	0.08997(3)
0.10	0.10	0.08	0.07930(70)	0.07977(23)	0.07986(14)
(2) Changing volatilities[5]					
0.10	0.09	0.09513	0.09510(3)	0.09512(1)	0.09513(0)
0.10	0.08	0.09055	0.09044(11)	0.09051(4)	0.09053(2)
0.10	0.07	0.08631	0.08605(26)	0.08622(9)	0.08626(5)
(3) Mean reversion and changing volatilities[6]					
0.10	0.09	0.08513	0.08510(3)	0.08512(1)	0.08513(0)
0.10	0.08	0.08055	0.08055(0)	0.08055(0)	0.08055(0)
0.10	0.07	0.07631	0.07626(5)	0.07629(2)	0.07630(1)

[1] In each case the periods are $t_1 = 1$ year, $t_2 = 2$ years.
[2] In all cases n is the same for year 1 and year 2 (i.e., $n_1 = n$, $n_2 = n$).
[3] All volatilities are annualized. The volatility over the first period $\sigma_{0,1}$ is exact, by construction.
[4] In parentheses we show $(\sigma_{0,2} - \hat{\sigma}_{0,2}) \times 100,000$. This is the approximation error. This is also the size of the error $(\sigma_{1,2} - \hat{\sigma}_{1,2})$.
[5] In this case we assume that X_t follows random walk with changing variance; hence, $\sigma_{0,2}^2 = (\sigma_{0,1}^2 + \sigma_{1,2}^2)/2$.
[6] In this case we assume that X_t follows a mean reverting process with changing conditional variance. In these examples, we choose $\sigma_{0,2} = \sqrt{(\sigma_{0,1}^2 + \sigma_{1,2}^2)/2 - 0.01}$.

calculation of X_2. In extreme cases, however, a change in the size of n_i can be made which can improve the accuracy.[P]

Theorem 1 allows us to approximate a process with given mean, variance, and covariance characteristics over the periods $(0, t_i)$ and (t_{i-1}, t_i), where $1 = 1, 2, \ldots, m$. We can therefore construct a process using all the dates $t_1, t_2 \ldots, t_m$. Successive application of Theorem 1 guarantees that the volatilities converge to their given values over each time period. The resulting binomial process can be used to evaluate multiple exerciseable options with a given number of exercise dates as in Breen (1991).

4. The Multivariate Case

There are two ways in which an option valuation problem may involve a multivariate probability distribution. The first is when the payoff on the option depends

[P] Simulations of the case where X_t is fitted over three periods, t_1, t_2, t_3, yield similar results. For example, assuming constant conditional volatility of 10 percent and mean reversion such that $\sigma_{0,2} = 0.09$, $\sigma_{0,3} = 0.08$, we found that for $n = 10$, $\hat{\sigma}_{0,3} = 0.07974(26)$; $n = 30$, $\hat{\sigma}_{0,3} = 0.07991(9)$; $n = 50$, $\hat{\sigma}_{0,3} = 0.07995(5)$. Similar results were obtained for the case of changing volatility and where the process has both mean reversion and volatility change.

on the outcome of two or more random variables. This is the case for the yen oil option discussed in the introduction. The payoff there depends on the dollar oil price and the yen/dollar exchange rate. The second motivation stems from the consideration of the value of an *American*-style option where the intermediate value of the option depends on the price of the underlying asset and other variables, such as the rate of interest. In this case we need to model the covariance of the asset price and the rate of interest, as well as the time series properties of the two variables. In this section we consider the general problem of approximating the multivariate distribution of two or more variables each of which has the properties (of mean reversion and changing volatility) discussed in Secc. 3.[q]

Initially, we consider the case of a single time period $(0, T)$. This is relevant for *European-style* options where the payoff depends on two or more variables. We wish to approximate the joint distribution of the variables $(X, Y, Z, \ldots)$ with a multivariate binomial distribution. Hence, we choose the conditional probabilities (of Y given X, and of Z given Y and X) given the covariance between the variables.[r] We confine ourselves to the case where there are three relevant random variables (X, Y, Z). The first step involves approximating the first random variable X_T with a binomial distribution using the method outlined in Sec. 3. We then build a vector of Y_T values using Eqs. (7) and (8), and using the conditional volatility of Y_T given X_T. We then choose the conditional probability of Y_T given the outcome X_T. This conditional probability is denoted $q(y_T \mid x_{T,r})$, where r is the node of the X binomial tree, and $x_{T,r}$ is the value of x_T at the rth node. In Appendix B we show that the appropriate value of $q(y_T)$ is[s]

$$q(y_T \mid x_T = x_{T,r}) = \frac{\alpha_y + \beta_y x_{T,r} - n_y \ln d_y}{n_y(\ln u_y - \ln d_y)}, \qquad (11)$$

where α_y and β_y are the coefficients from the simple regression of Y_T on X_T, and n_x and n_y are the number of binomial stages in the x and y approximations, respectively.

In the next stage, we build a vector of Z_T using the conditional volatility of Z_T given X_T and Y_T. We then choose the probability of Z given an X and a Y node. This conditional probability is shown in Appendix B [Eq. (B.3)] to be

$$q(z_T \mid x_T = x_{T,r}, y_T = y_{T,s}) = \frac{\alpha_z + \beta_z x_{T,r} + \gamma_z y_{T,s} - n_z \ln d_z}{n_z(\ln u_z - \ln d_z)}, \qquad (12)$$

where α_z, β_z, and γ_z are the coefficients from the multiple regression of Z_T on X_T and Y_T. If the conditional probabilities are chosen by Eqs. (11) and (12), it follows

---

[q]Boyle (1988) develops a technique for the multivariate case based on a trinomial distribution. Our method differs from Boyle's in that we use a simpler binomial distribution. Our method allows us to solve the general problem of constructing a multivariate process with arbitrary cross-sectional variances and covariances, and conditional variances and covariances over time.
[r]An alternative method suggested by Amin (1991) models the covariance using factor loadings on two independent random variables. A problem with this approach is that the number of Y and Z nodes increases rapidly with the n size making the computation of complex option prices difficult.
[s]This follows from Eq. (B.3) WITH $z = y$ and $\gamma_z = 0$.

by an extension of Theorem 1 that the volatilities of Y and Z converge, as n_y and n_z increase, to their values σ_y and σ_z.

We now consider the accuracy of the covariance and correlation between the approximated X and Y variables. We have

Theorem 2. *Suppose the up and down movements of two correlated random variables, X and Y, are chosen by Eqs. (7) and (8) and that the conditional probabilities are given by Eq. (11). Then the approximated value of the conditional covariance is equal to its true value. Also, when there are three or more correlated variables, for example X, Y, and Z, where X is the first fitted variable, the covariance of the pairs (X, Y) and (X, Z) are exact and that of (Y, Z) limits to its true value.*

Proof. See Appendix A. □

Note that, in the case of just two variables X and Y, this cross-sectional property is an *exact* one and holds for all values of n and not just for the limiting values. The approximation in the case of three or more variables is due to the fact that Y has only an approximately correct variance which limits to its true value as n increases. Aslo, the covariance property holds in the multiperiod case for all the conditional covariances. However, since our method approximates the variances of the variables, the resulting correlation is only asymptotically correct.

4.1. *The multiperiod, multivariate case*

We now turn to the general problem of a multiperiod, multivariate process. In this case we may be concerned with say two variables (the dollar price of oil and the yen/dollar exchange rate) at a series of dates $(t_1, t_2, \ldots)$. Our general problem is to construct a multiperiod, multivariate process over periods $(0, t_1), (t_1, t_2), \ldots$. For simplicity, we illustrate the general methodology with an example of a two-period $[(0, t_1), (t_1, t_2)]$, bivariate (X, Y) case. This case is sufficient, for example, if we wish to value a compound option whose payoff depends on two variables.

Suppose that (X_1, X_2) and (Y_1, Y_2) are multivariate lognormally distributed with volatilities $(\sigma_{0,1,x}, \sigma_{0,2,x}, \sigma_{1,2,x})$ and $(\sigma_{0,1,y}, \sigma_{0,2,y}, \sigma_{1,2,y})$. Also, assume that the correlation between x_1 and y_1 is $\rho_{0,1,x,y}$. The conditional correlation between x_2 and y_2 is denoted $\rho_{1,2,x,y}$. Note that, for the joint lognormal distribution, the conditional correlation is nonstochastic. However, in this method there is no need to restrict the conditional correlation to be the same at each point in time.

The steps in the computation for the general case are:

1. Compute the up and down movements for X_1 and Y_1, and X_2 and Y_2, independently using the methodology of Sec. 3. Specifically, the vector of Y_1 is constructed using the conditional volatility of Y_1 given X_1, X_2 requires the conditional volatility $\sigma_{1,2,x}$, and Y_2 requires the conditional volatility of Y_2 given both Y_1 and X_2.

2. Compute the conditional probability of an up movement in Y_1 given a value of X_1 using Eq. (11). We denote these probabilities as $q(y_1 \mid x_1 = x_{1,r})$ since they are of Y_1 given that X_1 is at a node r at time t_1.

3. Compute the conditional probability of an up movement in X_2 given a value of X_1 using the methodology of Sec. 3 and Eq. (10) in particular. These probabilities are denoted $q(x_2 \mid x_1 = x_{1,r})$.

4. Compute the conditional probability of Y_2 given both Y_1 and X_2. We denote this probability as $q(y_2 \mid y_{1,r}, x_2 = x_{2,s})$. This probability must satisfy both the time-series (Y_2 on Y_1) and cross-sectional (Y_2 on X_2) properties of Y_2. In other words, it must satisfy the volatilities ($\sigma_{0,1,y}, \sigma_{0,2,y}, \sigma_{1,2,y}$) and the conditional correlation $\rho_{1,2,x,y}$. In Appendix B, we show that in this case

$$q(y_2) = \frac{a_2 + b_2 y_{1,r} + c_2 x_{2,s} - [(N_1 - r) \ln u_{2,y} + r \ln d_{2,y}] - n_2 \ln d_{2,y}}{n_2 [\ln u_{2,y} - \ln d_{2,y}]} , \quad (13)$$

where the probability $q(y_2) = q(y_2 \mid y_1 = y_{1,r}, x_2 = x_{2,s})$, and where a_2, b_2, and c_2 are the multiple regression coefficients from the regression of y_2 on y_1 and x_2.

Again, an extension of Theorem 1 can be used to show that when $q(y_2)$ and $q(x_2)$ are chosen in this manner, both the variances and the correlations of the multivariate process converge to their given values. However, the probabilities in Eqs. (10), (11), and (13) could be greater than one or less than zero when the n size is finite. As we see now, this is particularly likely when the correlation between the variables is high.

It is important now to consider the limitations on the accuracy of the multivariate method when the natural limits are placed on the conditional probabilities in Eqs. (10), (11), and (13). In Table 2, we show the results of simulations with and without a nonnegativity constraint placed on the conditional probabilities. For simplicity, we show just the case of a single time period $(0, t_1)$ with $t_1 = 1$ year. Even when the conditional probability is constrained, the simulations show that the accuracy of the approximation is not adversely affected by the nonnegativity constraint for correlations under 0.8. For high values of ρ (e.g., $\rho = 0.9$) some inaccuracy is apparent, but this reduces for large n values.[t]

5. Applications of the Methodology to Option Pricing

In applying the methodology to the pricing of options we need to approximate the *risk neutral* distribution of the asset prices. For the problems considered here, this distribution is joint lognormal where the means of the variables are their respective forward prices and the volatilities are the same as those of the actual distribution. It follows from Theorem 1 that if the u_i and d_i are chosen as in Eqs. (7) and (8) with the additional requirement that the mean equals the forward price, then the distribution converges to the required risk neutral distribution. In this section, we price multiple exerciseable options on two variables using these approximating distributions. However, before applying the method we need to establish that the option payoffs and prices also converge as the n size increases.

---

[t]In fact the inaccuracy can be eliminated by appropriate choice of n size for the dependent variable.

Table 2. Accuracy of the binomial approximation in a multivariate case with and without nonnegativity constraint on probabilities.

	Volatility of approximated process, $\hat{\sigma}_{0,1,y}$ [1]			
	Binomial stages, $n_{1,x} = n_{1,y} = 10$		Binomial stages, $n_{1,x} = n_{1,y} = 20$	
Correlation[2] $\rho_{0,1,x,y}$	Estimated $\hat{\sigma}_{0,1,y}$	Approximation error[3]	Estimated $\hat{\sigma}_{0,1,y}$	Approximation error
0.0	0.1000	0	0.1000	0
0.1	0.0999	1	0.1000	0
0.2	0.0998	2	0.0999	1
0.3	0.0995	5	0.0998	2
0.4	0.0992	8	0.0986	4
0.5	0.0987	13	0.0984	6
0.6	0.0982	18	0.0991	9
0.7	0.0975	25	0.0988	12
0.8	0.0960	40	0.0984	16
0.8[4]	(0.0967)	(33)	(0.0984)	(16)
0.9	0.0878	122	0.0961	39
0.9[4]	(0.0959)	(41)	(0.0979)	(21)

[1] In all cases, each of the volatilities (conditional and unconditional) for X and Y is 10 percent.
[2] $t_1 = 1$ year.
[3] The approximation error is shown $(\sigma_{0,1,y} - \hat{\sigma}_{0,1,y}) \times 100,000$.
[4] The number in parentheses below the constrained estimated volatility $\hat{\sigma}_{0,1,y}$ is the corresponding unconstrained volatility estimate.

In the case of a finite n size, we propose the option payoffs and values only as approximations to their limiting values. The method makes no use of the (unknown) true mean of the underlying asset and does not assume a complete market in which no-arbitrage option values are determined. We show that the option values we compute converge to the prices in the complete market, continuous-time economy.[u]

[u]The prices which our finite n-size economy values converge to are thus the no-arbitrage prices. Our method is in the spirit of Rajasingham (1990) who shows that it is not necessary for the finite economies to be complete when the economies are used as approximations. Our task, therefore, differs somewhat from that of Cox, Ross, and Rubinstein (1979) and Duffie (1988), who assume knowledge of the asset's true mean and show that the no-arbitrage, finite economy prices converge to the continuous economy limit. It is interesting to note that Amin (1990) and He (1990) both suggest an alternative three-state approach to the valuation of options on multivariate processes which allows a complete market, no-arbitrage valuation in the finite economy. In our bivariate binomial case, the finite economy is incomplete; however, our approximated values can still be shown to converge to the continuous-time limiting values.

We will first discuss and prove convergence in the case of a European-style call option on a single asset: the Black–Scholes case. We then extend the proof to the case of American-style options. Our procedure in the univariate case is as follows. If the option has a maturity t, the asset price on which the option is written has a lognormally distributed price X_t with volatility σ. We construct a binomially distributed variable $X_t(n)$ with binomial probability $q = 0.5$, an expected value $E[X_t(n)] = F_{0,t}$: the asset forward price, and u and d chosen by the formula in Eqs. (7) and (8). It follows from Lemma 1 that $X_t(n)$ converges in distribution to the risk neutral density of X_t: that is, to a distribution with a mean $F_{0,t}$ and volatility σ. We now establish:

Theorem 3. *Define $X_t(n)$ using Eqs. (7) and (8) with $E[X_t(n)]F_{0,t}$. Then*

(a) *$X_t(n)$ converges in distribution to X_t, where X_t is lognormal with mean $F_{0,t}$ and logarithmic variance $\sigma^2 t$.*

(b) *For a European-style call option with payoff function $g(X_t) = \max[X_t - K, 0]$, $g[X_t(n)] \to g(X_t)$.*

(c) *$E[g(X_t(n))] \to E[g(X_t)] = C_0/B_{0,t}$ is the Black–Scholes value of the call option.*

Proof.

(a) Follows directly from Lemma 1.

(b) Follows since $g(\cdot)$ is a continuous function; and Bilingsley (1986), Theorem 25.7, Corollary 1.

(c) Follows from the fact that the sequence $g[X_t(n)]$ is uniformly integral, given the lognormal distribution; and Bilingsley (1986), Theorem 25.12.   □

Parts (b) and (c) of Theorem 3 follow quite closely the analysis of Duffie (1998, pp. 244–248). An important implication of the theorem is that hedge ratios and other risk parameters, such as the deltas, thetas, vegas, and gammas (i.e., the derivatives of C_0), also converge as the n size increases. This allows the approximation of prices and hedge ratios with the binomial method.

5.1. *American-style options*

We now extend the discussion of convergence to the case of American-style options. Consider first a call option that is exerciseable twice, at time t_1 and at time t_2. Let the exercise price at time t_1 be K_1, and at time t_2 be K_2. The payoff at time t_2, if the option is not exercised at t_1, is

$$g(X_2) = \max[X_2 - K_2, 0].$$

The value of the option, if alive at time t_1, is given by the Black–Scholes function

$$C_1 = C_1(X_1).$$

Just prior to the exercise decision at time t_1, the value of the option can be written

$$g(X_1) = \max[C_1, X_1 - K_1].$$

The value at time 0 of the claim is

$$C_0 = B_{0,1}E[g(X_1)],$$

where the expectation is again taken with respect to the risk neutral density.[v] We now approximate C_0 using a binomial distrubution for X_1 and X_2.

To approximate C_0, we first construct $X_1(n)$ and $X_2(n)$ using Eqs. (7) and (8) with $E(X_1) = F_{0,1}$ and $E(X_2) = F_{1,2}$. We then compute $g(X_2(n))$ and $C_1(n)$ in the same way as in the case of the European-style option above. It then follows:

Corollary 1. *The payoff and value of a twice exerciseable option converge to their continuous limits as n increases.*

Proof. From Theorem 3, it follows that $C_1(n)$ converges to the Black–Scholes value C_1. We know also the payoff (if exercised) at t_1, $X_1(n) \to X_1$. Hence $g(X_1(n)) \to g(X_1)$. Finally, since $g(X_1)$ is again uniformly integral it follows that the sequence $E[g(X_1(n))]$ converges to $E[g(X_1)]$, and $C_0(n)$ converges to C_0. □

The above extension of Theorem 3 shows that our approximation converges for a twice exerciseable option. By induction it must follow that an m times exerciseable option can be approximated in a similar way choosing the u and d proportions according to Eqs. (7) and (8). Hence, in the limit, as m increases the values converge to the continuously exerciseable American option price.[w]

The proof in Theorem 3 and Corollary 1 extend to the multivariate case, as shown for example in He (1990). Here, what is required again is that the option payoff is uniformly integral where the underlying state space is multivariate.

5.2. *An example: the valuation of a twice exerciseable, quality option*

Consider the following quality option contract. The option is to acquire either asset X or asset Y. The holder has the additional option to exercise at time t_2 or time t_1. If he exercises at t_2, he can buy either asset X at a price $K_{2,x}$ or asset Y at a price $K_{2,y}$. Alternatively, he can exercise his option at time t_1 and buy asset X for $K_{1,x}$ or asset Y for $K_{1,y}$. This is a quality option with an *American* feature.[x]

---

[v]This follows form Geske and Johnson (1984).

[w]Strictly, this argument require that the strike price of the option is a continuous function of time. Otherwise, it is possible, for example, that the option payoff could be positive if exercised at an irrational date and zero otherwise. An alternative rigorous proof of convergence for American options is provided by Amin and Khanna (1994).

[x]An example of a quality option is the option available in most bond futures contracts. When we use the term "American" feature we mean that it can be exercised (at least once) before the final maturity.

The payoffs on the option are as follows:
At time t_2,

$$g(X_2, Y_2) = \max[X_2 - K_{2,x}, Y_2 - K_{2,y}, 0],$$

$$\text{if option not exercised at } t_1,$$

$$= 0, \qquad \text{if option exercised at } t_1.$$

At time t_1 the value of the option is

$$g(X_1, Y_1) = \max[X_1 - K_{1,x}, Y_1 - K_{1,y}, C_1],$$

where C_1 is the value of the option at time t_1, if it is not exercised at time t_1. In order to value this option, we make the following assumptions: (1) interest rates in the economy are nonstochastic, (2) $t_1 = 1$ and $t_2 = 2$, (3) the relevant prices of the assets X_1, X_2, Y_1, and Y_2 are joint lognormally distributed, and (4) a preference-free valuation relationship exists for the valuation of the option.

The fourth assumption implies that the option can be valued using a *risk neutral* density function where the means of the stochastic variables are their respective forward prices. This implies that the means of the variables are

$$E(X_1) = F_{0,1,x}, \quad E(Y_1) = F_{0,1,y},$$

$$E(X_2) = F_{0,2,x}, \quad E(Y_2) = F_{0,2,y},$$

where $F_{0,t}$ refers to the forward price of the asset for delivery at time t. We approximate the required distributions with a binomial distribution which has n stages from 0 to t_1 and from t_1 to t_2.

In Table 3, we show the input data for the option valuation. Note that the unconditional volatility for asset X_t is less than its (constant) conditional volatility, indicating mean reversion. Also the conditional volatility of Y_t is nonconstant. For convenience we choose the forward prices of X_t and Y_t to be constant, equal to each other, and equal to 1. The strike prices are also 1 at time t_2. At time t_1, the strike prices are assumed to be lower (0.96) in order to allow the possibility of early exercise, and to make the valuation problem more interesting.

In Table 3, we illustrate the convergence of the option price as the number of binomial stages (both n_1 and n_2 for X and Y) increases. Note that reasonably accurate answers are obtainable with $n = 8$ even when the correlation between the variables is high. This is due to the fact that with two variables and two periods, there are $(n + 1)^4$ states. Table 3 also shows that this *quality option* declines in value as the assumed correlation between the underlying assets increases: with $n = 30$, the value is 8.85 percent in the uncorrelated case and only 7.06 percent when $\rho = 0.8$.

This example illustrates the rapid convergence of the method. Even for relatively small values of n and highly correlated variables, the errors of the approximated process are relatively small. The resulting option prices are a accurate and can be computed rapidly given the small values of n.

Table 3. A twice exerciseable option: sensitivity analysis.

Correlation Coefficient[2], ρ	Number of binomial stages[1], n				
	4	8	12	16	30
0.0	0.0881	0.0884	0.0884	0.0885	0.0885
0.2	0.0845	0.0849	0.0850	0.0850	0.0849
0.4	0.0802	0.0808	0.0810	0.0810	0.0810
0.6	0.0750	0.0759	0.0761	0.0763	0.0764
0.8	0.0677	0.0694	0.0700	0.0702	0.0706

Time[3]	$t_1 = 1$ year	$t_2 = 2$ years
Forward prices[4]	$F_{0,1,x} = F_{0,1,y} = 1$	$F_{0,2,x} = F_{0,2,y} = 1$
Zero bond prices[5]	$B_{0,1} = 0.9$	$B_{0,2} = 0.81$
Volatility[6]:		
Conditional (x)	$\sigma_{0,1,x} = 0.1$	$\sigma_{1,2,x} = 0.1$
Conditional (y)	$\sigma_{0,1,y} = 0.08$	$\sigma_{1,2,y} = 0.07$
Unconditional[7] (x)		$\sigma_{0,2,x} = 0.09$
Unconditional[8] (y)		$\sigma_{0,2,y} = 0.07517$
Strike prices	$K_{1,x} = K_{1,y} = 0.96$	$K_{2,x} = K_{2,y} = 1$

[1] The n size applies to n_1 for X and Y, and to n_2 for both X and Y.
[2] This is the value of ρ both for time t_1 and for time t_1 to t_2.
[3] The following input data for the twice exerciseable quality option on two assets is assumed for all option price calculations.
[4] The fact that forward prices are the same for delivery at years 1 and 2 implies that the assets pay dividends.
[5] These prices imply a constant rate of interest of 11.11 percent per year.
[6] These volatilities are all quoted on an annualized basis.
[7] This lower unconditional volatility implies mean reversion in the X_t process.
[8] $\sigma_{0,2,y} = \sqrt{(\sigma_{0,1,y}^2 + \sigma_{1,2,y}^2)/2}$. Hence, y follows a random walk with changing variance.

6. Conclusions

In this article, we have shown that binomial distributions with changing probability parameters can be constructed to approximate the joint distribution of the asset price at various (exercise) dates. The method can be extended, using state-dependent probabilities, to approximate the covariance characteristics in the multivariate case, where there is more than one state variable determining the option payoff. Tests of the accuracy of the multivariate binomial approximation for variables with changing variance and covariance characteristics show it to be quite accurate even with a small number of binomial stages. The accuracy also extends to cases where the correlation between the variables is high. The method offers a fast and efficient computational method for multiple exerciseable options prices which can be extended to American-style option prices using the Geske and Johnson (1984) approximation technique. This would facilitate a generalization of the work of Breen (1991). Alternatively, it can be used to value a compound option whose payoff depends on two or more variables.

The principal feature of our methodology is its computational efficiency compared with alternative option pricing methods using numerical integrations. This advantage is clear-cut for American options where exercise is possible on two or more dates; our approach is less computationally intensive. This is because in our method, the number of nodes increases only linearly with the number of exercise dates. For instance, numerical integration along the lines of Geske and Johnson (1984) would involve n^3 instead of $(3n + 1)$ nodes in our binomial method in the thrice-exerciseable case. For options involving two or more state variables, and/or more exercise dates, the comparative efficiency of our methodology is even more significant. For example, if Amin's (1991) orthogonal factor method is used with three state variables, $(n + 1)^3$ nodes are generated for the third variable, whereas each variable has just $n + 1$ nodes using our method. This leads to a dramatic reduction in the number of requited integrations. For some options our method can make the difference between feasible and nonfeasible computation. These computational advantages become even more important when hedge ratios and other risk management sensitivity parameters are required in addition to the option values.

Appendix A. Proof of Lemma 1 and Theorem 2

Proof of Lemma 1. (a) For a given N_i, the mean of the approximated process is

$$\frac{\hat{E}(X_i)}{X_0} = \Pi_{l=1}^{i}[q(x_l)u_i + \{1 - q(x_l)\}d_i]^{n_l} .$$

Since, by the condition of the theorem, $q(x_l) \to 0.5$ as $n_l \to \infty$, $l = 1, 2, \ldots, i$,

$$\lim_{\substack{n_l \to \infty \\ l=1,2,\ldots,i}} \frac{\hat{E}(X_i)}{X_0} = [0.5(u_i + d_i)]^{N_i} .$$

From Eq. (8), however,

$$u_i + d_i = 2\left[\frac{E(X_i)}{X_0}\right]^{\frac{1}{N_i}} .$$

Hence,

$$\lim_{\substack{n_l \to \infty \\ l=1,2,\ldots,i}} \frac{\hat{E}(X_i)}{X_0} = \frac{E(X_i)}{X_0} .$$

(b) The conditional variance can be written in the limit, since $q(x_i) \to 0.5$, when $n_i \to \infty$,[y] as

$$\lim_{n_i \to \infty}[\hat{\sigma}_{i-1,i}(n_i)]^2 = \frac{1}{t_i - t_{i-1}} n_i(0.5)^2(\ln u_i - \ln d_i)^2$$

$$= \frac{0.25 n_i}{t_i - t_{i-1}}\left[4\sigma_{i-1,i}^2 \frac{(t_i - t_{i-1})}{n_i}\right] ,$$

that is, $\lim_{n_i \to \infty}[\hat{\sigma}_{i-1,i}(n_i)]^2 = \sigma_{i-1,i}^2$.   □

[y] See Cox and Rubinstein (1985), p. 201.

Proof of Theorem 2. In the case of just two variables (X, Y) it is sufficient to establish the theorem for the special case of $n_x = 1$. For $n_x = 1$, the means of the two variables X and Y, where Y is chosen to be conditional on X, are given by

$$\mu_x = (\ln u_x + \ln d_x)/2\,,$$

$$\mu_y = q \ln u_y + (1 - q) \ln d_y\,,$$

$$q = (q_0 + q_1)/2\,,$$

where $q_r = q(y|x = x_r)$ as defined by Eq. (11).

The conditional covariance between X and Y is given by

$$\hat{\sigma}_{xy} = \left(\frac{1}{2}\right)^2 (q_0 - q_1)(\ln u_x - \ln d_x)(\ln u_y - \ln d_y)\,.$$

But from Eq. (11), q_0 and q_1 are given by

$$q_r = \frac{\alpha + \beta[r \ln d_x + (1 - r) \ln u_x] - \ln d_y}{\ln u_y - \ln d_y}\,, \quad r = 0, 1$$

so that the approximated covariance $\hat{\sigma}_{xy}$ can be written as

$$\hat{\sigma}_{xy} = \left(\frac{1}{2}\right)^2 \beta(\ln u_x - \ln d_x)^2 = \beta\sigma_x^2\,,$$

that is, $\hat{\sigma}_{xy} = \sigma_{xy}$. $\qquad\qquad\square$

A similar argument can be used for the case of three or more variables. For instance, consider the case of three variables X, Y, and Z, where X is the first fitted variable. First, the proof for the case of two variables applies directly to the covariances between X and Y, and X and Z. Next, since in the limit, the binomial probability of Y converges to 0.5, the covariance of Z with Y converges to its true value.

Appendix B. The Binomial Probability in the Case of a Multivariate Lognormal Stochastic Process

B.1. *The general problem*

The method used in the article can be applied to construct a binomial tree as a discrete time approximation for any multivariate lognormal distribution. We first consider the general problem of approximating variables with a given covariance structure. We then apply this general method to the problem in the text. To see the specific details of the method consider the case of a trivariate lognormal distribution of three variables X, Y, and Z with the following variance-covariance matrix (between the logarithms of the variables):[z]

$$\Omega = \begin{pmatrix} \sigma_x^2 & \sigma_{x,y} & \sigma_{x,z} \\ \sigma_{y,x} & \sigma_y^2 & \sigma_{y,x} \\ \sigma_{z,x} & \sigma_{z,y} & \sigma_z^2 \end{pmatrix}$$

For notational convenience we use lowercase letters to denote natural logarithms (i.e., $x = \ln x$, $y = \ln Y$, $z = \ln Z$).

Since x, y, and z are normally distributed, the multiple regression

$$z = \alpha_z + \beta_z x + \gamma_z y + \varepsilon \,,$$

where

$$\beta_z = \frac{\sigma_{z,x}\sigma_y^2 - \sigma_{y,z}\sigma_{x,y}}{\sigma_x^2\sigma_y^2 - \sigma_{x,y}^2} \,,$$

$$\gamma_z = \frac{\sigma_{y,z}\sigma_x^2 - \sigma_{x,y}\sigma_{z,x}}{\sigma_x^2\sigma_y^2 - \sigma_{x,y}^2} \,,$$

and

$$\alpha_z = E(z) - \beta_z E(x) - \gamma_z E(y) \,,$$

is linear, and the conditional expectation of z is

$$E(z \mid x, y) = \alpha_z + \beta_z x + \gamma_z y \,. \tag{B.1}$$

First, we construct separate binomial trees for the variables x, y, z, using the method described in the text. We then choose the conditional probability of an up movement in z so that Eq. (B.1) is satisfied each node. Given that z is a binomial process this implies that

$$\alpha_z + \beta_z x_r + \gamma_z y_s = n_z\{q(z)\ln u_z + [1 - q(z)]\ln d_z\} \,, \tag{B.2}$$

where $q(z) = q(z \mid x = x_r, y = y_s)$ is the probability of an up movement in z given that x is at node r and y is at node s of their respective binomial distributions. In Eq. (B.2) n_z is the number of stages in the binomial process of z, and u_z and d_z are the up and down movements of z. Solving Eq. (B.2) we find

$$q(z) = \frac{\alpha_z + \beta_z x_r + \gamma_z y_s - n_z \ln d_z}{n_z(\ln u_z - \ln d_z)} \,. \tag{B.3}$$

---

[z]The method is readily generalized to the n-variable case. However, the notation is complex.

B.2. *The time series problem with two time periods*

In the text we consider the problem of constructing a process for X_t with given volatility characteristics at two specified points in time $ti - 1$ and t_i. In this case we construct binomial trees of $x = x_i$ and $x = x_{i-1}$. The resulting values of X_{i-1} and X_i are illustrated in Fig. 1 in the text. The difference in this case is that x_i is a time-series (i.e. a cumulative) variable. In this case, the conditional expectation of x_i given node r at time t_{i-1} is

$$a_i + b_i x_{i-1} = n_i \{ q(x_i) \ln u_i + [1 - q(x_i)] \ln d_i \}$$
$$+ (N_{i-1} - r) \ln u_i + r \ln d_i . \tag{B.4}$$

In Eq. (B.4) the first term represents the drift from time t_{i-1} to time t_i. The second term is the expected value of x_i, given that the variable is in state r at time t_{i-1}, if the drift from time t_{i-1} to t_i is zero.[z']

Solving Eq. (B.4) for $q(x_i)$ we have in this case

$$q(x_i) = \frac{\alpha_i + \beta_i x_{i-1} - [(N_{i-1} - r) \ln u_i + r \ln d_i] - n_i \ln d_i}{n_i (\ln u_i - \ln d_i)} ,$$

where $q(x_i) = q(x_i | x_{i-1} = x_{i-1}, r)$ is the probability of an up movement in x_i given that x_{i-1} is at node r.

B.3. *The multivariate time-series problem*

In the text we consider the example where there are two relevant variables (X_t, Y_t) and two time periods $(i - 1, i)$. In this case we suggest first constructing the binomial process for (X_{i-1}, X_i) and the relevant conditional probabilities $q(x_i)$ using the methods described above. The remaining problem is to compute the conditional probability of Y_2 given both Y_1 and X_2. This must reflect both the time-series properties of the Y_t process and the conditional correlation of the two variables $(\rho_{1,2,x,y})$.

This is a multivariate problem involving three variables (Y_i, Y_{i-1}, X_i). Hence we can use the general case in Sec. 1 of Appendix B with $z = Y_i, x = Y_{i-1}$, and $y = X_i$. Again, we recognize that y_i is a time-series variable. In this case, therefore, the conditional expectation of y_i, given node r of y_{i-1} and node s of x_i, is

$$a_i + b_i y_{i-1} + c_i x_i = n_i \{ q(y_i) \ln u_{i,y} + [1 - q(y_i)] \ln d_{i,y} \}$$
$$+ (N_{i-1} - r) \ln u_{i,y} + r \ln d_{i,y} . \tag{B.5}$$

Solving Eq. (B.5) for $q(y_i)$ we have

$$q(y_i) = \frac{a_i + b_i y_{i-1} + c_i x_i - [(N_{i-1} - r) \ln u_{i,y} + r \ln d_{i,y}] - n_i \ln d_{i,y}}{n_i (\ln u_{i,y} - \ln d_{i,y})} ,$$

[z'] Note that, in the special case where the volatility is constant over time $u_z = u_x$ and Eq. (B.4) becomes

$$a_i + b_i x_{i-1} = n_i \{ q(x_i) \ln u_i + [1 - q(x_i)] \ln d_i \} + x_{i-1} .$$

In this case the conditional expectation of x_i is the drift plus x_{i-1}.

where $q(y_i) = q(y_i|y_{i-1} = y_{i-1,r}, x_i = x_{i,s})$ is the conditional probability of an up movement in y_i given that y_{i-1} is at node r and x_i is at node s.

References

K. I. Amin, "A simplified Discrete Time Approach for the Pricing of Derivative Securities with Stochastic Interest Rates", working paper, University of Michigan, Ann Arbor, 1990.

K. I. Amin, "On the Computation of Continuous Time Option Prices Using Discrete Approximations", *Journal of Financial and Quantitative Analysis* **26** (1991) 477–495.

K. I. Amin and J. N. Bodurtha Jr., "Discrete Time Valuation of American Options with Stochastic Interest Rates", *Review of Financial Studies* **8** (1995) 193–234.

K. I. Amin and A. Khanna, "Convergence of American Option Values from Discrete to Continuous-Time Financial Models", *Mathematical Finance* **4** (1994) 289–304.

P. Billingsley, *Probability and Measure*, 2nd ed. John Wiley and Sons, New York, 1986.

R. Breen, "The Accelerated Binomial Option Pricing Model", *Journal of Financial and Quantitative Analysis* **26** (1991) 153–164.

P. P. Boyle, "A Lattice Framework for Options Pricing when there are Two State Variables", *Journal of Financial and Quantitative Analysis* **23** (1988) 1–12.

P. P. Boyle, J. Evnine and S. Gibbs, "Numerical Evaluation of Multivariate Contingent Claims", *Review of Financial Studies* **2** (1989) 241–250.

D. S. Bunch and H. Johnson, "A Simple and Numerically Efficient Valuation Method for American Puts Using a Modified Geske–Johnson Approach", *Journal of Finance* **47** (1992) 809–816.

J. C. Cox and M. Rubinstein, *Options Markets*, Prentice-Hall, Englewood Cliffs, New Jersey, 1985.

J. C. Cox, S. A. Ross and M. Rubinstein, "Option Pricing: A Simplified Approach", *Journal of Financial Economics* **7** (1979) 229–263.

D. Duffie, *Security Markets*, Academic Press, San Diego, Calif., 1988.

R. Geske, "The Valuation of Compound Options", *Journal of Financial Economics* **7** (1979) 63–81.

R. Geske and H. Johnson, "The American Put Valued Analytically", *Journal of Finance* **39** (1984) 1511–1542.

H. He, "Convergence from Discrete- to Continuous-Time Contingent Claims Prices", *Review of Financial Studies* **3** (1990) 523–546.

D. B. Nelson and K. Ramaswamy, "Simple Binomial Processes as Diffusion Approximations in Financial Models", *Review of Financial Studies* **3** (1990) 393–430.

A. Rajasingham, "Convergence of Discrete Time Event Trees to Continuous Time Economies and the Pricing of Contingent Claims", Technical Report, International Monetary Fund, Washington, D.C., 1990.

R. C. Stapleton and M. G. Subrahmanyam, "The Valuation of Options on Portfolios", *Proceedings of the 15th Annual Conference of the European Finance Association*, Madrid, Spain, 1988.

R. M. Stulz, "Options on the Minimum or the Maximum of Two Risky Assets: Analysis and Applications", *Journal of Financial Economics* **10** (1982) 161–185.

Reprinted from Int. J. Theor. Appl. Finance
1(3) (1998) 349–376
© World Scientific Publishing Company

DERIVING CLOSED-FORM SOLUTIONS FOR GAUSSIAN PRICING MODELS: A SYSTEMATIC TIME-DOMAIN APPROACH

ALEXANDER LEVIN

The Dime Bancorp, Inc., Treasury Department,
589 5th Ave., New York, NY 10017, USA
E-mail: levina@dime.com

Received 6 April 1998

A systematic time-domain approach is presented to the derivation of closed-form solutions for interest-rate contingent assets. A financial system "asset – interest rate market" is assumed to follow an any-factor system of linear stochastic differential equations and some piece-wise defined algebraic equations for the payoffs. Closed-form solutions are expressed through the first two statistical moments of the state variables that are proven to satisfy a deterministic linear system of ordinary differential equations.

A number of examples are given to illustrate the method's effectiveness. With no restrictions on the number of factors, solutions are derived for randomly amortizing loans and deposits; any European-style swaptions, caps, and floors; conversion options; Asian-style options, etc. A two-factor arbitrage-free Gaussian term structure is introduced and analyzed.

Keywords: Linear stochastic systems, statistical moments, state variables, closed-form analytics, Gaussian term structure models, arbitrage-free conditions.

JEL classification codes: C490, E430, E490.

Finding closed-form solutions for stochastic pricing of interest-rate contingent, dynamic assets has been a target of numerous studies. Among many other advantages, these approaches usually enable constructing "analytical" term structures where the underlying stochastic process is consistent with the observed market and the entire yield curve can be defined at any instance of time, given the observed factors. When pricing complex path-dependent or multi-factor claims, such a model can successfully complement the Monte-Carlo scheme, which simulates by itself only pre-defined model's factors, not the entire term structure.[a] In addition, the use of exact solutions (when they exist) reduces processing time and establishes a perfect

---

[a]For example, the derivation of a long rate as a function of the short rate (the factor) would require knowledge of the complete set of future short rate paths and their probabilities that is simply unavailable to the Monte-Carlo or any other "forward-looking" methods.

pricing benchmark to control accuracy of and help testing complex numerical systems and algorithms.

Although the importance of the problem is well assessed by academicians and financial engineers, the industry has not been able to fully appreciate results beyond the scope of the term structure modeling itself. The existence of closed-form pricing solutions under a Gaussian term structure is well established, see Jamshidian [6,7], Beaglehole and Tenney [1], Hull and White [9,10], Duffie [4], Heath, Jarrow and Morton [8], Brenner and Jarrow [3], Kennedy [13], Brace and Musiela [2], Ritchken and Sankarasubramanian [15], and Franchot [5]. However, most published results are presented as illustrative examples typically involving discount bonds or simple options on them in a one- (rarely two-) factor environment.

Our method is different from most others in that it is a systematic time-domain approach structurally neutral to the model size. The financial system "security–interest rate market" is assumed to satisfy an any-order system of linear differential and some piece-wise defined algebraic equations. The closed-form solution requires an analytic or numeric integration of a system of deterministic linear differential equations, which are explicitly written for the expectations and covariances of all needed random variables. We will be able to identify quite extensive and complicated asset classes that can be effectively priced under the proposed scheme including randomly amortizing loans and deposits; any European-style swaptions, caps, and floors; conversion options; Asian-style options, etc.

The number of factors or equations does not restrict the proposed analytical framework. In fact, all the examples of closed-form solutions have been derived with no assumptions about the model size. We also demonstrate how an arbitrage-free two-factor Gaussian term structure can be constructed. This model enables a stochastic yield curve twist partially independent from the short rate process. The model's parameters are time-independent and sought fitting three points of the observed volatility term structure.

While our approach will speed up processing of many interest-rate contingent assets, it can also serve as a paradigm for model- or asset-specific derivations.

1. Finding Conditional Statistics via the Linear System Theory

1.1. *An example of a linear system: the Hull–White interest rate model*

A good example of a linear, analytically tractable system is the Hull–White model [9]:

$$dr(t) = a(t)[\theta(t) - r(t)]dt + \sigma(t)dz(t), \qquad (1.1)$$

$$dy(t) = -r(t)dt, \qquad (1.2)$$

where $r(t)$ denotes the spot short rate, $\theta(t)$ is the arbitrage-free "long-term" equilibrium, $y(t)$ is natural logarithm of the discount factor applied to \$1 to be received in time t, $z(t)$ is a standard Brownian motion, $a(t)$, $\sigma(t)$ are the mean reversion and volatility, respectively. Notice that the model presented here is in the form of linear continuous differential equations and has to be complemented by initial conditions for $r(0)$ and $y(0) = 0$.

Today's price $P_t(0)$ of a t-maturity discount bond is computed by the known econometric formula where E^{af} denotes mathematical expectation with respect to an arbitrage-free interest rate dynamics:

$$P_t(0) = E^{af}(e^{y(t)}) = \exp\left\{ E^{af}[y(t)] + \frac{1}{2}\text{Var}[y(t)] \right\}. \tag{1.3}$$

The last equality in (1.3) is valid because all the variables in systems (1.1) and (1.2) are normally distributed. We, therefore, have to know the mean and variance of $y(t)$ conditional upon initial conditions for models (1.1) and (1.2). In general, we may have to know averages and covariances for other variables that belong to the system's state space, at any point of time, t. The next section explains how to do it with the use of the linear system theory.

1.2. *The linear system theory: finding conditional statistics*

The method we propose employs closed deterministic forms for all needed first and second statistical moments and is essentially based on the following known mathematical statement that is valid for a general (time-dependent) linear system of stochastic differential equations:

$$dX(t) = [A(t)X(t) + C(t)]dt + B(t)dZ(t), \ X(0) = X_0, \tag{1.4}$$

where $X(t)$ is a vector of n unknowns, $Z(t)$ is a vector of m Brownian motions having a matrix of covariances Q_Z {that is, $Q_Z = \lim_{\delta \to 0} E[Z(t+\delta)Z^T(t+\delta) - Z(t)Z^T(t)]/\delta$}, $C(t)$ is a free term vector, $A(t)$ is a square matrix (n by n), $B(t)$ is a rectangular matrix (n by m).

Theorem (see I. Karatzas and S. Shreve [12], p. 355).[b] *At any instance of time t, vector $X(t)$ has the following conditional average $E[X(t)]$ and conditional autocovariance matrix $Q_X(t, \tau)$ (where $\tau \geq t$):*

$$E[X(t)] = \Phi(t, 0)X_0 + \int_0^t \Phi(t, \tau)C(\tau)d\tau, \tag{1.5}$$

$$Q_X(t, \tau) = E\{[X(t) - EX(t)][X(\tau) - EX(\tau)]^T\} = Q_X(t, t)\Phi^T(\tau, t), \tag{1.6}$$

where

$$Q_X(t, t) = \int_0^t \Phi(t, \tau)B(\tau)Q_Z B^T(\tau)\Phi^T(t, \tau)d\tau \tag{1.7}$$

---

[b]In I. Karatzas and S. Shreve [12], $Q_Z = I$, and Φ matrix depends on one time argument only, such that $\Phi(t)\Phi^{-1}(\tau)$ means the same as $\Phi(t, \tau)$ in our notations.

and $\Phi(t, \tau)$ is the fundamental solution of the homogeneous matrix differential equation

$$\frac{d\Phi(t, \tau)}{dt} = A(t)\Phi(t, \tau) \tag{1.8}$$

with the unity initial conditions, $\Phi(\tau, \tau) = I$.

Alternatively, vector of means $E[X(t)]$ can be computed as the solution to the homogeneous version of Eq. (1.4),

$$\frac{d}{dt}E[X(t)] = A(t)E[X(t)] + C(t), \quad E[X(0)] = X_0, \tag{1.9}$$

and matrix of covariances $Q_X(t, t)$ can be found as the solution to the following Lyapunov linear matrix differential equation:

$$\frac{dQ_X(t, t)}{dt} = A(t)Q_X(t, t) + Q_X(t, t)A^T(t) + B(t)Q_Z B^T(t) \tag{1.10}$$

with the zero initial conditions: $Q_X(0, 0) = 0$.

Matrix Φ is often written in the exponential form $\Phi(t, \tau) = \exp[\int_\tau^t A(\lambda)d\lambda]$ and can be built from the Jordan transformation of matrix $\int_\tau^t A(\lambda)d\lambda$ as shown in Appendix A. Thus, one can derive the major conditional statistics for process (1.4). Note that the conditional covariance matrix $Q_X(t, \tau)$ always depends on two time arguments, not on their difference, even if A, B, and C are time-invariant (only unconditional (long-term) statistics for system (1.4) can be stationary). We can also comment here that the differential Eqs. (1.9) and (1.10) are preferred for large-scale or time-dependent models since they do not involve matrix Φ and require less computational and algorithmic efforts than explicit formulae (1.5) and (1.7).

Let us apply the Theorem to the systems (1.1) and (1.2), and, for simplicity, assume that Eq. (1.1) is time-invariant ($a = $ const, $\sigma = $ const). Computing the integrals in (1.5) and (1.7), we get

$$E[r(t)] = e^{-at}\left[r(0) + a\int_0^t e^{a\tau}\theta(\tau)d\tau\right] \tag{1.11a}$$

$$E[y(t)] = -\frac{r(0)}{a}(1 - e^{-at}) - \int_0^t [1 - e^{a(\tau - t)}]\theta(\tau)d\tau \tag{1.11b}$$

$$\text{Cov}[r(t), r(\tau)] = \frac{\sigma^2}{2a}(1 - e^{-2at})e^{-a(\tau - t)} \tag{1.11c}$$

$$\text{Cov}[y(t), r(\tau)] = -\frac{\sigma^2}{2a^2}(1 - e^{-at})^2 e^{-a(\tau - t)} \tag{1.11d}$$

$$\text{Cov}[r(t), y(\tau)] = \frac{\sigma^2}{2a^2}(1 - e^{-2at})e^{-a(\tau - t)} + \frac{\sigma^2}{a^2}(e^{-at} - 1) \tag{1.11e}$$

$$\text{Cov}[y(t), y(\tau)] = \frac{\sigma^2}{a^2}\left[t - \frac{2}{a}(1 - e^{-at}) + \frac{1}{2a}(1 - e^{-2at})\right]$$

$$- \frac{1 - e^{-a(\tau-t)}}{a}\text{Cov}[r(t), y(t)]. \tag{1.11f}$$

Differential Eqs. (1.9) and (1.10) for the general, time-dependent case will take the following short forms:

$$\frac{dE[r(t)]}{dt} = -a(t)\{E[r(t)] - \theta(t)\}$$

$$\frac{dE[y(t)]}{dt} = -E[r(t)] \tag{1.12}$$

$$\frac{d\text{Var}[r(t)]}{dt} = -2a(t)\text{Var}[r(t)] + \sigma^2(t)$$

$$\frac{d\text{Var}[y(t)]}{dt} = -2\text{Cov}[r(t), y(t)] \tag{1.13}$$

$$\frac{d\text{Cov}[r(t), y(t)]}{dt} = -\text{Var}[r(t)] - a(t)\text{Cov}[r(t), y(t)]$$

1.3. *Conditional statistics for quasi-linear systems*

The pure linear system (1.4) is not the only one with the first and the second order statistical moments satisfying some ordinary linear differential equations. Let us generalize the model keeping its linear kernel, but allowing for random, generally factor-dependent volatilities:

$$dX(t) = [A(t)X(t) + C(t)]dt + B(t, X)dZ(t), \quad X(0) = X_0, \tag{1.14}$$

where $B(t, X)Q_Z B^T(t, X) = \alpha(t) + L(t, X) + Q(t, X)$, $\alpha(t)$ is a free random matrix term, independent of X, $L(t, X)$ is a linear matrix function of X, i.e. $L_{ij}(t, X) = X^T l_{ij}(t)$, $Q(t, X)$ is a quadratic matrix function of X, i.e. $Q_{ij}(t, X) = X^T q_{ij}(t)X$.

In the above definitions, $l_{ij}(t)$ are random vectors and $q_{ij}(t)$ are random n-by-n matrices, both independent of X, $i, j = 1, \ldots, n$. Certainly, Eq. (1.14) covers the constant volatility (Gaussian) case ($L = 0$ and $Q = 0$), CIR-like models ($\alpha = 0$ and $Q = 0$), the factor-proportional volatility case ($\alpha = 0$ and $L = 0$), as well as the pure stochastic volatility case. We assume that the model is constructed such that the unique strong solution exists. The following statement that generalizes the Theorem is proven in Appendix B (herein $\bar{\alpha}$ denotes expectation of α, etc.).

Proposition 1.1. *Matrix of covariances $Q_X(t, t)$ of system (1.14) is the solution to the following linear matrix differential equation:*

$$\frac{dQ_X(t, t)}{dt} = A(t)Q_X(t, t) + Q_X(t, t)A^T(t) + K(t), \tag{1.15}$$

where $Q_X(0,0) = 0$, and $K(t) \equiv K\{t, E[X(t)], Q_X(t,t)\}$ is linear in elements of Q_X:

$$K_{ij}(t) = \bar{\alpha}_{ij}(t) + E[X^T(t)]\bar{l}_{ij}(t) + E[X^T(t)]\bar{q}_{ij}(t)E[X(t)]$$

$$+ \sum_{k,m=1}^{n} [Q_X]_{km}[\bar{q}_{ij}(t)]_{km}, \quad i,j = 1,\ldots,n,$$

vector of means still satisfies Eq. (1.9) and relationship (1.6) sill holds true.

For example, consider a generalized quasi-linear single-factor process,

$$dr(t) = a[\theta(t) - r(t)]dt + \sqrt{\alpha + lr(t) + qr^2(t)}dz(t). \tag{1.16}$$

We assume that $a > 0$ and $\text{sign}[\theta(t) - r] = \text{sign}[l + 2qr]$ for any $t > 0$ and real r satisfying equation $\alpha + lr + qr^2 = 0$ (a boundary reflection criterion). The system (1.15) will have a form similar to (1.13) except the $\sigma^2(t)$ term in the equation for $\text{Var}[r(t)]$ is now replaced by $\alpha + lE[r(t)] + q[Er(t)]^2 + q\text{Var}[r(t)]$. In particular, we note that this equation is stable (providing a bound short-rate variance for bound short-rate mean) if and only if $2a > q$. Thus, the first two statistical moments can be analytically constructed for some quasi-linear models although the probability distributions of the system's state variables are not going to be normal.

2. Constructing Arbitrage-Free Gaussian Term Structures

2.1. *Local arbitrage-free conditions for Gaussian models*

Although Gaussian term structures are not globally arbitrage-free (Jamshidian [7]) by the definition (the interest rates are not bound), the absence of arbitrage in a local sense often satisfies their users. In a local no-arbitrage term structure, every default-free bond is expected to return the short rate $r(t)$. Applying this constraint to a t-maturity discount bond we can equate its price $P_t(0)$ to the cumulative discount factor known from today's forward curve:

$$P_t(0) = \exp\left\{E[y(t)] + \frac{1}{2}\text{Var}[y(t)]\right\} = \exp\left[-\int_0^t f(\tau)d\tau\right], \tag{2.1}$$

where $f(t)$ is the short rate forward in t years, and hereinafter index "af" is omitted for brevity. Taking natural logarithm from the both sides of equality (2.1), we have

$$E[y(t)] + \frac{1}{2}\text{Var}[y(t)] = -\int_0^t f(\lambda)d\lambda. \tag{2.2}$$

Taking the first time derivative from the both sides of (2.2) and changing the order of the E (or Var) and the d/dt operators one gets an equivalent condition:

$$E[r(t)] = f(t) - \text{Cov}[r(t), y(t)]. \tag{2.2'}$$

It is seen from (2.2') that there exists a unique relationship between the short rate's drift (the expectation term) and the model's volatility (the covariance term). Heath, Jarrow and Morton [8] have derived this relationship, for generic, no-arbitrage term structure models.

At any future instance of time t, the entire yield curve is implied by Eqs. (1.1) and (1.2) and the factors. Indeed, price $P_T(t)$ for a T-maturity discount bond will equal

$$P_T(t) = \exp[-r_T(t)T] = E|_t \exp\left[-\int_t^{t+T} r(\tau)d\tau\right] = E|_t \exp[y(t+T)-y(t)], \quad (2.3)$$

where the $E|_t$ sign denotes mathematical expectation conditional upon information at time t. To simplify the derivations, we can reset $y(t)$ to zero and start the short rate process at time t yet following Eqs. (1.1) and (1.2). This transformation excludes $y(t)$ from (2.3) leading to

$$-r_T(t)T = E|_{t,y(t)=0}\, y(t+T) + \frac{1}{2}\mathrm{Var}|_{t,y(t)=0}\, y(t+T). \quad (2.4)$$

Thus, conditions (2.2) [or (2.2')] and (2.4) exclude arbitrage opportunities, in the local sense, for an any-factor Gaussian term structure model.

2.2. *Arbitrage-free conditions for the Hull–White model*

Substituting the covariance term (1.11e) written for $t = \tau$ into condition (2.2') yields

$$E[r(t)] = f(t) + \frac{\sigma^2}{2a^2}(1 - e^{-at})^2. \quad (2.5)$$

Equating the right-hand sides of formulae (11.1a) and (2.5) and resolving thus obtained integral equation for the drift function $\theta(t)$, we immediately derive the following condition:

$$\theta(t) = f(t) + \frac{1}{a}\frac{df}{dt} + \frac{\sigma^2}{2a^2}(1 - e^{-2at}). \quad (2.6)$$

To establish the inter-rate relationship, we note that the second term in the right-hand side of (2.4) is simply given by formula (1.11f) with $t = \tau = T$, whereas the first term is computed by a modification of formula (1.11b) using t as the starting point and $t+T$ as the ending one:

$$E|_{t,y(t)=0}\, y(t+T) = -\frac{r(t)}{a}(1 - e^{-aT}) - \int_t^{t+T} [1 - e^{a(\tau-t-T)}]\theta(\tau)d\tau.$$

Substituting the required conditional statistics and expression (2.6) for $\theta(t)$ in formula (2.4) we eventually derive the Hull–White formula presented here in slightly different notations:

$$r_T(t) = B(T)r(t) - A(t,T), \quad (2.7)$$

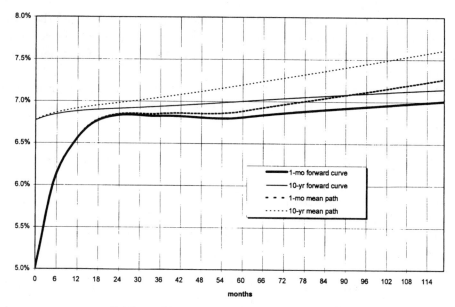

Exhibit 1. Mean-rate curves versus forward curves.

where $A(t,T) = f(t)B(T) - f_T(t) - \frac{\sigma^2}{4a^3T}(1 - e^{-aT})^2(1 - e^{-2at})$, $B(T) = (1 - e^{-aT})/aT$, and $f_T(t)$ is the forward rate for a T-maturity discount bond, at instance t. Taking expectations from the both sides of (2.7) and using (2.5), we derive the mean path for the T-maturity rate:

$$E[r_T(t)] = f_T(t) + \frac{\sigma^2}{4a^3T}(1 - e^{-aT})[2(1 - e^{-at})^2 + (1 - e^{-2at})(1 - e^{-aT})]. \quad (2.8)$$

Note the difference between the mean rates determined by formulae (2.5) and (2.8) and the corresponding forward rates (Exhibit 1). This difference is always positive and caused by a non-linear, positively convex nature of the discounting operation. Often called "convexity adjustment", this term can be approximated by $\frac{1}{2}TB_T^2\sigma^2t$ for small t; as $t \to \infty$ it approaches $B_T^2\sigma^2T/4a$.

Volatility term structure is entirely defined by the $B(T)$ function. Since all rates are perfectly correlated, we list below the trivial second-order moments to complete this case study:

$$\text{Var}(r_T) = B^2(T)\text{Var}(r), \quad \text{Cov}(r_Ty) = B(T)\text{Cov}(ry), \quad \text{Cov}(rr_T) = B(T)\text{Var}(r). \quad (2.9)$$

2.3. Introduction of a second factor into the interest rate model

One-factor stationary models have at least two limitations when practically applied. First, the yield curve cannot twist and assets sensitive to its slope cannot be priced accurately. Second, volatility term structure implied by these models is such that the long rate volatility cannot exceed the short rate volatility. We consider here a straightforward, easy-to-implement two-factor generalization of the Hull–White model that creates enough variability of the term structure including its potential twist and a fit of the observed volatility term structure without explicit time dependency of the model parameters. Let us introduce two pairs of state variables (x_1, y_1) and (x_2, y_2) that satisfy equations similar to (1.1) and (1.2):

$$dx_i(t) = a_i[\theta_i(t) - x_i(t)]dt + \sigma_i dz_i(t), \qquad i = 1, 2, \tag{2.10}$$

$$dy_i(t) = -x_i(t)dt, \qquad i = 1, 2, \tag{2.11}$$

where $\theta_2(t)$ can be set to zero, two standard Brownian motion increments, $dz_1(t)$ and $dz_2(t)$ have a correlation of ρ.[c]

We define the short rate $r = x_1 + x_2$, therefore, $y = y_1 + y_2$. Matrix A of system (2.10), (2.11) and matrix $\Phi(t, \tau)$ of its fundamental solutions are block-diagonal matrices such that each of the two diagonal blocks has a 2-by-2 dimension and corresponds to either (x_1, y_1) or (x_2, y_2), see the bottom of Appendix A. Financial meanings of variables x_1 and x_2 are not relevant for finding neither the arbitrage-free conditions nor the implied term structure analytics. Indeed, (2.10) is simply the canonical form for any time-invariant two-factor linear model with all the eigenvalues of matrix A being real.

Since Eqs. (2.10) and (2.11) are identical to (1.1) and (1.2) with the accuracy of notations, formulae (1.11) will be certainly valid if (r, y) is formally replaced by (x_i, y_i), and (a_i, σ_i) is used instead of (a, σ). The mixed covariances not listed in formulae (1.11) are given below for $t = \tau$ [if $t \neq \tau$, use formulae (2.12) below, statement (1.6), and the obtained matrix $\Phi(t, \tau)$]:

$$\text{Cov}[x_1(t), x_2(t)] = \rho \frac{\sigma_1 \sigma_2}{a_1 + a_2}[1 - e^{-(a_1 + a_2)t}] \tag{2.12a}$$

$$\text{Cov}[x_i(t), y_j(t)] = \rho \frac{\sigma_1 \sigma_2}{a_j}\left[\frac{1 - e^{-(a_1 + a_2)t}}{a_1 + a_2} - \frac{1 - e^{-a_i t}}{a_i}\right], \quad i, j = 1, 2 \tag{2.12b}$$

$$\text{Cov}[y_1(t), y_2(t)] = \rho \frac{\sigma_1 \sigma_2}{a_1 a_2}\left[t - \frac{1 - e^{-a_1 t}}{a_1} - \frac{1 - e^{-a_2 t}}{a_2} + \frac{1 - e^{-(a_1 + a_2)t}}{a_1 + a_2}\right] \tag{2.12c}$$

As in the Hull–White case, arbitrage-free condition (2.2') represents an integral equation for the unknown $\theta(t)$ function. Resolving it in a similar fashion we get

$$\theta(t) = \left[1 + \frac{1}{a_1}\frac{d}{dt}\right]\{E[r(t)] + L(t)\}, \tag{2.13}$$

where $L(t) = -x_1(0)e^{-a_1 t}/a_1 - x_2(0)e^{-a_2 t}/a_2$, and

$$E[r(t)] = f(t) + \frac{1}{2}\sum_{i=1}^{2}\frac{\sigma_i^2}{a_i^2}(1 - e^{-a_i t})^2 + \rho\frac{\sigma_1 \sigma_2}{a_1 a_2}(1 - e^{-a_1 t})(1 - e^{-a_2 t}) \tag{2.14}$$

---

[c]Hull and White [9] and Duffie [4] have considered similar models assuming $\rho = 0$. For this special case, the long-rate volatility cannot exceed the short-rate volatility. The model we consider is mathematically identical to that of Hull and White [10] and written in the canonical form.

(note again the convexity adjustment term). As the arbitrage-free condition (2.2')
for the short rate has a universal nature, the same is true for condition (2.4) keeping
in mind that the statistics are now conditional upon observations of all four state
variables of the model, at time t. The required expressions for the conditional
expectation and variance can be easily written from formulae (1.11) and (2.12) with
the following formal substitutions reflecting the initial condition $y(t) = 0$: $t = T$
for the variance; time instances t and $t + T$ should be taken as the starting and the
ending points, correspondingly, when using formula (1.11b) for the expectation. The
derivation of the following statement is purely technical and omitted for brevity:

Proposition 2.1. *The T-maturity rate of the two-factor term structure* (2.10)
and (2.11) *is linear in factors,*

$$r_T(t) = B_1(T)x_1(t) + B_2(T)x_2(t) - A(t,T),\qquad(2.15)$$

and has the following expectation:

$$E[r_T(t)] = f_T(t) + \frac{1}{2}\sum_{i=1}^{2}\frac{\sigma_i^2}{a_i^2}[2B_i(T)(1 - e^{-a_i t}) - B_i(2T)(1 - e^{-2a_i t})]$$

$$+\rho\frac{\sigma_1\sigma_2}{a_1 a_2}\{B_1(T)(1 - e^{-a_1 t}) + B_2(T)(1 - e^{-a_2 t})$$

$$-B_\Sigma(T)[1 - e^{-(a_1+a_2)t}]\},\qquad(2.16)$$

where $B_i(T) = (1 - e^{-a_i T})/a_i T,\ i = 1,2;\ B_\Sigma(T) = [1 - e^{-(a_1+a_2)T}]/(a_1 + a_2)T$

$$A(t,T) = B_1(T)f(t) - f_T(t) - x_2(0)e^{-a_2 t}[B_1(T) - B_2(T)]$$

$$+\sum_{i=1}^{2}\frac{\sigma_i^2}{2a_i^2}[B_1(T)(1 - e^{-a_i t})^2 - 2B_i(T)(1 - e^{-a_i t}) + B_i(2T)(1 - e^{-2a_i t})]$$

$$+\rho\frac{\sigma_1\sigma_2}{a_1 a_2}\{-B_1(T)(1 - e^{-a_1 t})e^{-a_2 t} - B_2(T)(1 - e^{-a_2 t})$$

$$+B_\Sigma(T)[1 - e^{-(a_1+a_2)t}]\}.\qquad(2.17)$$

The first two terms in the right-hand side of formula (2.15) define volatility term
structure:

$$\text{Var}[r_T(t)] = B_1^2(T)\text{Var}[x_1(t)] + B_2^2(T)\text{Var}[x_2(t)]$$

$$+2B_1(T)B_2(T)\text{Cov}[x_1(t), x_2(T)],\qquad(2.18)$$

$$\text{Cov}[r_T(t), r(t)] = B_1(T)\text{Var}[x_1(t)] + B_2(T)\text{Var}[x_2(t)]$$

$$+[B_1(T) + B_2(T)]\text{Cov}[x_1(t), x_2(t)].\qquad(2.19)$$

An important practical question arising here is how to choose the five model's parameters, a's, σ's and ρ in order to fit the observed volatility term structure. Dividing formulae (2.18) and (2.19) by t and considering the limits as $t \to 0$ we can find today's instantaneous volatilities and correlations between the rates. Equating them to the market observations yields to the following algebraic constraints on the model's coefficients:

$$\sigma^2 = \sigma_1^2 + \sigma_2^2 + 2\rho\sigma_1\sigma_2$$

$$\sigma_T^2 = B_1^2(T)\sigma_1^2 + B_2^2(T)\sigma_2^2 + 2B_1(T)B_2(T)\rho\sigma_1\sigma_2 \qquad (2.20)$$

$$\text{Cov}[r, r_T] = B_1(T)\sigma_1^2 + B_2(T)\sigma_2^2 + [B_1(T) + B_2(T)]\rho\sigma_1\sigma_2$$

Appendix C presents a set of explicit formulae that derive the model's parameters given volatility and correlation for additional two points (besides the short one) of the yield curve. For $\rho < 0$, long rates can be more volatile than the short rate (including a "humped" shape) and may not perfectly correlate with it. Clearly, the Hull–White model cannot replicate such structures. On the other hand, some volatility term structures cannot be fitted.

Since the factors x_1 and x_2 have no financial meaning, we propose to transform the abstract models (2.10) and (2.11) into an equivalent form introducing a new r-independent "slope" variable, $v = x_1 + \beta x_2$ with $\beta = -\sigma_1(\sigma_1 + \rho\sigma_2)/\sigma_2(\rho\sigma_1 + \sigma_2) \neq 1$ (assuming $\rho\sigma_1 + \sigma_2 \neq 0$):

$$dr(t) = \left[a_1\theta(t) + \frac{a_1\beta - a_2}{1 - \beta}r(t) - \frac{a_1 - a_2}{1 - \beta}v(t)\right] dt + \sigma dz_r,$$

$$dv(t) = \left[a_1\theta(t) + \frac{a_1 - a_2}{1 - \beta}\beta r(t) + \frac{a_2\beta - a_1}{1 - \beta}v(t)\right] dt + \sigma_v dz_v, \qquad (2.21)$$

where the short rate volatility, σ, is defined in formula (2.20) above, $\sigma_v^2 = \sigma_1^2 + \beta^2\sigma_2^2 + 2\rho\beta\sigma_1\sigma_2$, and two Brownian motion increments, dz_r and dz_v are now uncorrelated, as are the factors r and v, at time $t = 0$ (i.e. $\text{Corr}[r(t), v(t)] \to 0$ as $t \to 0$).

Factor $v(t)$ determines a term structure move net of that propagated by the short rate:

$$r_T(t) = \frac{B_2(T) - \beta B_1(T)}{1 - \beta}r(t) + \frac{B_1(T) - B_2(T)}{1 - \beta}v(t) - A(t, T). \qquad (2.22)$$

The mean path for $v(t)$ is defined as $E[v(t)] = E[r(t)] + (v_0 - r_0)e^{-a_2t}$, where the initial value r_0 is observed, and the choice of v_0 is not relevant for the model (one can even set $v_0 = 0$).[d] This seemingly paradoxical statement holds true because both the mean paths (for any rate) and the volatility term structure functions $B_1(T)$, $B_2(T)$ are, in fact, v_0-independent.

---

[d]The only "undesired" choice would be $v_0 = r_0$, therefore $x_2(0) = 0$, which degenerates the second factor of the model.

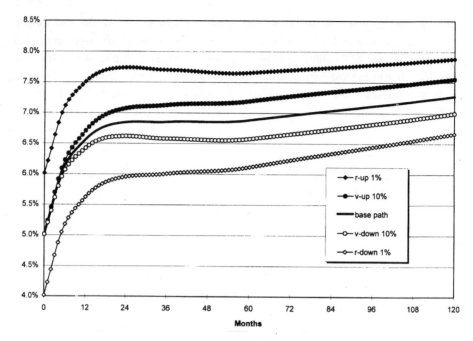

Exhibit 2. Short rate mean paths in the 2-factor model.

Let us assume, for example, that $\sigma = 0.7\%$ and volatility term structure is specified by two points of the curve, an intermediate point ($T = 3$ years, $\sigma_3 = 0.8\%$, $\rho_3 = 80\%$), and a long point ($T = 10$ years, $\sigma_{10} = 0.9\%$, $\rho_{10} = 60\%$). Writing the system of Eqs. (2.20) for the chosen maturity points and resolving for the model's parameters as explained in Appendix C we get: $a_1 = 0.046(yr^{-1})$, $a_2 = 0.457(yr^{-1})$, $\sigma_1 = 1.40\%$, $\sigma_2 = 1.23\%$, $\rho = -0.866$. Then, the implied parameters for the "r-v"-model are $\sigma_v = 24.7\%$, $\beta = -19.06$. Basic r- and v-scenarios are plotted on the Exhibit 2; all others are their linear combinations.

2.4. *Modeling coupon rates*

F. Jamshidian [7] has proven that, under a one-factor Gaussian model, the option price on a coupon-bearing bond equals the sum of the option prices on the constituent zeros. This result essentially uses proportionality in the random deviations of different rates and, therefore, is not valid beyond one-factor model. In order to price options on coupon-bearing bonds or swaps, one has to link the coupon rates to the term structure model's factors.

Even if the spot rates are linear in factors, the coupon rates will be not. A simple explanation for this non-linear relationship is that, to compute a coupon rate, one has to weigh the constituent spot rates by present values, which are themselves exponential in rates. In this section, we introduce a possible linearization approach that would enable us to derive closed-form solutions when coupon-bearing bonds, swaps or options on them are involved. We denote $P_T^{Ann}(t) = \int_t^{t+T} e^{-r_\tau(t)\tau}d\tau$ the value at time t of an annuity continuously paying \$1 per annum during the T year period that starts at t. Then, the rate on the continuously coupon paying T-maturity bonds at time t, $r_T^*(t)$ is defined from the following simple relation:

$$r_T^*(t)P_T^{Ann}(t) + P_T(t) = 1, \qquad (2.23)$$

where, as above, $P_T(t) = e^{-r_T(t)T}$ is the value of the T-maturity discount bond, at time t. Let us assume that a linear term structure is disturbed by n centered factors, $x_i(t)$, $i = 1, \ldots, n : r_\tau(t) = \bar{r}_\tau(t) + \sum_{i=1}^n B_{i\tau}x_i(t)$, where $\bar{r}_\tau$ denotes $E(r_\tau)$. Representing the τ-maturity coupon rate in a similar form, $r_\tau^*(t) = \bar{r}_\tau^*(t) + \sum_{i=1}^n B_{i\tau}^*x_i(t)$, substituting in (2.23), equating the free terms and the first partial derivatives with respect to the factors we get:

$$\bar{r}_T^*(t) = \frac{1 - \exp[-T\bar{r}_T(t)]}{\int_t^T \exp[-\tau\bar{r}_\tau(t)]d\tau};$$

$$\qquad (2.24)$$

$$B_{iT}^*(t) = \frac{TB_{iT}\exp[-T\bar{r}_T(t)] + \bar{r}_T^*(t)\int_t^T \tau B_{i\tau}\exp[-\tau\bar{r}_\tau(t)]d\tau}{\int_t^T \exp[-\tau\bar{r}_\tau(t)]d\tau}.$$

Volatility term structure function $B_{iT}^*(t)$ of coupon rates appeared to be time-variant because it is sensitive to the expected changes in the yield curve's shape. This finding does not present any problem in deriving closed-form solutions, as our approach is free of any constraints on time dependency. Numerical experiments support the proposed linear approximation. Thus, for the 2-factor model introduced above, with a 90%-confidence, the 10-year and the 30-year coupon rates can be reconstructed with the accuracy of 1–2 basis points and 3–4 basis points, correspondingly.

3. A Price Premium Formula for Dynamic Assets

We have considered "analytical" arbitrage-free interest rate models that, by the definition, provide correct pricing for simple Treasury bonds. This approach has created a framework for pricing more complicated securities having interest-rate-contingent payoffs. For an instrument paying interest at a $c(t)$ rate, having a $B(t)$ balance, which retires at a $\lambda(t)$ speed, price $P(t)$ certainly satisfies the following equation with an arbitrary time horizon T:

$$P(0) = E\int_0^T B(t)[c(t) + \lambda(t)]e^{y(t)}dt + E[P(T)e^{y(T)}], \qquad (3.1)$$

where

$$\frac{dB(t)}{dt} = -\lambda(t)B(t), \quad \text{therefore,} \quad B(t) = B(0)\exp\left(-\int_0^t \lambda(\tau)d\tau\right). \qquad (3.2)$$

System (3.1), (3.2) involves an additional state variable, balance B. To eliminate one, the principal cashflow component is integrated by parts with the use of the terminal balance, $B(T)$:

$$\int_0^T B(t)\lambda(t)e^{y(t)}\,dt = B(0)\left\{1 - \int_0^T r(t)e^{y'(t)}\,dt\right\} - B(T)e^{y(T)}, \qquad (3.3)$$

where y' is the natural logarithm of an artificial discount factor,

$$dy' = -[r(t) + \lambda(t)]dt, \quad y'(0) = 0. \qquad (3.4)$$

Substituting (3.2) and (3.3) in (3.1), we derive the following compact formula for the relative premium (or discount) $p = (P - B)/B$:

$$p(0) = \int_0^T E\{[c(t) - r(t)]e^{y'(t)}\,dt\} + E[p(T)e^{y'(T)}]. \qquad (3.5)$$

The obtained result states that the premium (discount) consists of two components. The first sums the differentials between the paid rate and the short discount rate. These differentials are discounted at an artificial rate, which is equal to the sum of the actual short rate and the rate of balance amortization. The second component is the expected terminal premium value also discounted at the above-mentioned artificial rate.

An equivalent interpretation of formulae (3.4) and (3.5) is as follows. If we transform the economy having shifted the discount rates by a time-varying, generally random rate of amortization $\lambda(t)$, formula (3.5) will be reduced to a constant-par asset's pricing formula. To illustrate the advantage of this transformation, consider a simple one-factor tree built for $r(t)$ in the arbitrage-free economy. The balance $B(t)$ is not known on the nodes of this tree and its straight use to price amortizing assets is not feasible. However, let us assume that λ is a function of r and t. For the "λ-shifted" tree, the standard pricing scheme will work perfectly for premiums rather than for prices.

It is easy to show that, if a part of the interest is not paid, but rather credited to the balance, it has to be subtracted from $\lambda(t)$ remaining included in $c(t)$. On the other hand, formula (3.5) has to be modified for "interest-only" assets. The model may still account for balance changes but ignores the principal cashflow:

$$p(0) = \int_0^T E[c(t)e^{y'(t)}]dt + E[p(T)e^{y'(T)}], \qquad (3.6)$$

where p now denotes P/B. For certainty, we will assume that a security pays principal and is priced in accordance to formula (3.5), unless version (3.6) is explicitly mentioned.

A choice of the time horizon T is certainly determined by the nature of the asset. For maturing assets, T can be a finite maturity length thus eliminating the terminal value; for European-style options, T is the expiration; for non-maturing deposits or consumer's loans, one can set either a large T or work with $T = \infty$. A strict mathematical assumption to be made for the latter case is the convergence on probability of $Ee^{y'(T)}$ as $T \to \infty$.

Another interesting comment we could make here is that (3.5) and (3.6) are the Feynman–Kac solutions (see Karatsas and Shreve [12]) to the following partial differential equations:

for principal-paying instruments,

$$[r(X,t) + \lambda(X,t)]p(X,t) = c(X,t) - r(X,t) + Dp(X,t) \qquad (3.7)$$

for interest-only instruments,

$$[r(X,t) + \lambda(X,t)][p(X,t) + 1] = c(X,t) + Dp(X,t), \qquad (3.7')$$

where X is a vector of factors and state variables having drift μ and autocovariance matrix V, and D is the well-known "expected instantaneous price return" operator: $Dp = p_t + p_x\mu + \frac{1}{2}\text{tr}(Vp_{xx})$. Thus, for the general linear model (1.4), $\mu = AX + C$, $V = BQ_ZB^T$. Certainly, PDEs (3.7) or (3.7') have to be complemented by appropriate boundary conditions for $p(X,t)$.

4. Derivation of Closed-Form Solutions

4.1. *A key transition: factoring out the discount factor*

Without any limitation on the factor number or the model size, we consider arbitrary normally distributed m-dimensional vector $\boldsymbol{x}$, a scalar y', and any measurable function $g(\boldsymbol{x}) : R^m \to R^1$. A proof of the following result is given in Appendix D:

Proposition 4.1.

$$E[g(\boldsymbol{x})e^{y'}] = E[g(\boldsymbol{u})]E(e^{y'}), \qquad (4.1)$$

where $\boldsymbol{u} = \boldsymbol{x} + \text{Cov}(\boldsymbol{x}, y')$.

For a generic amortizing asset, $E[e^{y'(t)}]$ is the price $P'_t(0)$ of the t-maturity discount bond, in the λ-shifted economy, i.e. $P'_t(0) = \exp\{E[y'(t)] + \frac{1}{2}\text{Var}[y'(t)]\}$, and $g(\boldsymbol{x})$ is the interest (or interest spread) paid, option payoff, or the terminal value. That is why, most of the closed-form solutions derived further are based on Proposition 4.1 with all the needed statistics being computed as shown in the first part. Hereinafter, we use bold letters to denote vectors or matrices when $m > 1$.

4.2. *Special (Jamshidian's) case: $m = 1$, $x = r(t)$ and $y' = y(t)$*

Under the assumption that the short rate process $r(t)$ follows an arbitrage-free one-factor Gaussian dynamics, Jamshidian [6] proves that $E[u(t)] \equiv \mu_u(t)$ has to be the short forward rate, $f(t)$. This statement, however, holds true regardless of the number of factors and immediately follows from the universal arbitrage-free condition (2.2'): $f(t) = E[r(t)] + \text{Cov}[r(t), y(t)]$, an exact expression for μ_u as implied by Proposition 4.1. It is also easy to see that the other multiplier, $E[e^{y(t)}]$, is the observed price $P_t(0)$ of a t-maturity discount bond.

4.3. *Special case: a linear payoff*

Let us assume that $g(x) = a + b^T x$, then (4.1) gives us

$$E[g(x)e^{y'}] = E(e^{y'})(a + b^T \mu_u). \qquad (4.2)$$

Note that, despite of the perfect linear symmetry, the volatility assumption will generally affect the result (unless $m = 1$ and $x = r$).

4.4. *Special case: a piece-wise defined payoff*

Most options have piece-wise defined exercise fashion. Let us assume that $m = 1$, $g(x) = g_i(x)$, for $x_i \leq x < x_{i+1}$, $i = 0, 1, \ldots, N - 1$. Then,

$$E[g(x)e^{y'}] = E(e^{y'}) \sum_{i=0}^{N-1} \int_{x_i}^{x_{i+1}} g_i(x)p_{[\mu_u, \sigma_u]}(x)dx, \qquad (4.3)$$

where $p_{[\mu_u, \sigma_u]}(\cdot)$ denotes the density function, for the normal distribution of u with parameters μ_u, σ_u. A list of functions $g_i(x)$ for which the integral in (4.3) can be computed in closed-form is quite extensive. We restrict illustrations with the following two most important cases.

Linear function. If $g_i(x) = a_i + b_i x$, then

$$\int_{x_i}^{x_{i+1}} g_i(x)p_{[\mu_u, \sigma_u]}(x)dx = (a_i + b_i\mu_u)[\Phi(d_{i+1}) - \Phi(d_i)] - b_i\sigma_u[\varphi(d_{i+1}) - \varphi(d_i)],$$
$$(4.4)$$

where $d_i = (x_i - \mu_u)/\sigma_u$, and $\Phi(\cdot)$ and $\varphi(\cdot)$ refer, respectively, to the cumulative and density functions, for the standard normal distribution.

Exponential function. If $g_i(x) = c_i e^{\gamma_i x}$, then

$$\int_{x_i}^{x_{i+1}} g_i(x)p_{[\mu_u, \sigma_u]}(x)dx = c_i[\Phi(h_{i+1}) - \Phi(h_i)] \exp\left(\mu_u\gamma_i + \frac{1}{2}\sigma_u^2\gamma_i^2\right), \qquad (4.5)$$

where $h_i = (x_i - \mu_u)/\sigma_u - \gamma_i\sigma_u$. It is seen that the last multiplier in the right-hand side of (4.5) is simply $E(e^{\gamma_i u})$.

Since the price is linear in payoffs, it will be computed by a linear combination of the above formulae if the payoffs are such.

4.5. *Summary of the method*

To price a dynamic security, we propose to describe it and the interest rate market by a system of linear stochastic and some generally piece-wise defined algebraic equations for the payoffs complemented by Eq. (3.4) for the natural logarithm of the artificial discount factor. Having constructed matrices A and B of the model we write down the Lyapunov matrix differential Eq. (1.10) for covariances along with vector differential Eq. (1.9) for averages. We integrate them either analytically (for small and time-independent models) or numerically concurrently computing the premium value (3.5) or (3.6) with the help of expectation hints (4.1)–(4.5). Since all the formulae are deterministic, we need only one computational run over a pre-defined pricing horizon.

5. Examples

In this section, we describe some of interest-rate contingent assets and demonstrate how to derive closed-form solutions. No assumptions will be made about the number of factors, but $m = 1$ holds in every case.

5.1. *European options on discount bond*

Let us consider a European-style call on a T-maturity discount bond with expiration at t (i.e. the bond itself matures at $t + T$) and strike K. The bond's price at the expiry is $\exp[-r_T(t)T]$, and the contract will have no payoffs until the expiration when it costs $\{\exp[-r_T(t)T] - K\}^+$, where, as usually in the mathematics of derivatives, the "+" superscript means that the negative values are ignored. Thus, the payoff is given in a piece-wise linear-exponential form and we can immediately combine formulae (4.2)–(4.5) with $x = r_T(t)$ and $u = r_T(t) + \text{Cov}[r_T(t), y(t)]$. The model's non-zero parameters are $N = 2$, $x_0 = -\infty$, $x_1 = -\frac{1}{T}\ln K$, $x_2 = \infty$, $a_0 = -K$, $c_0 = 1$, and $\gamma_0 = -T$. Therefore, $d_0 = h_0 = -\infty$, $d_2 = h_2 = \infty$, $d_1 = -(\frac{1}{T}\ln K + \mu_u)/\sigma_u$, $h_1 = d_1 + T\sigma_u$. It yields to $p(0) = P_t(0)[\exp(-T\mu_u + \frac{1}{2}T^2\sigma_u^2)\Phi(h_1) - K\Phi(d_1)]$. Reading again Proposition 4.1 "from right to left" we recognize $P_t(0)\exp(-T\mu_u + \frac{1}{2}T^2\sigma_u^2) = P_t(0)E(e^{-uT})$ as $E[e^{y(t)}e^{-r_T(t)T}]$, i.e. today's price $P_{t+T}(0)$ for a discount bond maturing at $t + T$. Therefore,

$$p(0) = P_{t+T}(0)\Phi(h_1) - KP_t(0)\Phi(d_1). \tag{5.1c}$$

Similarly, for the European put,

$$p(0) = -P_{t+T}(0)[1 - \Phi(h_1)] + KP_t(0)[1 - \Phi(d_1)]. \tag{5.1p}$$

Thus, we have obtained familiar option pricing structures that are valid for an any-factor model. For options on coupon-bearing bonds, refer to the next section.

5.2. Swaps, swaptions, caps, floors

Let us consider a T-maturity continuous swap paying the difference between the short rate, r and a strike rate, r^s and employ the underlying pricing formula (3.6) in which $c = r - r^s$. Without loss of generality, we assume that $\lambda = 0$ (non-amortizing swap) keeping in mind that the λ-shift of discount rates leads to this case. It is important to mention that there is no theoretical difference between pricing swaps and fixed-rate coupon-bearing bonds. Indeed, a swap can be thought of as an exchange of two bonds, a fixed-rate bond and a perfect floater.[e] Since the latter costs a par, the swap's price will equal the premium (discount) on the fixed-rate bond. We also assume that the swap (coupon) rates are linearized in factors as recommended in Sec. 2.4.

Swaps. For the "at-the-money" (ATM) swap, the strike rate r^s is chosen to produce a zero price. It is therefore equal to the today's rate $r_T^*(0, t)$ on a T-maturity, continuously paying bond to be delivered in t years, at par. This rate can be found from the following equation:

$$P_t(0) = P_{t+T}(0) + r_T^*(0, t) P_T^{Ann}(0, t), \qquad (5.2)$$

where $P_T^{Ann}(0, t) = E \int_t^{t+T} e^{y(\tau)} d\tau = \int_t^{t+T} e^{-r_\tau(0)\tau} d\tau$ is today's (known) value of a forward annuity continuously paying \$1 per annum during the T year period that starts in t years. Let us comment here that, in practice, one does not need to solve (5.2) for r_T^* if $t = 0$ as the coupon curve is usually known, rather the spot rate and the forward rate curves have to be constructed. For $t > 0$, not all the coupon rates can be observable, and they are reconstructed solving (5.2).

Let us consider now either out-of-the-money (OTM), swap, $r^s > r_T^*$, or in-the-money (ITM) swap, $r^s < r_T^*$. Combining (3.6) and (5.2) we derive a purely deterministic formula:

$$p(0) = \int_t^{t+T} E\{[r(\tau) - r^s]e^{y(\tau)}\} dt = E \int_t^{t+T} [r(\tau)e^{y(\tau)}] dt - r^s \int_t^{t+T} E e^{y(\tau)} dt$$

$$= \dots \text{first integral is taken by parts} \dots$$

$$= P_t(0) - P_{t+T}(0) - r^s P_T^{Ann}(0, t) = [r_T^*(0, t) - r^s] P_T^{Ann}(0, t) \qquad (5.3)$$

Future price of swaps. A choice of the interest rate model (including volatility assumptions) is irrelevant for the pricing swaps. However, the swap's future price

---

[e]The frequency of this adjustment ("coupon reset") most often matches maturity of the underlying rate. Thus, a semi-annually resetting floater will be typically indexed to the six-month rate, etc. Since the short rate r specifies the floating leg in our model, we have to assume that the swap resets continuously. This assumption does not restrict our approach, rather it sets the floating leg's value to par.

$p(t)$ is certainly going to be random. We rewrite formula (5.3) for some future point of time t keeping notations for $P_T^{Ann}(t)$ and $r_T^*(t)$ of part II:

$$p(t) = [r_T^*(t) - r^s]P_T^{Ann}(t)$$

$$= [r_T^*(t) - r^s]E|_t \int_t^{t+T} e^{y(\tau)-y(t)}d\tau$$

$$= [r_T^*(t) - r^s]e^{-y(t)}E|_t \int_t^{t+T} e^{y(\tau)}d\tau,$$

where $E|_t$ is the expectation conditional upon the information at time t (hence, we factor $e^{-y(t)}$ out of this operator). Therefore, the today's value of the same forward swap can be presented as

$$p(0) = E[e^{y(t)}p(t)]$$

$$= E\left\{[r_T^*(t) - r^s]E|_t \int_t^{t+T} e^{y(\tau)}d\tau\right\}$$

$$= E\int_t^{t+T} [r_T^*(t) - r^s]e^{y(\tau)}d\tau. \qquad (5.4)$$

The last formal transition holds true because (A) the external unconditional expectation E superposes the internal conditional expectation $E|_t$, and (B) the rate spread $r_T^*(t) - r^s$ actually does not depend on τ. Comparing (5.4) with the generic pricing formula (3.6) we see that one can formally replace the swap's actual floating rate, $r(\tau)$, $t \le \tau \le t+T$, with the random fixed rate $r_T^*(t)$ on a T-maturity coupon bond traded at instance t. This conclusion has a clear financial sense as both legs for an ATM swap cost a par.

European swaptions. Since the price of annuity is always positive, an option expiring at time t on a "pay fixed" T-maturity swap will be exercised if and only if $r_T^*(t) > r^s$. It leads to an obvious modification of formula (5.4):

$$p(0) = E\int_t^{t+T} [r_T^*(t) - r^s]^+ e^{y(\tau)}d\tau. \qquad (5.5)$$

We have obtained a piece-wise linear pricing model and can use formulae (4.3) and (4.4), in which $x = r_T^*(t)$ and $u = r_T^*(t) + \text{Cov}[r_T^*(t), y(\tau)]$. The other model's non-zero parameters are:

$$N = 2, \; x_0 = -\infty, \; x_1 = r^s, \; x_2 = \infty, \; a_1 = -r^s, \; b_1 = 1,$$

$$d_0 = -\infty, \; d_1 = (r_s - \mu_u)/\sigma_u, \; d_2 = \infty.$$

Hence,

$$p(0) = \int_t^{t+T} P_\tau(0)\{(\mu_u - r^s)[1 - \Phi(d_1)] + \sigma_u\varphi(d_1)\}d\tau. \qquad (5.6)$$

(Parameters μ_u and d_1 are functions of τ, the integration variable.)

Caps, floors, collars. A cap, floor, or collar is a portfolio of swaptions, and we can immediately utilize the results from the preceding section. Alternatively, we can think of them as swaps with piece-wise linear payoffs. Thus, for the kth caplet $[t_k, t_{k+1})$, the terminal payoff is a piece-wise linear function:

$$p(t_{k+1}) = [r_T^*(t_k) - r^s]^+ (t_{k+1} - t_k) . \qquad (5.7)$$

The advantage of this specification is that it can easily accommodate the case when the caplet length $t_{k+1} - t_k$ does not coincide with the maturity T of the underlying rate. For example, one can price a cap or floor indexed to 5- or 10-year Treasury rate, but being reset monthly or quarterly. The option pricing model structure is very similar to that of a swaption except for having only one payoff at t_{k+1} instead of the continuous cashflow integrated in (5.6). Therefore, the caplet's value is going to be

$$p(0) = (t_{k+1} - t_k) P_{t_{k+1}}(0) \{ (\mu_u - r^s)[1 - \Phi(d_1)] + \sigma_u \varphi(d_1) \} , \qquad (5.8)$$

where $u = r_T^*(t_k) + \text{Cov}[r_T^*(t_k), y(t_{k+1})]$, and d_1 is as defined above.

Options on coupon-bearing bonds. An option on a swap costs as much as the same-type option on a coupon bond, which pays the same strike rate r^s and is struck at par. If the strike price is not a par, one can approximate the bond option value computing first its yield to maturity and then use the one as r^s in the appropriate swaption formula. A possible error arising here is caused by an implicit assumption that the yields on the same-maturity bonds having different coupon rates should perfectly correlate with one another. In fact, two same-maturity bonds paying differing coupons will have slightly different payoff term structures. However, for most coupons and prices observed on the fixed income market, this assumption sounds reasonable for most cases and the errors are going to be insignificant. Rare exceptions include options on very long bonds struck at a substantial premium or discount.

5.3. Non-maturing deposits

We assume that a bank manages its deposit rates to stabilize $\lambda(t)$ (the rate of redemption or growth) and perfectly achieves this goal. To do so, the bank must adjust the paid rate c properly. In a simple case we can use the following first-order linear model:

$$\frac{dc'}{dt} = -a[c'(t) - \beta r(t) - \alpha] , \qquad (5.9)$$

where $\beta \le 1$ and α specify the long-term steady state. Under the linear model (5.9), the payoff rate $c = c'$ will be a Gaussian variable itself, and we can employ the main pricing formula (3.5) complemented by the computational hint (4.2) with $x = c'(\tau) - r(\tau)$ and $u = x + \text{Cov}(x, y')$:

$$p(0) = \int_0^\infty P_\tau'(0) \mu_u(\tau) d\tau . \qquad (5.10)$$

One can complicate the model simulating a typically conservative style of the bank's deposit management. Namely, if $c'(t)$ computed from (5.9) is greater than $\beta r(t) + \alpha$, only a k-portion ($0 \le k \le 1$) of the "correctly computed" spread variable $x = c' - \beta r - \alpha$ is actually paid:

$$c = x + \beta r + \alpha \text{ if } x < 0,$$

$$c = kx + \beta r + \alpha, \text{ otherwise}.$$

(5.11)

Introducing a piece-wise linear function, $g(x) = x$ if $x < 0$, $g(x) = kx$ otherwise, and using the main pricing formula (3.5) we get

$$p(0) = E \int_0^\infty [c(\tau) - r(\tau)] e^{y'(\tau)} d\tau$$

$$= E \int_0^\infty [\alpha + (\beta - 1) r(\tau)] e^{y'(\tau)} d\tau + E \int_0^\infty g[x(\tau)] e^{y'(\tau)} d\tau$$

$$\equiv E_1 + E_2.$$

Then, expectation E_1 is the Jamshidian's special case whereas E_2 is computed using statement (4.4) for a piece-wise linear payoff with $u(\tau) = x(\tau) + \text{Cov}[x(\tau), y'(\tau)]$. Parameters used in (4.4) are $N = 2$, $x_0 = d_0 = -\infty$, $x_1 = 0$, $d_1 = -\mu_u/\sigma_u - \sigma_u$, $x_2 = d_2 = \infty$, $b_0 = 1$, $b_1 = k$. The final result is

$$p(0) = \int_0^\infty \{\alpha + (\beta - 1) f(\tau) + \mu_u[k + \Phi(d_1)(1 - k)] - \sigma_u \varphi(d_1)(1 - k)\} P_\tau'(0) d\tau. \quad (5.12)$$

5.4. *Loans and mortgages*

They represent a major yet the most general class of assets (random prepayments, coupon adjustments to a random index). Our analytical approach faces some limitations as the redemption rate λ is required to be a state variable of some linear system. In contrast, many analysts think of the prepayment rate as of an S-like function of the spread between the gross coupon rate and some refinancing rate (driven by r_T^* with T to be somewhat between 5 and 10 years). This function is saturated by the zero (or another minimum) level from the lower side, and by credit limitations for refinancing, from the upper side. A possible local linearization to be used in the context of our approach is $\lambda = \lambda_0 + k(c - r_T^* - \theta_s)$, where λ_0 is the zero-spread redemption rate, k is the sensitivity to the spread, θ_s accounts for servicing and the typical Mortgage–Treasury spread. Such a prepayment model satisfactorily describes some loans and mortgages within limited deviations of the refinancing spread.

Additional factor- or time-dependent amortization rates can simulate seasoning and seasonality (let us recall that our approach has no limitations on the time dependency). Life-time caps for adjustable-rate loans are modeled as piece-wise linear interest payoffs discussed above.

5.5. *The conversion option*

Pricing of the conversion option involves two stochastic processes, for the bond and the stock. We present here the derivations for a simplified ("conceptual") model when one can convert, at expiration, a T-maturity zero-coupon bond (such as LYONs) into a non-dividend stock, the price of which $S(t)$ (conversion-ratio-adjusted) follows an arbitrage-free lognormal process:

$$dS(t) = r(t)S(t)dt + S(t)\sigma_S dz_S(t), \quad S(0) = S_0, \qquad (5.13)$$

where $r(t)$ is the same short rate that underlies a Gaussian interest rate term structure. We assume that the standard Brownian motion $dz_S(t)$ has some known correlation with those disturbing the interest rates. We introduce a new Gaussian variable, $s = \ln(S/S_0)$:

$$ds(t) = \left[r(t) - \frac{1}{2}\sigma_S^2\right]dt + \sigma_S dz_S(t), \quad s(0) = 0. \qquad (5.14)$$

We assume that, following the linear system theory paradigm of Part I, one has derived the expectations and covariances for the state variables, r_T, y, and s, at the time of expiration t. The conversion option value is described in terms of a piece-wise linear-exponential payoff, structurally similar to the option on a discount bond:

$$P(0) = E\{e^{y(t)}[S_0 e^{s(t)} - e^{-T r_T(t)}]^+\} = E\{e^{y(t)+s(t)}[S_0 - e^{-s(t)-T r_T(t)}]^+\}. \quad (5.15)$$

We introduce new variables, $y' = y(t) + s(t)$, $x = -s(t) - T r_T(t)$, define $u = x + \text{Cov}(x, y')$, and use formulae (4.3)–(4.5) with $N = 2$, $x_0 = -\infty$, $x_1 = \ln S_0$, $x_2 = \infty$, $a_0 = S_0$, $c_0 = -1$, $\gamma_0 = 1$, therefore, $d_0 = h_0 = -\infty$, $d_1 = (\ln S_0 - \mu_u)/\sigma_u$, $h_1 = d_1 + \sigma_u$, $d_2 = h_2 = \infty$. It immediately follows from equations (1.2) for $y(t)$ and (5.14) for $s(t)$ that $e^{y'(t)}$ is a matringale. Then, the conversion option value is derived as

$$p(0) = S_0 \Phi(d_1) - E(e^u)\Phi(h_1) \qquad (5.16)$$

where $E(e^u)$ becomes known from the statistics of $s(t)$, $y(t)$ and $r_T(t)$.

5.6. *A term-structure-extended Black–Scholes formula (R. Merton's problem)*

In particular, if $r_T = \text{const}$, (5.16) appears to be a *term-structure-extended* Black–Scholes formula for European call option on a stock struck at $K = \exp(-r_T T)$ and priced under a volatile short rate process $r(t)$. This problem has been considered by R. Merton [16]. It is relatively easy to prove that $E(e^u) = K P_t(0)$ in this case, and formulae for d_1 and h_1 are those from the classical Black–Scholes with the stock volatility σ_S being now replaced by a *term-structure-adjusted* stock volatility $\sigma_{S(\text{adj})}$ as follows:

$$\sigma_{S(\text{adj})}^2 t = \sigma_S^2 t + \text{Var}[y(t)] - 2\text{Cov}[y(t), y'(t)]. \qquad (5.17)$$

Note that the adjusted stock volatility depends on the actual stock volatility, the short rate volatility and the correlation between two processes. Calculations show that, for 1 year to expiration, the required volatility adjustment can be as large as 2–4% (of the stock volatility) if the short rate has a strong correlation (positive or negative) with the stock. Interestingly, the Black–Scholes formula with a constant short rate does not yield the same exact result as term-structure-extended one, even if the short rate is uncorrelated with the stock.

5.7. *Asian-style (arithmetic) options and lookback options*

The Asian (arithmetic) style of derivatives involves the average value $x^{\mathrm{ave}}(t)$ of some underlying variable $x(t)$, over the past period of time $[t-T,t]$. Assuming that $x(t)$ is Gaussian we complement the state space with a new Gaussian variable, $\xi(t)$, such that

$$d\xi(t)/dt = x(t), \qquad x^{\mathrm{ave}}(t) = [\xi(t) - \xi(t-T)]/T. \tag{5.18}$$

Equations (5.18) are linear, with a delayed argument. The same model's property is also typical for lookback options. To build closed-form solutions, covariance matrix $Q_X(t,\tau)$ has to be computed for $t \neq \tau$.

6. Conclusion

We have presented an analytical framework that allows for stochastic pricing of interest-rate contingent dynamic assets via purely deterministic formulae. The asset and the interest rate market are assumed to follow a system of linear stochastic differential equations and some piece-wise defined algebraic equations. The method derives the price directly from the expectations and covariances of the state variables. To use this approach, one needs to find analytic or numeric solution to a system of deterministic linear differential equations for the statistics of those state variables.

We have demonstrated how to use these statistics for developing arbitrage-free one- and two-factor linear term structures. Then, we established a computational paradigm for derivations of closed-form pricing solutions for interest-rate contingent assets. The range of examples we have given is much wider than what is typically considered including interest rate swaps and swaptions, randomly amortizing loans and deposits, conversion options, Asian-style options. All the pricing formulae obtained do not assume any particular number of factors or variables.

Appendix A

Matrix $\Phi(t,\tau)$ for the one-factor system (1.1) and (1.2). The original Hull–White model is presentable in the general form (1.4), where $A(t) =$

$$\begin{array}{|c|c|}\hline -a & 0 \\\hline -1 & 0 \\\hline\end{array}, \quad B(t) = \begin{array}{|c|}\hline 1 \\\hline 0 \\\hline\end{array}, \quad Q_Z = \sigma.$$

To find matrix Φ of fundamental solution for the system, we use the following algebraic fact: if J is the Jordan form for matrix $\int_\tau^t A(\lambda)d\lambda$, and P is the matrix of the Jordan transformation (that is, $\int_\tau^t A(\lambda)d\lambda = PJP^{-1}$), then $\Phi(t,\tau) = Pe^J P^{-1}$. If the Jordan form is diagonal, $J = \text{diag}[\lambda_k]_{k=1}^n$ (the most common case), then $e^J = \text{diag}[\exp(\lambda_k)]_{k=1}^n$. Therefore, finding matrix Φ is a simple 3-step procedure.

Step 1.   Finding eigen-values and eigen-vectors of matrix $\int_\tau^t A(\lambda)d\lambda =$
$$\begin{bmatrix} -a(t-\tau) & 0 \\ -(t-\tau) & 0 \end{bmatrix}:$$

$$\lambda_1 = -a(t-\tau), \quad X_1 = \begin{bmatrix} a \\ 1 \end{bmatrix}; \qquad \lambda_2 = 0, \quad X_2 = \begin{bmatrix} 0 \\ 1 \end{bmatrix};$$

Step 2. From Step 1, $P = \begin{bmatrix} a & 0 \\ 1 & 1 \end{bmatrix}$, therefore, $P^{-1} = \begin{bmatrix} 1/a & 0 \\ -1/a & 1 \end{bmatrix}$.

Step 3. $\Phi(t,\tau) = \exp(\int_\tau^t A(\lambda)d\lambda) = Pe^J P^{-1} = P \begin{bmatrix} e^{-a(t-\tau)} & 0 \\ 0 & 1 \end{bmatrix} P^{-1} =$

$$\begin{bmatrix} e^{-a(t-\tau)} & 0 \\ \frac{1}{a}[e^{-a(t,\tau)} - 1] & 1 \end{bmatrix}.$$

Matrix $\Phi(t,\tau)$ for the two-factor systems (2.10) and (2.11). Matrix A of system (2.10) and (2.11), and matrix $\Phi(t,\tau)$ of its fundamental solutions are block-diagonal matrices such that each diagonal block corresponds to either (x_1, y_1) or (x_2, y_2) as they were found above:

$$A = \text{diag}\left\{ \begin{bmatrix} -a_i & 0 \\ -1 & 0 \end{bmatrix} \right\}_{i=1,2} ; \quad \Phi(t,\tau) = \text{diag}\left\{ \begin{bmatrix} e^{-a_i(t-\tau)} & 0 \\ \frac{1}{a_i}[e^{-a_i(t,\tau)} - 1] & 1 \end{bmatrix} \right\}_{i=1,2} .$$

Appendix B

Proof of Proposition 1.1. First, we state that the direct application of the mathematical expectation operator E to both sides of Eq. (1.14) leads to differential Eq. (1.9) for the mean vector. Here we employ the fact that $X(t)$ is independent of $dZ(t)$. Second, there exists an analytical form of solution for (1.14),

$$X(t) = \Phi(t,0)X_0 + \int_0^t \Phi(t,\tau)C(\tau)d\tau + \int_0^t \Phi(t,\tau)B[\tau, X(\tau)]dZ(\tau). \qquad \text{(B.1)}$$

The first two terms comprise $E[X(t)]$. Therefore,

$$Q_X(t,t) = E\left\{ \left[\int_0^t \Phi(t,\tau)B[\tau, X(\tau)]dZ(\tau) \right] \left[\int_0^t \Phi(t,\tau)B[\tau, X(\tau)]dZ(\tau) \right]^T \right\} .$$

To figure out this integral, we divide the $[0, t]$ time interval into a large number N of small steps $0, h, \ldots, Nh$, and replace a continuous Brownian motion's increment vector $dZ(t)$ with serially uncorrelated discrete time series $dZ_k, k = 0, 1, \ldots, N$, such that $E[dZ_i dZ_j^T] = \delta_{ij} h Q_Z$, where δ_{ij} is the Kroneker's symbol ($\delta_{ij} = 1$ if $i = j$, 0 otherwise):

$$Q_X(t, t) = \lim_{N \to \infty} E \left\{ \sum_{k=0}^{N} \Phi(t, t - kh) B_k dZ_k \sum_{k=0}^{N} dZ_k^T B_k^T \Phi^T(t, t - kh) \right\},$$

where B_k denotes $B[t - kh, X(t - kh)]$, a random variable. For any $i > j$, dZ_i is independent of dZ_j, B_i and B_j. Since dZ_i is centered, all those "cross-products" have zero expectations, therefore

$$Q_X(t, t) = \lim_{N \to \infty} E \left\{ \sum_{k=0}^{N} \Phi(t, t - kh) B_k dZ_k dZ_k^T B_k^T \Phi^T(t, t - kh) \right\}.$$

Let us consider the expectation of $B_k dZ_k dZ_k^T B_k^T$. The (i, j)th element of this matrix equals

$$\sum_{l,m} E[(B_k)_{il}(B_k)_{jm}(dZ_k)_l(dZ_k)_m] = h \sum_{l,m} E[(B_k)_{il}(B_k)_{jm}](Q_Z)_{lm}$$

$$= h \sum_{l,m} E[(B_k)_{il}(Q_Z)_{lm}(B_k)_{jm}]$$

since B_k are dZ_k independent. Therefore $E[B_k dZ_k dZ_k^T B_k^T] = h E[B_k Q_Z B_k^T] = h K_k$, where K_k denotes $K(t - kh)$, a function defined in Proposition 1.1. It leads to

$$Q_X(t, t) = \lim_{N \to \infty} h E \left[\sum_{k=0}^{N} \Phi(t, t - kh) K_k \Phi^T(t, t - kh) \right]$$

$$= \int_0^t \Phi(t, \tau) K(\tau) \Phi^T(t, \tau) d\tau. \tag{B.2}$$

Note that (B.2) is not yet the final formula to compute $Q_X(t, t)$ because matrix $K(t)$ generally also depends on the second statistical moments. To prove (1.15), simply differentiate (B.2) with respect to t and account for differential Eq. (1.8) for matrix $\Phi(t, \tau)$.

Appendix C

We find an algebraic solution to system (2.20) thus computing all five parameters, a's, σ's, and ρ, used in the two-factor interest rate model. We assume known: the short rate volatility, σ, volatilities σ_{T_1} and σ_{T_2} for two other maturity points of the term structure, T_1 and T_2, and correlations of those rates with the short rate, ρ_{T_1} and ρ_{T_2}. Then, using the inter-rate relationship (2.22) and denoting $B_r = (B_2 - \beta B_1)/(1-\beta)$, $B_v = (B_1 - B_2)/(1-\beta)$, we get:

$$\sigma_{T_i}^2 = B_r^2(T_i)\sigma^2 + B_v^2(T_i)\sigma_v^2, \qquad i = 1,2 \tag{C.1}$$

$$\rho_{T_i}\sigma_{T_i} = \sigma B_r(T_i), \qquad i = 1,2 \tag{C.2}$$

Combining (C.1) and (C.2) for each i, we have $\sigma_{T_i}^2(1-\rho_{T_i}^2) = B_v^2(T_i)\sigma_v^2$, therefore,

$$\frac{B_v^2(T_1)}{B_v^2(T_2)} = \frac{\sigma_{T_1}^2(1-\rho_{T_1}^2)}{\sigma_{T_2}^2(1-\rho_{T_2}^2)} \equiv \gamma^2, \tag{C.3}$$

a given parameter.

From the definitions, $B_v = B_1 - B_r$, therefore

$$\frac{B_v^2(T_1)}{B_v^2(T_2)} = \frac{[B_1(T_1) - B_r(T_1)]^2}{[B_1(T_2) - B_r(T_2)]^2} = \dots \text{ using (C.3)} \dots = \frac{[\rho_{T_1}\sigma_{T_1} - \sigma B_1(T_1)]^2}{[\rho_{T_2}\sigma_{T_2} - \sigma B_1(T_2)]^2}. \tag{C.4}$$

Equating the right-hand sides of (C.3) and (C.4) we arrive at a transcendent equation for one unknown only, a_1:

$$\frac{[\rho_{T_1}\sigma_{T_1} - \sigma B_1(T_1)]^2}{[\rho_{T_2}\sigma_{T_2} - \sigma B_1(T_2)]^2} = \gamma^2. \tag{C.5}$$

In the domain of positive numbers, Eq. (C.5) may have either two solutions (use one for a_1 and another for a_2), or none. In the latter case, the two-factor time-invariant term structure cannot fit the given set of inputs.

After a's are computed, σ_v and β can be found as follows:

$$\beta = \frac{\rho_{T_i}\sigma_{T_i} - \sigma B_2(T_i)}{\rho_{T_i}\sigma_{T_i} - \sigma B_1(T_i)}, \qquad \sigma_v^2 = \sigma_{T_i}^2(1-\rho_{T_i}^2)/B_v^2(T_i), \tag{C.6}$$

where any i, 1 or 2, can be used. Parameters of the (x_1, x_2)-model are then calculated as

$$\sigma_1^2 = (\sigma_v^2 + \beta^2\sigma_r^2)/(1-\beta)^2, \quad \sigma_2^2 = (\sigma_v^2 + \sigma_r^2)/(1-\beta)^2,$$

and

$$\rho = -(\sigma_v^2 + \beta\sigma_r^2)/\sigma_1\sigma_2(1-\beta)^2.$$

Appendix D

Proof of Proposition 4.1. Two correlated normally distributed variables, vector $\boldsymbol{x}$ with the mean vector $\boldsymbol{\mu}_x$ and the autocovariance matrix $\boldsymbol{V}$, and scalar y' always satisfy the following exact linear regression:

$$y' = \boldsymbol{\beta}^T\boldsymbol{x} + \varepsilon, \tag{D.1}$$

where $\beta = V^{-1}\text{Cov}(x, y')$ and ε is also normally distributed and independent of x. Therefore,

$$E[g(x)e^{y'}] = E(e^{\varepsilon})E[g(x)e^{\beta^T x}]$$

$$= E(e^{\varepsilon})\frac{1}{\sqrt{(2\pi)^m|V|}}$$

$$\times \int_{R^m} g(x)\exp\left[-\frac{1}{2}(x-\mu_x)^T V^{-1}(x-\mu_x) + \beta^T x\right]dx.$$

It is easy to check that the quadratic function in the brackets has the following canonical form:

$$-\frac{1}{2}(x-\mu_x)^T V^{-1}(x-\mu_x) + \beta^T x = -\frac{1}{2}(x-\mu_u)^T V^{-1}(x-\mu_u) + \eta,$$

where $\mu_u = \mu_x + \sigma_x\beta \equiv \mu_x + \text{Cov}(x, y')$ and $\eta = \beta^T\mu_x + \frac{1}{2}\beta^T V\beta$. Hence, $e^{\eta} = E(e^{\beta^T x})$ and

$$E[g(x)e^{y'}] = E(e^{\varepsilon})E[e^{\beta^T x}]\frac{1}{\sqrt{(2\pi)^m|V|}}$$

$$\times \int_{R^m} g(x)\exp\left[-\frac{1}{2}(x-\mu_u)^T V^{-1}(x-\mu_u)\right]dx. \qquad \text{(D.2)}$$

Since x are ε independent, $E(e^{\varepsilon})E(e^{\beta^T x}) = E(e^{\beta^T x+\varepsilon}) = E(e^{y'})$ due to regression (D.1), whereas the remaining multiplier in formula (D.2) is recognized as $E[g(u)]$.

Acknowledgements

The author is thankful to John Hull, Marco Avellaneda, Raphael Douady, Owen Walsh, an anonymous referee for helpful comments, and the participants of Mathematical Finance Seminar at NYU/Courant Institute.

References

[1] D. R. Beaglehole and M. S. Tenney, *General solutions of some interest rate-contingent claim pricing equations*, J. Fixed Income **4** (1991).

[2] A. Brace and M. Musiela, *A multifactor Gauss Markov implementation of Heath, Jarrow, and Morton*, Math. Finance **4** (July 1994) 259–283.

[3] R. Brenner and R. Jarrow, *A simple formula for options on discount bonds*, in Advances in Futures and Options Research **6** (1993) 45–51.

[4] D. Duffie, *Dynamic Asset Pricing Theory*, Second Ed., Princeton Univ. Press, Princeton, NJ (1996).

[5] A. Franchot, *Factor models of domestic and foreign interest rates with stochastic volatilities*, Math. Finance **5** (April 1995) 167–185.

[6] F. Jamshidian, *An exact bond option pricing formula*, J. Finance **44** (March 1989) 205–209.

[7] F. Jamshidian, *Bond and option evaluation in the Gaussian interest rate model*, Research in Finance **9** (1991) 131–170.

[8] Heath, D. R. Jarrow, and A. J. Morton, *Bond pricing and the term structure of interest rates: a new methodology for contingent claims valuation*, Econometrica **60** (1992) 77–105.

[9] J. C. Hull and A. D. White, *Pricing interest rate derivative securities*, Rev. Financial Studies **3** (1990) 573–592.

[10] J. C. Hull and A. D. White, *Numerical procedures for implementing term structure models II: two-factor models*, J. Derivatives **2**(2) (Winter 1994) 37–48.

[11] J. C. Hull, *Options, Futures and Other Derivative Securities*, Prentice Hall, Englewood Cliffs, NJ (1993).

[12] I. Karatzas and E. S. Shreve, *Brownian Motion and Stochastic Calculus*, Springer-Verlag New York (1991).

[13] D. P. Kennedy, *The term structure of interest rates as a Gaussian random field*, Math. Finance **4** (July 1994) 247–258.

[14] A. Levin, *Linear system theory in stochastic pricing models*, in *Yield Curve Dynamics*, ed. R. Ryan, Glenlake Publ., Chicago, IL (1997) 123–146.

[15] P. Ritchken and L. Sankarasubramanian, *Volatility structure of forward rates and the dynamics of the term structure*, Math. Finance **5** (Jan. 1995) 55–72.

[16] R. C. Merton, *Theory of rational option pricing*, Bell J. Economics and Management Science **4** (Spring 1973), pp. 141–183.

MODELS FOR ESTIMATING THE STRUCTURE
OF INTEREST RATES FROM OBSERVATIONS
OF YIELD CURVES[*]

K. O. KORTANEK

Department of Management Sciences
College of Business Administration
Program in Applied Mathematical and Computational Sciences
University of Iowa
Iowa City, IA 52242, USA

V. G. MEDVEDEV

Department of Optimal Control Methods
Faculty of Applied Mathematics and Informatics
Byelorussian State University
F. Skorina pr. 4, Republic Belarus

Contents

[*]Partial financial support from the Donald E. and Felicity A. Sodaro Revocable Trust is gratefully acknowledged. We also acknowledge the Office of Vice President of Research for providing computer equipment during Dr. Medvedev's visit to Iowa City and the College of Business Administration for providing office space (7/4/97–6/15/98). This is a revision of the 25 September 1997 Technical Report.

We present a **dynamical systems** approach for modelling the term structure of interest rates based on a linear differential equation under uncertainty. In contrast to a stochastic process we introduce impulse or point-impulse perturbations on either (a), the spot (shortest-term, risk neutral) interest rate as the unknown function, or (b), its integral, namely the yield function, or both simultaneously. Parameters are estimated by minimizing the maximum absolute value of the measurement errors. Termed the Optimal Observation Problem, *OOP*, it defines our *norm of uncertainty*, in contrast to the *expectation operator* for a stochastic process. Beyond the learning period (the current time), the solved-for spot rate function becomes the forecast of the unobservable function in a future period, while its integral should approximate the yield function well.

Non-arbitrage is addressed by providing a necessary condition expressed as constraints in the OOP, under which non-arbitrage is guaranteed. The property of mean-reversion is also preserved, and functional estimates are provided for the market price of risk. Analogous concepts to "drift" and "volatility" are treated in a manner that provides a criterion for the choice of perturbation to employ in a given real situation. Additional constraints, if necessary, guarantee non-negative short and forward rates, a property not automatically fulfilled in the stochastic case. We test the approach empirically with daily Treasury yield curve rates data, mainly for discount bonds having 3- to 6-month maturities over observation periods of up to one year. Computational results are reported for many numerical experiments together with some financial interpretations, see http://kwel.biz.uiowa.edu/

1. Introduction: Approach for Estimating Uncertainty in Dynamical Systems

In the early stages of mathematical modelling of control problems involving state variables that can be changed by means of control variables it was generally assumed that exact and complete state measurements could be obtained. It was soon learned that the optimal behavior of a dynamical system under uncertainty, DSU, cannot be effectively obtained without having an observation system capable of identifying unknown parameters of the underlying dynamical system.

Generally, the problem of estimation occurring in non-deterministic systems has been investigated by means of many stochastic models beginning with the papers of Wiener [30] and Kalman [13]. Earlier in the 1970's, nonstochastic observation models ("minimax", "guaranteed") under uncertainty appeared in [17, 18, 7]. In [8] a new approach for optimization of linear dynamical systems under uncertainty was presented based on the earlier fundamental papers of Gabasov and Kirillova.

This approach to treating uncertainty is in contrast to other qualitative approaches, for example, based on stochastic differential equations. In the latter case certain mathematical assumptions are made about the underlying stochastic processes which may be difficult to verify in real situations, for example, in the financial derivatives and assets markets.

Needed are effective computational algorithms for solving problems where estimates of a control mechanism are available based upon sensor observations. There are at least two major extremal problems that must be addressed in order to solve such classes of optimal observation problems. Termed the Program Problem, this model must be solved in order to generate estimates for the underlying linear dynamical system. This problem is solved after all observations are made, and the solution is helpful in understanding the structure of the underlying linear dynamical system. But in practice real time estimates are needed of the dynamical system, so another type of optimal observation problem arises termed the Positional Problem. Solving this problem provides estimates of the unknown system parameters in a real-time regime while providing the additional benefit of being able to construct parameter estimates adaptively and continuously as new information from the sensor is received.

During the last 7 years computational methods have been developed that take advantage of the structure of both of these types of problems by employing semi-infinite programming (SIP) algorithms. SIP models are a next level of extension of ordinary linear programming and have finitely many decision variables appearing in infinitely many constraining inequalities, see [11, 16]. SIP methods incorporate many properties of the underlying linear control system, such as the classical Cauchy Fundamental matrix. The main algorithmic construction updates the current SIP optimal solution by taking the previous position as an efficient advanced start. Special rules are based on higher order derivatives and the necessary SIP-optimality conditions. Previous uses of ordinary linear programming are extended to more general classes of measurable functions as input disturbances, see [22–25].

In this paper we apply the program/positional treatment of DSU observations to the modelling the shortest term riskless rate of interest, the *spot rate*, by linear dynamical systems with unknown parameters under nonstochastic uncertainty, $LDSU$. As data, spot rates are unobservable, and so we employ observations from Treasury yield curves, namely, observed yields of bonds with either common maturity or differing maturities. Included in the outputs are the estimated yields to maturity (based on observed yields to maturity), sometimes referred to as the learning phase of the process. Observed data are discrete, but stability in the dynamic models is enhanced if we build piecewise continuous extensions of the discrete data. We consider two such extensions: piecewise constant and piecewise linear. We consider *one-factor* dynamical systems where perturbations occur only in the spot rate differential equation, and present the results of numerical experiments in tabular and graphical form. We also generate yield estimates from *bilevel* models, where now perturbations are present both for the spot rate and for the piecewise continuous extension of the observed yields.

The basic numerical outputs of the approach are (1), estimated prices of discount bonds, (2), the term structure of interest rates, and (3), the predicted future prices of discount bonds.

We review the basic systems approach from the viewpoint of linear dynamic systems under uncertainty with perturbations, $LDSU$.

1.1. *Min-Max observation problem under uncertainty with perturbations*

$$\dot{x} = Ax + Dw(t), \quad x(0) = x_0, \quad \forall t, \ t_a \leq t \leq t_c$$

$$x_0 \in X_0 = \{x \in \mathbf{R}^n \,|\, d_* \leq x \leq d^* \text{ and } g_* \leq Gx \leq g^*\};$$

$$w(t) \in W(t) = \{w(t) \in \mathbf{R}^l \,|\, w_* \leq w(t) \leq w*\}; \tag{1}$$

$$D \in \mathbf{R}^{n \times l}; \ w_*, w^* \in \mathbf{R}^l; \ d^*, \ d^* \in \mathbf{R}^n; \ g_*, \ g^* \in \mathbf{R}^m, \ G \in \mathbf{R}^{m \times n}.$$

Associated with (1) is the Cauchy Fundamental matrix F having the following constructive properties:

$$\dot{F} = AF, \quad F(0) = E, \quad F(t-s) = F(t-p)F(p-s), \quad F(t+s) = F(t)F(s),$$

$$F^{-1}(t) = F(-t).$$

Throughout this paper our common technique will be to construct the form of a solution to system (1) by using the Cauchy matrix as follows:

$$x(t) = F(t)x_0 + \int_{t_a}^{t} F(t)F(-\tau)Dw(\tau)\,d\tau\,. \tag{2}$$

An estimate, $\hat{x}(t)$, of the state $x(t)$ of the system (1) is obtained from measurements made with a sensor system of the following integral from which is used throughout with varying limits of integration:

$$y(t) = \int_{t_a}^{t} x(\tau)d\tau + z(t)\,, \quad \forall t\,, \quad t_a \le t \le t_b\,, \quad \text{where } t_b \le t_c \tag{3}$$

which gives inexact and incomplete information about current state of system (1), where $z(t)$ is an *unknown* piecewise continuous measurement error function, noting that in general t_a may be a function of t.

Let (1), (3) generate a signal $y^*(t)$, $t \in T$ with some measurement error $z^*(t)$, $t \in T$. A measure of the precision of our estimate is L_∞-norm measure of the error, wherein we seek the function $\hat{x}(\cdot)$ according to the following minimax *observation problem*:

$$\min_{(x,w(\cdot))\in X_0 X W(\cdot)} \max_{t_a \le t \le t_b} |z^*(t)|\,. \tag{4}$$

This problem leads to an **infinite linear program**. First, substitute (2) into (3) to obtain:

$$z^*(t) = y^*(t) - x_0 \int_{t_a}^{t} F(\tau)d\tau - \int_{t_a}^{t} F(\tau)\left[\int_{t_a}^{\tau} F(-s)Dw(s)ds\right]\,, \quad \forall t \in [t_a,\ t_b]\,.$$

The linear structure of (4) is seen by rewriting it in the usual equivalent, double inequality form:

$$\min_{(x,w(\cdot))\in X_0 \times W(\cdot)} v \quad \text{subject to}$$

$$y^*(t) \le x \int_{t_a}^{t} F(\tau)d\tau - \int_{t_a}^{t} F(\tau)\left[\int_{t_a}^{\tau} F(-s)Dw(s)ds\right] + v\,,$$

$$x \int_{t_a}^{t} F(\tau)d\tau - \int_{t_a}^{t} F(\tau)\left[\int_{t_a}^{\tau} F(-s)Dw(s)ds\right] - v \le y^*(t)\,, \tag{5}$$

$$x \in X_0\,, \quad v \ge 0\,, \quad \text{and } w(t) \in W(t)\,, \quad \forall t \in [t_a,\ t_b]\,.$$

Our recurring application throughout this paper begins with an optimal solution, (x^0, w^0, v^0), to (5). We obtain an estimate, $\hat{x}(\cdot)$, of the state $x(\cdot)$ of the system (1) by:

$$\hat{x}(t) = x^0 \int_{t_a}^{t} F(\tau)d\tau - \int_{t_a}^{t} F(\tau)\left[\int_{t_a}^{\tau} F(-s)Dw^0(s)ds\right]\,, \quad \forall t\,, \quad t_a \le t \le t_c\,.$$

This estimate gives the minimum of the maximum absolute value v^0 of the measurement error $z(\cdot)$. Problem (5) is an infinite linear programming problem. A Program Algorithm and Positional Algorithm for (4) have been developed by the second author, and we apply these algorithms in our numerical study.

Our basic procedure is to first specify an $LDSU$ model for the underlying price behavior, and then specify a class (or classes) of perturbations. Related to the concept of *random walk* we introduce the methodology with an *uncertainty walk* next.

2. A Simple Model for Asset Prices with Nonstochastic Uncertainty

2.1. *A description of an uncertainty walk*

It is often stated that asset prices must move randomly because of the *efficient market hypothesis*. Generally this hypothesis has two meanings:

1. the past history is fully reflected in the present price, which does not have any additional information, and
2. markets respond immediately to any new information about an asset.

Hence, the modelling of asset prices really is concerned with the modelling of new information which affects the price. In this paper our response to the two assumptions above is to introduce a class of linear dynamic systems under uncertainty for unanticipated changes in the asset price.

First, we note that the *absolute* change in the asset price is not by itself a useful quantity; a change of $1 is more significant when the asset price is $10 than when it is $500. Therefore, we associate the *investor's return* which is the change in asset price plus dividends divided by the asset price, a well-established measure of return.

Assume at time t the asset price is S, and consider a small subsequent time interval, dt, during which S changes to $S + dS$. The standard way of modeling the return $\frac{dS}{S}$ is to decompose the return into two parts. First there is the predictable, deterministic, and anticipated return equivalent to a risk-free return in a government bond or certificate of deposit from a bank. This component of return takes the form

$$\mu dt,$$

where μ is a measure of the average rate of growth of the asset price. In simple models μ is constant, but it can be a function of S and t.

The second component of $\frac{dS}{S}$ models the uncertain change in the asset price due to unanticipated external effects. We shall assume that upper and lower bounds for uncertain changes in the price are known, denoted respectively as functions of time, t, $w^*(t)$ and $w_*(t)$. Later we will postulate properties for these bounding functions, but now merely indicate their dependence on time. Therefore, the second component in the return of the asset price reveals its uncertain structure through

$$\frac{w(t)}{S}dt, \quad w_*(t) \leq w(t) \leq w^*(t),$$

where $w(\cdot)$ is an integrable function.

Combining these two components yields the following **differential equation under uncertainty (DEU)**:

$$\frac{dS}{S} = \mu dt + \frac{w(t)}{S} dt, \quad w_*(t) \le w(t) \le w^*(t), \tag{6}$$

which is the mathematical representation of the recipe we adopt for generating asset prices.

We rewrite (6) in a form that enhances the use of the Cauchy formula, a transformation that will be used throughout this paper in different linear dynamical system contexts.

$$\frac{dS}{dt} = \mu S + w(t), \quad w_*(t) \le w(t) \le w^*(t). \tag{7}$$

Assuming that μ is constant, the Cauchy formula yields,

$$S(t) = S_0 e^{\mu(t-t_0)} + \int_{t_0}^{t} e^{\mu(t-\tau)} w(\tau) d\tau, \quad t \ge t_0; \tag{8}$$

$$w_*(\tau) \le w(\tau) \le w^*(\tau), \quad \tau \in [t_0, t],$$

where S_0 is the value of the asset at $t = t_0$. Figure 2.1 illustrates these components of a discrete uncertainty walk. Point B corresponds to the deterministic case, while point A corresponds to the maximal asset price achieved under current conditions of the market. Next, point D corresponds to the minimal possible asset price, while point C describes the realized current asset price.

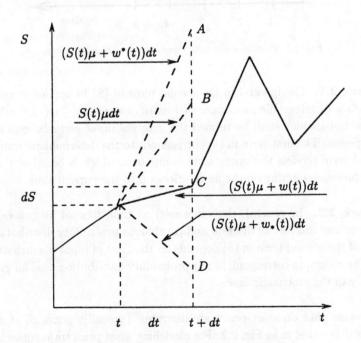

Fig. 2.1. A discrete *uncertainty walk* for asset prices.

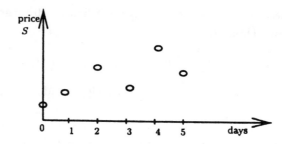

Fig. 2.2. A discrete scatter of observed daily prices.

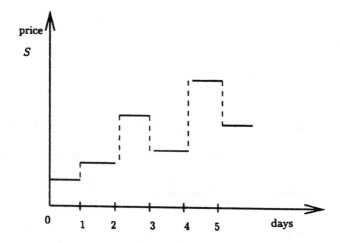

Fig. 2.3. Piecewise constant representations of asset prices.

Remark 2.1. The integral in the second term in (8) to applies to an *a priori* specified class of integrable functions, $w(\cdot)$ satisfying, $w_*(\tau) \leq w(\tau) \leq w^*(\tau)$, $\tau \in [t_0, \ t]$, functions which shall be termed the *external input perturbations* acting on the asset prices. The first term in (8) corresponds to the deterministic return, while the second term models the uncertainty component which is based of the *history of the perturbations* acting on the asset prices up to the current time.

Remark 2.2. In general, the value $w(t)$ at time t could be generated by a sample from any mixture of discrete and continuous probability distributions. The structure of the second term in (8) depends on the class of input perturbations, and these can be chosen to correspond to any probability distribution used for generating uncertainty in the stochastic case.

Observable data on asset prices are discrete. Typically, prices $\hat{S}_i$ at day i are discrete as illustrated in in Fig. 2.2. For modelling asset price trajectories by means of a differential Eq. (7), where t varies over a continuum, we need to represent the

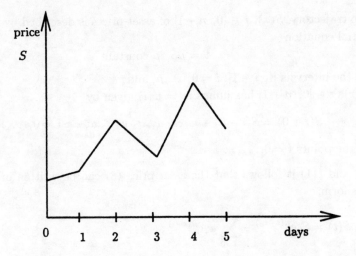

Fig. 2.4. Piecewise linear representations of asset prices.

discrete information as a function of t. It is clear from (8) that the class of input perturbations defines a family of trajectories for the asset price. In the following section we describe how to choose a class of input perturbations in a way that depends on the structure of the asset price itself. We focus on two families of asset price trajectories, piecewise constant, see Fig. 2.3 and piecewise linear, see Fig. 2.4, although much more general families are admissible.

2.2. *Modelling asset prices with input perturbations*

Perturbations acting on asset prices and must have a structure compatible with an accurate determination of the trajectory of asset prices. We consider perturbations which generate piecewise constant and piecewise linear asset price trajectories, respectively.

2.2.1. *Point impulse perturbations for piecewise constant asset prices*

We assume asset prices yare fixed within each day, namely

$$\hat{S}(t) = \hat{S}_i, \quad t \in [i, i+1[, \quad i = 0, \ldots, n. \tag{9}$$

Figure 2.3 illustrates the jumps in asset prices at the fixed points, $0, 1, \ldots, n$. The class of delta functions (point impulse) is particularly convenient for obtaining these jumps through the second term of (8). These are defined as follows.

Definition 2.1. Point-impulse perturbations have the following properties. $\{w(\tau), \tau \in [0, n+1]\}$ acting on (6) are the following ones.

(1) $w(\tau) = 0, \tau \in [0, \ n+1] \setminus \{0, \ldots, n\}$;

(2) the trajectory, $S(\tau)$, $\tau \in [0, \ n+1]$ of asset prices is described by the differential equation

$$\dot{S} = \mu S, \ \mu \ \text{constant} \tag{10}$$

on the intervals $]i, \ i+1[$, $i = 0, \dots, n$, and

(3) $S(\tau)$, $\tau \in [0, \ n+1]$ has jumps $w_i = w(i)$ given by

$$S(i+0) = S(i-0) + w_i, \quad w_{i*} \le w_i \le w_i^*, \quad i = 0, \dots, n, \tag{11}$$

at the points $i = 0, \dots, n$.

From (10) and (11) it follows that the asset price (8) can be written in piecewise continuous form:

$$S(t) = S_0 e^{\mu(t-t_0)} + \sum_{i=0}^{j} w_i e^{\mu(t-i)},$$

$$t \in [j, \ j+1], \quad j = 0, \dots, n; \quad w_{i*} \le w_i \le w_i^*, \quad i = 0, \dots, n.$$

2.2.2. *Impulse perturbations for piecewise linear asset prices*

We now assume that the asset price is a piecewise linear function of time. Figure 2.1 is simply obtained from Figure 2.1 by connecting neighboring points by line segments.

In this case, analogous to (9) we obtain piecewise linear changes in asset prices:

$$\hat{S}(t) = S_i + \frac{t - t_i}{t_{i+1} - t_i}(S_{i+1} - S_i), \quad t \in [i, \ i+1], \quad i = \overline{0, n-1}. \tag{12}$$

The trajectory of the asset price is continuous, and while there are various classes of functions for describing perturbations of this type, we choose the following impulse type.

Definition 2.2. Impulse perturbations: $\{w(\tau), \ \tau \in [0, \ n+1]\}$ acting on (6) are simply piecewise constant functions, namely:

$$w(\tau) = w_i, \quad w_{i*} \le w_i \le w_i^*, \quad \tau \in [i, \ i+1[, \quad i = 0, \dots, n-1. \tag{13}$$

By means of (13) we obtain the following specialization of (8), where $S(t)$ is continuous:

$$S(t) = S_0 e^{\mu(t-t_0)} + \sum_{i=0}^{j-1} w_i \int_i^{i+1} e^{\mu(t-\tau)} d\tau + w_j \int_j^t e^{\mu(t-\tau)} d\tau$$

$$= S_0 e^{\mu(t-t_0)} + \sum_{i=0}^{j-1} \frac{w_i}{\mu} \left(e^{\mu(t-i)} - e^{\mu(t-(i+1))} \right) + \frac{w_j}{\mu} \left(e^{\mu(t-j)} - 1 \right),$$

$$t \in [j, \ j+1], \quad j = \overline{0, n-1}; \quad w_{i*} \le w_i \le w_i^*, \quad i = \overline{0, n-1}.$$

We have described two basic ways of constructing assset price trajectories over a continuum from discrete observations, using respectively point-impulse and impulse perturbations. We next describe how to estimate these input pertubations using using "real" observations.

2.3. *Estimating input perturbations and asset price trajectories*

Our approach uses previous observations of the asset price during a predetermined special period termed the *period of asset price observations*.

We begin with (1), known values of lower and upper bounds for the parameter, μ, μ_* and μ^*, respectively, and (2) known values of lower and upper bounds on input disturbances, w, w_* and w^*, respectively. Assume observations $\hat{S}(t)$ are taken over on an interval $[t_0, T]$, which are modelled by either (11) or (12) with initial state $S_0 = S(t_0)$.

The *error of estimation of the asset price* is the function:

$$
\begin{aligned}
\epsilon(t, \mu, w(\cdot)) &= \hat{S}(t) - S(t) \\
&= \hat{S}(t) - S_0 e^{\mu(t-t_0)} - \int_{t_0}^{t} e^{\mu(t-\tau)} w(\tau) d\tau, \quad t \in [t_0, T].
\end{aligned} \tag{14}
$$

Based upon *minimax* $L_{+\infty}$, (1)–(4), we obtain (biextremal) optimization problem for computing estimated values of μ, and $w(\cdot)$:

$$
\min_{\mu, w(\cdot)} \max_{t \in [t_0, T]} \epsilon(t, \mu, w(\cdot)). \tag{15}
$$

Using the classes of input perturbations described above, problem (15) becomes a special nonlinear semi-infinite programming problem, that has occurred in earlier dynamic control contexts, e.g., [17]; see [11] for a more general statement and description of its mathematical properties.

Let $(\mu^0, w^0(\cdot))$ be a solution of Problem (15). Then the function,

$$
S^0(t) = S_0 e^{\mu^0(t-t_0)} + \int_{t_0}^{t} e^{\mu^0(t-\tau)} w^0(\tau) d\tau, \quad t \in [t_0, T], \tag{16}
$$

is an estimate of the asset price over the entire observation period. The optimal solution, $(\mu^0, w^0(\cdot))$, will be used to predict future values of asset prices beyond the observation period, as well for updating new bounds for the perturbation functions. The result shall be an automatic updating of asset pprice trajectories as new information becomes available.

In the following sections we implement our approach for modelling uncertainty for the purpose of estimating the term structure of interest rates under various models of observation.

3. Building Linear Dynamic Systems from Stochastic Differential Interest Rate Models

We introduce the fundamental modelling construction of dynamical differential equations systems with unknown parameters under nonstochastic uncertainty that correspond to stochastic differential equations models assumed to govern the interest rates of shortest duration, namely the *spot rates.*

We illustrate the approach with the classical model of Vasicek, [28] where the standard Brownian motion, Z, underlies the stochastic differential equation, and where α, β, and σ are parameters, see [6, 14, 9, 29, 2]:

$$dr = (\alpha + \beta r)dt + \sigma dZ, \tag{17}$$

where r is the shortest term riskless rate of interest, the *spot rate,* and the parameters are employed to capture shifts and volatility of the this rate, [6].

We associate to (17) the following *linear differential equation under uncertainty, LDEU:*

$$\dot{r} = \beta r + \alpha + \sigma w(t), \quad r(0) = r_0 \in R_0 = \{r \in R,\ r_* \le r \le r^*\};$$
$$w_* \le w(t) \le w^*, \quad t \in [0, t^*], \tag{18}$$

where $w(\cdot) = (w(t), t \in)$ is an unknown piecewise continuous function. We shall term (18) the **analog** of the given stochastic system.

Assume that the values of $r(t)$, $t \in [0, t^*]$ can be measured with some piecewise continuous error function $z(t)$ defined in an observation model as follows:[a]

$$y(t) = r(t) + z(t). \tag{19}$$

Compute the estimates of r_0 and $w(\cdot)$ according to minimax estimation:

$$\min_{x \in R_0,\, w_* \le w(\cdot) \le w^*} \ \max_{t \in [0, t^*]} |z(t)|. \tag{20}$$

Following our method, we obtain the following **linear infinite problem:**

$$\min \quad v$$

$$\text{subject to } y(t) + \frac{\alpha}{\beta}(1 - e^{\beta t}) \le e^{\beta t} x + \sigma \int_0^t e^{\beta(t-\tau)} w(\tau) d\tau + v,$$

$$e^{\beta t} x + \sigma \int_0^t e^{\beta(t-\tau)} w(\tau) d\tau - v \le y(t) + \frac{\alpha}{\beta}(1 - e^{\beta t}), \tag{21}$$

$$v \ge 0, \ r_* \le x \le r^*, \text{ and } w_{i*} \le w(t) \le w_i^*, \ \forall t \in [0, t^*].$$

Definition 3.1. Throughout this paper T_i shall denote the specific time interval, $[t_{i-1},\ t_i]$. In addition we shall use the following notation for other intervals for integer i, j, $j \le i$.

$$T_{i,j} = [t_{i-j},\ t_i] \text{ and } T^i = [t_0,\ t_i].$$

---

[a]This observation model is introduced as a simple illustration for constructing the minimax extremal estimation problem. A general approach will be described in the next section.

For a positive integer N define

$$T^N = \bigcup_{i=1}^N T_i = [0, t_N], \text{ and } \tilde{T}^N = \bigcup_{i=0}^N \{t_i\}. \tag{22}$$

We approximate the unknown piecewise continuous perturbation function $w(\cdot)$ by impulse perturbations, (13) using the definition of T^N, (22). This choice reduces problem (21) to the following linear semi-infinite programming problem:

$$\min v \quad \text{subject to}$$

$$y(t) + \frac{\alpha}{\beta}(1 - e^{\beta t}) \le e^{\beta t} x + \sum_{i=1}^{j-1} \int_{t_{i-1}}^{t_i} e^{\beta(t-\tau)} \sigma w_i d\tau + \int_{t_{j-1}}^t e^{\beta(t-\tau)} \sigma w_j d\tau + v,$$
$$\tag{23}$$
$$e^{\beta t} x + \sum_{i=1}^{j-1} \int_{t_{i-1}}^{t_i} e^{\beta(t-\tau)} \sigma w_i d\tau + \int_{t_{j-1}}^t e^{\beta(t-\tau)} \sigma w_j d\tau - v \le y(t) + \frac{\alpha}{\beta}(1 - e^{\beta t}),$$

$$v \ge 0, \quad r_* \le x \le r^*, \text{ and } w_{i*} \le w_i \le w_i^*, \quad \forall t \in T_i, \quad j = \overline{1, N}.$$

The stochastic system (17) is referred to as *one factor* continuous time stochastic system. We shall consider several important one factor models in Sec. 4.2 and construct the *LDEU* model for each of them.

Remark 3.1. Anticipating actual market experience we can obtain the values of a measurement signal once a day. For example, the values of Treasury yield curve rates are updated daily, usually by 5:30 PM Eastern time. So we can assume, for example, that a measurement signal (19) is a piecewise-constant function. If we choose the perturbation function class, $w(\cdot)$, according to (13),where points t_i correspond to business days, then we obtain linear semi-infinite problem (23) instead of the infinite problem (21).

4. Estimation of the Term Structure of Interest Rates Under Observation of Yield Curves

As reviewed in the Introduction, a new approach for optimization of linear dynamic systems under uncertainty was suggested in [8]. This approach lead to the development of the so-called *program algorithm* together with *the positional (real-time) algorithm* for solving the optimal observation problem under a nonstochastic model of uncertainty, see in [22–24]. In this paper we apply these optimization algorithms to the problem of estimating the spot riskless rate under the condition of having observations from the yield curves.

4.1. *A basis for defining admissible sets of parameters and spot rate functions*

As is well known in the literature, the term structure of interest rates is the relationship between the yields on default-free, discount bonds and their maturities.

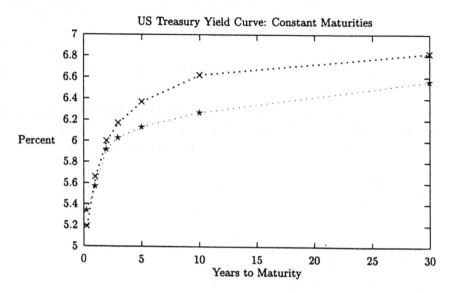

Fig. 4.1. × for 8/15/96 and ⋆ for 8/14/97.

A default-free discount bond maturing at time T, $T \geq 0$, is a security that will pay one dollar at T and nothing at any other time. Let $P(t,T)$ denote the price of this bond at time $t, 0 \leq t \leq T$. By definition, $P(T,T) = 1$. For $t \leq T$, the *yield to maturity* $R(t,T)$ prevailing at time t is the internal rate of return at time t on a bond with maturity date T:

$$R(t,T) = -\frac{\log P(t,T)}{T-t} \tag{24}$$

$$P(t,T) = e^{-(T-t)R(t,T)} . \tag{25}$$

The following subset of $\mathbf{R}^3$,

$$\{(t,T,R(t,T)) \,|\, T \geq t\},$$

defines the *term structure*.

The *implied forward rates curve*, $\{(s, f(t,s)) \,|\, s \geq \text{fixed } t\}$, is defined by

$$P(t,T) = e^{-\int_t^T f(t,s)ds} , \tag{26}$$

where yet another reference for (26) is [1, (8.3)], which in the literature is usually accompanied by the equivalent expression for the *forward rate for instantaneous borrowing*,

$$f(t,T) = -\frac{\partial}{\partial T} \log P(t,T) , \tag{27}$$

e.g., [10, (1)], [1, (8.4)], and our [15, (37)].

Let us define the instantaneous short (*spot*) rate as

$$r(t) = \lim_{T \to t} R(t, T), \tag{28}$$

from which it follows that (see [3, page 132]),

$$r(t) = -\frac{\partial}{\partial T} \log P(t, t) = f(t, t). \tag{29}$$

see also [10, (3)] and [1, page 172]. Simply differentiating (26) above with respect to t yields,

$$
\begin{aligned}
P_t(t, T) &= \left(e^{-\int_t^T f(t,s)ds} \right)'_t = e^{-\int_t^T f(t,s)ds} \left(-\int_t^T f(t, s)ds \right)'_t \\
&= P(t, T) \left(f(t, t) - \int_t^T f_t(t, s)ds \right) \\
&= P(t, T) \left(r(t) - \int_t^T f_t(t, s)ds \right).
\end{aligned}
\tag{30}
$$

Hence, we obtain

$$P_t(t, T) = \left(r(t) - \int_t^T f_t(t, s)ds \right) P(t, T). \tag{31}$$

The point of departure for constructing analogous linear dynamic systems under uncertainty is the following continuous time stochastic process extension of (17):

$$dr = (\alpha + \beta r)dt + \sigma r^\theta dB_t, \tag{32}$$

where B_t is the standard Brownian motion and α, β, σ and θ are parameters.

In order to construct various types of linear dynamical systems corresponding to different models of the term structure of interest rates we specify admissible parameters and spot rate functions next.

Definition 4.1. Let T^N be chosen according to (22). Introduce

$$
\begin{aligned}
\Omega = \{ \omega \in \mathbf{R}^{N'} &: g_j(\omega) = 0, j = \overline{1, l}; \quad g_j(\omega) \leq 0, \quad j = \overline{l+1, m}; \\
\omega_{*i} \leq \omega &\leq \omega_i^*, \quad i = \overline{1, L} \}, \text{ where } N' \text{ and } N \text{are integers}, \quad N' \geq N.
\end{aligned}
\tag{33}
$$

We shall term Ω the **a priori distribution of unknown parameters**. The functions $g_j(\omega), j = \overline{1, m}$ are twice continuously differentiable, with some number of them, l, generating equality constraints and others generating inequality constraints.

Assumption 4.1. Let $\Re(t)$ be a family of functions defined by

$$\Re(t) = \{h(t; \omega), \quad \omega \in \Omega\}, \quad t \in T^N, \tag{34}$$

where

(1) $h(t,\omega)$ *is twice continuously differentiable for every* $t \in T^N$;
(2) $h(t,\omega)$ *is piecewise continuous for every* $\omega \in \Omega$;
(3) $h(t,\omega)$ *is twice continuously differentiable inside intervals of continuity for every* $\omega \in \Omega$.

Assume that the behavior of the spot rate, $r(\cdot|\omega)$, *(for each fixed* $\omega \in \Omega$*) can be described by the following relations:*

$$r(t|\omega) = h(t,\omega), \quad t \in [t_0, \ t_N], \quad \omega \in \Omega. \tag{35}$$

The form of $h(\cdot,\cdot)$ depends on the type of model used for the term structure of interest rates; for example, the type of differential equation governing the spot rate and the class of the perturbations acting on the model. The set Ω describes the unknown parameters of the *LDEU*, such as the initial state of the spot rate and the coefficients in the *LDEU* together with the values of the perturbations acting on the model.

In the stochastic theory of the term structure there are several expectation type hypotheses for pricing a bond, see, for example, [12, Theories of the Term Structure, p. 86] and [1, Chapter 8]. At first we consider one of them, termed the *unbiased expectations hypothesis, UEH*, and later in Sec. 4.3 we shall consider the more modern case.

The *UEH* assumes that forward rates $f(t,T)$ and the expectation of future short rates, $E(r_T)$ are equal. There is an analog of this assumption occurring in our modelling of uncertainty. Given our parameter set (33) the analog is

$$f(t,T) = r(T|\omega^0) \text{ for all } t, \quad T \geq t \geq 0, \quad \omega^0 \in \Omega, \tag{36}$$

where ω^0 is an estimate of the unknown parameters in the *LDEU* obtained by solving the optimal observation problem. From (38) and (26) it follows that

$$P(t,T) = e^{-\int_t^T r(s|\omega^0)ds}, P(T,T) = 1, \quad \text{or} \tag{37}$$

$$dP_t(t,T) = r(t|\omega^0)P(T,T), \quad P(T,T) = 1. \tag{38}$$

This expression for $P(t,T)$ also appears as [31, 2.5] and [26, 7.23'], where the "short rate" is simply the force of interest used in the theory of compound interest.

Combining (31) with (38) under our analog of the *UEH*, it follows that $f_t(t,s) = 0$, namely, the forward rate doesn't depend on current time t.[b]

---

[b]While in this case we assume that the forward rate $f(t,s)$ doesn't depend on current time t, it really does because of actual one factor modelling to be introduced later in Sec. 4.2. This follows from the fact that our parameter vector of uncertainties, ω, depends on t, i.e.,

$$f(t_1,T) \neq f(t_2,T), \quad \text{for } t_1 \neq t_2;$$

see Sec. 7, Item 10.

We shall assume that the price of a discount bond is determined by the current assessment of spot rate trajectory over the term of the bond, for example, see [(A. 2)][28].

Assumption 4.2. *The price $P(t,T)$, $t < T$, of a discount bond is determined by the assessment at time t of the segment $\{r(\tau|\omega), t \leq \tau \leq T\}$ of the parameterized short rate curve over the term of the bond.*

Clearly for fixed T, Ω, and $h(\cdot, \cdot)$, we can construct a family of trajectories of prices and a family of trajectories of yields of a discount bond from the implied forward rates curve, (26).

A basic question addressed throughout this paper is how to choose a function $r^0(t) = h(t, \omega^0)$, $t \in T^N$ (22), $\omega^0 \in \Omega$, for the purpose of modelling the behavior of a real market. For example, one fruitful answer is to estimate the values of ω^0 by taking observations of Treasury yield curve rates: Bills, Notes, or Bonds.

4.2. *Programmed estimation of the spot riskless rate: one-factor models*

In the sense of [8] there exist *programmed estimates* and *positional estimates*. Programmed estimates are calculated at the conclusion of the process of taking observations. Positional estimates are continuously calculated during process of observation as soon as information from a sensor (observation system) is received. In this section we describe how to obtain the programmed estimate of the spot rate.

Bonds with common common maturities

Assume that we have observed values of the yield to maturity over the interval T^M (22), namely $\hat{R}(t, t+\mathcal{T}), t \in T^M = [t_0, t_M]$, for some given **time-to-maturity** $\mathcal{T}$ during M days.

Remark 4.1. Since interval T_i is the ith day of the observation period, the current moment of is the last day of the observation period, namely t_M. Under this interpretation we use the piecewise constant form of the yield to maturity,

$$\hat{R}(t, t+\mathcal{T}) = R_i^{(\mathcal{T})}, \quad t \in [t_{i-1}, t_i[, \; i = \overline{1, M}, \tag{39}$$

or the piecewise linear form,

$$\hat{R}(t, t+\mathcal{T}) = R_{i-1}^{(\mathcal{T})} + \frac{t - t_{i-1}}{t_i - t_{i-1}} \left(R_i^{(\mathcal{T})} - R_{i-1}^{(\mathcal{T})} \right), \quad t \in T_i, \; i = \overline{1, M}. \tag{40}$$

for describing the observed yield function. Here $R_i^{(\mathcal{T})} i = \overline{0, M}$, denotes the observed Treasury yield with time-to-maturity $\mathcal{T}$ corresponding to the ith day of observation.

Definition 4.2. By the Ω-**based yield** we shall mean the averaged integral

$$p(t, \omega | \mathcal{T}) = \frac{1}{\mathcal{T}} \int_{t}^{t+\mathcal{T}} h(\tau, \omega) \, d\tau \,, \quad t \in T^M, \ \omega \in \Omega \,. \tag{41}$$

By the interval of definition of the spot rate we shall mean T^N, where $N = M + \mathcal{T}$. The **estimation error** is the difference,

$$\varepsilon(t, \omega) = p(t, \omega | \mathcal{T}) - \hat{R}(t, t + \mathcal{T}) \,, \quad t \in T^M, \ \omega \in \Omega \,. \tag{42}$$

We compute the estimate ω^0 of unknown parameter ω by minimizing over $\omega \in \Omega$, the maximum absolute value of the function of estimation errors $\varepsilon(t, \omega)$ on the interval T^M. This leads to the following problem:

$$\min_{\omega \in \Omega} \max_{t \in T^M} |\varepsilon(t, \omega)| \,. \tag{43}$$

Problem (43) may be written as the following nonlinear semi-infinite programming problem, see [11]:

$$\min \quad v \quad \text{subject to}$$

$$p(t, \omega | \mathcal{T}) - v_{\mathcal{T}} \leq \hat{R}(t, t + \mathcal{T}) \,, \quad \hat{R}(t, t + \mathcal{T}) \leq p(t, \omega | \mathcal{T}) + v_{\mathcal{T}} \quad t \in T^M \,; \tag{44}$$

$$\omega \in \Omega \,, \quad v_{\mathcal{T}} \geq 0 \,.$$

We shall call problem (44) the $\mathcal{T}$-**programmed problem of spot rate estimation**.

Let $(\omega^0, v_{\mathcal{T}}^0)$ be the solution of problem (44). Then the function

$$r_{\mathcal{T}}(t) = h(t; \omega^0) \,, \quad t \in T^M, \ N = M + \mathcal{T} \tag{45}$$

will be *an estimate of the short rate* over the observation period, T^M. But over the future period $T^N \backslash T^M$ (45) become a **forecast** of the short rate. This forecast is an extra benefit from solving (44).

We shall call the function $r_{\mathcal{T}}(\cdot)$ defined by (45) $\mathcal{T}$-estimate of the short-term riskless rate. It follows from (26) that the function

$$P(t, s) = e^{- \int_{t}^{s} r_{\mathcal{T}}(\tau) \, d\tau} \,, \quad t \in [0, \, t_M], \ s \in [t, \, t_N] \,, \tag{46}$$

will be an estimate of the price at time t of a discount bond maturing at time s.

Remark 4.2. Note that the function $r_{\mathcal{T}}(t) = h(t, \omega^0)$, while an estimate in the interval $[t_0, \, t_M]$, becomes the forecast in the future interval (the time after the current time t_M), namely, the forecast interval $[t_M, \, t_N]$.

Definition 4.3. If the function $h(t, \omega^0)$ is defined over $t \in [t_N, t_{M^{\mathcal{P}}}]$, where $M^{\mathcal{P}}$ is an integer satisfying $M + \mathcal{T} < M^{\mathcal{P}}$, then the forecasted price of a discount bond maturing at time s is given by:

$$\tilde{P}(t, s) = e^{- \int_{t}^{s} r_{\mathcal{T}}(\tau) \, d\tau} \,, \quad t \in [t_M, \, t_{M^{\mathcal{P}}}] \quad s \in [t, \, t_{M^{\mathcal{P}}}] \,, \tag{47}$$

and the forecasted yield is

$$\tilde{R}(t,s) = \frac{1}{s-t} \int_t^s r_T(\tau) \, d\tau \,, \quad t \in [t_M, \, t_{M^P}], \; s \in [t, \, t_{M^P}]. \tag{48}$$

at time t of discount bond maturing at time s.

Bonds with differing maturities

Assume that we can observe the values of yields to maturity $\hat{R}(t, t+\mathcal{T}(j)), j = \overline{1,L}$, for different time-to-maturities $\mathcal{T}(j)$.

It follows from (38) that when the spot rate is independent of t, then (41) is extended in this case to:

$$R(t, t+\mathcal{T}(j)) = \frac{1}{\mathcal{T}(j)} \int_t^{t+\mathcal{T}(j)} h(\tau, \omega) \, d\tau \,, \quad j = \overline{1,L}. \tag{49}$$

From (44) we obtain the following nonlinear optimization problem:

$$\min \; \sum_{j=1}^L v_j$$

$$p(t,\omega|\mathcal{T}(j)) - v_j \leq \hat{R}(t, t+\mathcal{T}(j)), \quad \hat{R}(t, t+\mathcal{T}(j)) \leq \; p(t,\omega|\mathcal{T}(j)) + v_j \,, \tag{50}$$

$$t \in T^M \,, \quad \omega \in \Omega \,, \quad v_j \geq 0, \; j = \overline{1,L}.$$

We shall call the problem (50) *the programmed problem of total STRR estimation.*

Let (ω^0, v^0), $v^0 = (v_1^0, \ldots, v_L^0)$ be the solution of the problem (50), and recall the definition of T^N. (22) It follows that the function,

$$r(t) = h(t, \omega^0), \quad t \in T^N, \quad N = M + \max_{j=\overline{1,L}} \mathcal{T}(j) \tag{51}$$

will be an estimate of the spot riskless rate. We shall call the function $r(\cdot)$ defined by (51) the *total estimate of the short-term riskless rate (STRR)*. Analogous to theoretical developments we are able to compute one spot rate based on observations on bond yields for differing maturities.

In this section we consider some models associated with the stochastic theory of the term structure of interest rates, in particular, the stochastic differential equation of Vasicek, (32).

4.2.1. *Analog of Vasicek model with impulse perturbations*

Assume that the spot rate is governed by the following linear dynamic system *LDEU* with unknown parameters and nonstochastic uncertainty:

$$\dot{r} = \alpha + \beta r(t) + w(t) \,, \quad \beta \neq 0, \quad r(t_0) = r_0 \,; \quad t \in T^N. \tag{52}$$

Assume that *a priori* information about the unknown parameters of the *LDEU* (52) takes the form

$$\alpha_* \le \alpha \le \alpha^*, \quad \beta_* \le \beta \le \beta^*, \quad r_* \le r_0 \le r^*, \tag{53}$$

and that the perturbations, $w(\cdot)$, are of the *impulse* form defined by the following constraints:

$$w(t) = w_i, \quad w_{*i} \le w_i \le w_i^*, \quad t \in T_i, \ i + \overline{1, N}. \tag{54}$$

It follows from (33)–(34) that for model (52)–(54) we have

$$\omega = (r_0, \alpha, \beta, w_i, \ i = \overline{1, N}); \ \omega_* = (r_*, \alpha_*, \beta_*, w_{*i}, \ i = \overline{1, N});$$

$$\omega^* = (r^*, \alpha^*, \beta^*, w_i^*, \ i = \overline{1, N}), \tag{55}$$

$$\Omega = \{\omega \in \mathbf{R}^{N+3} : \omega_* \le \omega \le \omega^*\}.$$

By means of the Cauchy formula the solution of the differential Eq. (52) has the form

$$r(t|\omega) = e^{\beta t} r_0 + \frac{\alpha}{\beta}(e^{\beta t} - 1) + \sum_{j=1}^{i-1} w_j \frac{e^{\beta t}}{\beta}(e^{-\beta t_{j-1}} - e^{-\beta t_j}) + w_i \frac{1}{\beta}(e^{\beta(t-t_{i-1})} - 1), \tag{56}$$

$$t \in T_i, \quad i = \overline{1, N}.$$

In this way we have defined a member of the family Ω, see Assumption 4.1, namely

$$h(t, \omega) = r(t|\omega), \quad t \in T^N, \ \omega \in \Omega. \tag{57}$$

We derive an explicit form of problem (44), by first specifying $p(t, \omega|\mathcal{T})$, $t \in T_i$, $i = \overline{1, M}$. We proceed as follows, recalling that $\mathcal{T}$ is the time-to-maturity.

$$p(t, \omega|\mathcal{T})$$

$$= \frac{1}{\mathcal{T}} \int_t^{t+\mathcal{T}} h(\tau, \omega) \, d\tau = \frac{1}{\mathcal{T}} \left[r_0 \int_t^{t+\mathcal{T}} e^{\beta \tau} \, d\tau + \frac{\alpha}{\beta} \int_t^{t+\mathcal{T}} (e^{\beta \tau} - 1) \, d\tau \right.$$

$$+ \int_{\mathcal{T}}^{t_i} \left(\sum_{k=1}^{i-1} w_k \frac{e^{\beta \tau}}{\beta}(e^{-\beta t_{k-1}} - e^{-\beta t_k}) + w_i \frac{e^{\beta(\tau - t_{i-1})} - 1}{\beta} \right) d\tau$$

$$+ \sum_{j=i+1}^{i+\mathcal{T}-1} \int_{t_{j-1}}^{t_j} \left(\sum_{k=1}^{j-1} w_k \frac{e^{\beta \tau}}{\beta}(e^{-\beta t_{k-1}} - e^{-\beta t_k}) + w_j \frac{e^{\beta(\tau - t_{j-1})} - 1}{\beta} \right) d\tau$$

$$
+ \int_{t_i+\mathcal{T}-1}^{t+\mathcal{T}} \left(\sum_{k=1}^{i+\mathcal{T}-1} w_k \frac{e^{\beta\tau}}{\beta} (e^{-\beta t_{k-1}} - e^{-\beta t_k}) + w_i + \tau \frac{e^{\beta(\tau - t_{i+\mathcal{T}-1})} - 1}{\beta} \right) d\tau \Bigg]
$$

$$
= \frac{1}{\mathcal{T}} \Bigg[\frac{e^{\beta(t+\mathcal{T})} - e^{\beta t}}{\beta} r_0 + \left(\frac{e^{\beta(t+\mathcal{T})} - e^{\beta t}}{\beta} - \frac{\mathcal{T}}{\beta} \right) \alpha
$$

$$
+ \sum_{k=1}^{i-1} \frac{e^{-\beta t_{k-1}} - e^{-\beta t_k}}{\beta} w_k \int_t^{t+\mathcal{T}} e^{\beta\tau}\, d\tau
$$

$$
+ \left(\frac{1}{\beta} \int_t^{t_i} (e^{\beta(\tau - t_{i-1})} - 1)\, d\tau + \frac{e^{-\beta t_{i-1}} - e^{-\beta t_i}}{\beta} \int_{t_i}^{t+\mathcal{T}} e^{\beta\tau}\, d\tau \right) w_i
$$

$$
+ \sum_{k=i+1}^{i+\mathcal{T}-1} \left(\frac{1}{\beta} \int_{t_{k-1}}^{t_k} (e^{\beta(\tau - t_{k-1})} - 1)\, d\tau + \frac{e^{-\beta t_{k-1}} - e^{-\beta t_k}}{\beta} \int_{t_k}^{t+\mathcal{T}} e^{\beta\tau}\, d\tau \right) w_k
$$

$$
+ \frac{1}{\beta} \int_{t_i+\mathcal{T}-1}^{t+\mathcal{T}} (e^{\beta(\tau - t_{i+\mathcal{T}-1})} - 1) w_{i+\mathcal{T}}\, d\tau \Bigg]
$$

$$
= \frac{e^{\beta(t+\mathcal{T})} - e^{\beta t}}{\mathcal{T}\beta} r_0 + \left(\frac{e^{\beta(t+\mathcal{T})} - e^{-\beta t}}{\mathcal{T}\beta^2} - \frac{1}{\beta} \right) \alpha
$$

$$
+ \sum_{k=1}^{i-1} \frac{e^{\beta(t - t_{k-1} + \mathcal{T})} - e^{\beta(t - t_k + \mathcal{T})} + e^{\beta(t - t_k)} - e^{\beta(t - t_{k-1})}}{\mathcal{T}\beta^2} w_k
$$

$$
+ \left(\frac{e^{\beta(t - t_{i-1} + \mathcal{T})} - e^{\beta(t - t_i + \mathcal{T})} + e^{\beta(t - t_{i-1})} + 1}{\mathcal{T}\beta^2} - \frac{t_i - t}{\mathcal{T}\beta} \right) w_i
$$

$$
+ \sum_{k=i+1}^{i+\mathcal{T}-1} \left(\frac{e^{\beta(t - t_{k-1} + \mathcal{T})} - e^{\beta(t - t_k + \mathcal{T})}}{\mathcal{T}\beta^2} - \frac{t_k - t_{k-1}}{\mathcal{T}\beta} \right) w_k
$$

$$
+ \left(\frac{e^{\beta(t - t_{i+\mathcal{T}-1} + \mathcal{T})} - 1}{\mathcal{T}\beta^2} - \frac{t - t_{i+\mathcal{T}-1} + \mathcal{T}}{\mathcal{T}\beta} \right) w_{i+\mathcal{T}}, \quad t \in T_i. \tag{58}
$$

$$
a_k^i(\beta, t|\mathcal{T})
$$

$$
= \begin{cases}
\dfrac{e^{\beta(t - t_{k-1} + \mathcal{T})} - e^{\beta(t - t_k + \mathcal{T})} + e^{\beta(t - t_k)} - e^{\beta(t - t_{k-1})}}{\mathcal{T}\beta^2} & \text{if } k < i, \\[2ex]
\dfrac{e^{\beta(t - t_{k-1} + \mathcal{T})} - e^{\beta(t - t_k + \mathcal{T})} - e^{\beta(t - t_{k-1})} + 1}{\mathcal{T}\beta^2} - \dfrac{t_k - t}{\mathcal{T}\beta} & \text{if } k = i, \\[2ex]
\dfrac{e^{\beta(t - t_{k-1} + \mathcal{T})} - e^{\beta(t - t_k + \mathcal{T})}}{\mathcal{T}\beta^2} - \dfrac{t_k - t_{k-1}}{\mathcal{T}\beta} & \text{if } i < k < i + \mathcal{T}, \\[2ex]
\dfrac{e^{\beta(t - t_{k-1} + \mathcal{T})} - 1}{\mathcal{T}\beta^2} - \dfrac{t - t_{k-1} + \mathcal{T}}{\mathcal{T}\beta} & \text{if } k = i + \mathcal{T}.
\end{cases}
$$

$$
\tag{59}
$$

Then from (44)–(59) we obtain the following nonlinear semi-infinite programming problem:

$$\min \quad v_{\mathcal{T}} \quad \text{subject to}$$

$$\frac{e^{\beta(t+\mathcal{T})} - e^{\beta t}}{\mathcal{T}\beta} r_0 + \left(\frac{e^{\beta(t+\mathcal{T})} - e^{\beta t}}{\mathcal{T}\beta^2} - \frac{1}{\beta} \right) \alpha$$

$$+ \sum_{k=1}^{i+\mathcal{T}} a_k^i(\beta, t|\mathcal{T}) W_k - v_{\mathcal{T}} \leq \hat{R}(t, t+\mathcal{T}),$$

$$\hat{R}(t, t+\mathcal{T}) \leq \frac{e^{\beta(t+\mathcal{T})} - e^{\beta t}}{\mathcal{T}\beta} r_0 \tag{60}$$

$$+ \left(\frac{e^{\beta(t+\mathcal{T})} - e^{\beta t}}{\mathcal{T}\beta^2} - \frac{1}{\beta} \right) \alpha + \sum_{k=1}^{i+\mathcal{T}} a_k^i(\beta, t|\mathcal{T}) W_k + v_{\mathcal{T}},$$

$$t \in T_i, i = \overline{1, M};$$

$$\alpha_* \leq \alpha \leq \alpha^*, \quad \beta_* \leq \beta \leq \beta^*, \quad r_* \leq r_0 \leq r^*, \quad v \geq 0; \quad w_* \leq w_i \leq w^*,$$

$$i = \overline{1, N}.$$

Let $(\omega_0, v_{\mathcal{T}}^0)$ be the solution of problem (60). Then the function

$$r_{\mathcal{T}}(t) = r(t|\omega^0), \quad t \in T^M,$$

will be the **T-forecast of the spot riskless rate**, respectively. From (56) it follows that for some $M^P > N$ we have

$$r_{\mathcal{T}}(t) = e^{\beta^0 t} r_0^0 + \frac{\alpha^0}{\beta^0} (e^{\beta^0 t} - 1) + \sum_{j=1}^{N} w_j^0 \frac{e^{\beta^0 t}}{\beta^0} (e^{-\beta^0 t_{j-1}} - e^{-\beta^0 t_j}),$$

$$t \in T_i, \quad i = \overline{N+1, M^P}.$$

Hence we obtain a forecast of yields by means of Eqs. (47), (48) for any points, t and s, $t \in [0, t_{M^P}]$, and $s \in [t, t_{M^P}]$.

As a rule, in the theory of the term structure of interest rates the price of the bond has the following form (where we supress the dependence on r):

$$P(t, s) = \exp\{A(t, s) + r(t)B(t, s)\}, \quad t \leq s, \tag{61}$$

with $A(0, 0) = 0$ and $B(0, 0) = 0$. From (26) and (56) we obtain (see Appendix A) the following forms of, $A(t, s)$ and $B(t, s)$:

$$B(t,s) = \frac{1 - e^{\beta(s-t)}}{\beta},$$

$$A(t,s) = \frac{\alpha}{\beta}[(s-t) + B(t,s)] + \left(\frac{e^{\beta s}}{\beta}(e^{\beta t_i} - e^{-\beta t}) + t_i - t\right)\frac{w_i}{\beta}$$

$$- \sum_{k=i+1}^{j-1} \left(\frac{e^{\beta s}}{\beta}(e^{-\beta t_{k-1}} - e^{-\beta t_k}) - (t_k - t_{k-1})\right)\frac{w_k}{\beta}$$

$$- \frac{w_i}{\beta_j}\left(\frac{e^{\beta(s-t_{j-1})} - 1}{\beta} - (s - t_{j-1})\right), \quad t \in T_i, \ s \in T_j, \text{ and } i \le j.$$

In Tables 1–5 to follow we use the following common legend:

- ITER – number of iterations for solving support problems
- NFUN – number of times of the objective and the constraint functions are evaluated during solving support problems
- NGRAD – number of times the gradients of the objective and the constraint functions are evaluated during the support problems
- NQP – number of QP subproblems solved during the support problems
- ACCSP – stopping tolerance for solving the support problems
- NSP – number of support problems solved
- NCONFP – maximum number of constraints in the generated support problems
- CPU – elapsed CPU using a PC PENTIUM 120, coding in BORLAND PASCAL 6.0
- NPROB – number of the problem.

Table 1. Analog of Vasicek model with impulse perturbations under (44) and (39).

NPROB	1	2	3	4	5
β_0	−0.1	−0.5	−1.0	−1.5	−5.0
r_0^0	0.4672776	0.130256	0.136899	0.132424	0.163735
α^0	1.193621	2.401215	4.190845	4.453236	9.8241185
β^0	−0.2326357	−0.474331	−0.832853	−0.885401	−1.960342
$-\frac{\alpha^0}{\beta^0}$	5.130859	5.062310	5.031914	5.029626	5.011431
v_T^0	0.045	0.045	0.045	0.045	0.045
ITER	30	36	29	30	37
NFUN	30	38	29	30	37
NQP	30	36	29	30	37
NGRAD	30	36	29	30	37
ACCSP	10^{-5}	10^{-5}	10^{-7}	10^{-7}	10^{-7}
NSP	3	2	2	2	2
NCONSP	174	125	125	125	125
CPU(sec)	376.8	415.43	325.78	370.04	385.05

Example 4.1. For testing our approach we considered the observed values of Treasury yield rates from 1 Mar 96 to 31 Mar 96 for a time-to-maturity of 3 months. We choose the model (39) for modelling values of observed yield. The program code was written on BORLAND PASCAL 6.0 and run on PC PENTIUM 120 computer. In this example $M = 31$, $T = 91$, $N = 122$. The following initial point given in (44) was selected:

$$\alpha_0 = 0, \quad \beta_0, \quad r_0 = 0, \quad v_T^0 = 0, \quad w_k = 0, \quad k = \overline{1, N}.$$

The numerical results for different initial points are given in Table 1.

We observe that the problem (50) has optimal solutions.

Figure 1 illustrates the obtained estimate $p(\cdot|\cdot)$ of yield, the observed yield $R(\cdot, \cdot)$ and the obtained estimate of the spot rate $r(\cdot)$ for NPROB $= 1$. Figure 2 illustrates the trust region for the yield for NPROB $= 1$. Figure 3 illustrates the forecast for the spot rate for NPROB $= 1$. Figure 4 illustrates the estimation error for yield for NPROB $= 1$.

4.2.2. *Analog of Vasicek model with point-impulse perturbations*

Assume that the short rate follows model (52)–(53). Recall (22) for $\tilde{T}^N$. According to Definition 2.1 point-impulse perturbations $\{w(t), t \in T^N\}$ acting on *LDEU* (52)–(53) are specified by:

(1) $w(t) = 0, \quad t \in T^N \setminus \tilde{T}^N$ (62)

(2) the trajectory $r(t), \ t \in T^N$ of the system (52)–(53) is described by the following differential equation

$$\dot{r} = \alpha + \beta r(t), \tag{63}$$

on the intervals $]t_{i-1}, t_i[, \ i = \overline{1, N}$ and

(3) the trajectory $r(t), \ t \in t^N$ has jumps

$$r(t_i + 0) = r(t_i - 0) + w(t_i), \quad w_{i*} \leq w_i \equiv w(t_i) \leq w_i^*, \tag{64}$$

at the points $t_i \in \tilde{T}^N$, $\gamma \leq 0$.

By means of the Cauchy formula the solution of the differential Eq. (52) with perturbations defined by (62)–(64) has the form

$$r(t|\omega) = e^{\beta t} r_0 + \frac{\alpha}{\beta}(e^{\beta t} - 1) + \sum_{j=1}^{i} e^{\beta(t - t_{j-1})} w_j, \quad t \in T_i, \quad i = \overline{1, N}, \tag{65}$$

where $w_i = w(t_{i-1})$, $t_{i-1} \in \tilde{T}$, $i = \overline{1, N}$. Define the function $p(t, \omega|\mathcal{T})$, $t \in T_i$, $i = \overline{1, M}$ by (41), (57). From (65) it follows that

$$\mathcal{T}p(t,\omega|\mathcal{T}) = \int_t^{t+\mathcal{T}} h(\tau,\omega)\,d\tau$$

$$= \left(r_0 \int_t^{t+\mathcal{T}} e^{\beta\tau}\,d\tau + \frac{\alpha}{\beta}\int_t^{t+\mathcal{T}} (e^{\beta\tau}-1)\,d\tau \right.$$

$$+ \sum_{k=1}^{i} \int_t^{t+\mathcal{T}} w_k e^{\beta(\tau-t_{k-1})}\,d\tau$$

$$+ \left. \sum_{k=i+1}^{i+T} \int_{t_{k-1}}^{t+\mathcal{T}} w_k e^{\beta(\tau-t_{k-1})}\,d\tau \right) \tag{66}$$

$$= \left(\frac{e^{\beta(t+\mathcal{T})}-e^{\beta t}}{\beta} r_0 + \left(\frac{e^{\beta(t+\mathcal{T})}-e^{\beta t}}{\beta^2} - \frac{\mathcal{T}}{\beta} \right)\alpha \right)$$

$$+ \sum_{k=1}^{i} \frac{e^{\beta(t-t_{k-1}+\mathcal{T})}-e^{\beta(t-t_{k-1})}}{\beta} w_k$$

$$+ \sum_{k=i+1}^{i+\mathcal{T}} \frac{e^{\beta(t-t_{k-1}+\mathcal{T})}-1}{\beta} w_k.$$

Let

$$a_k^i(\beta,t|\mathcal{T}) = \begin{cases} \dfrac{e^{\beta(t-t_{k-1}+\mathcal{T})}-e^{\beta(t-t_{k-1})}}{\beta\mathcal{T}} & \text{if } k \le i, \\[3mm] \dfrac{e^{\beta(t-t_{k-1}+\mathcal{T})}-1}{\beta\mathcal{T}} & \text{if } i < k \le i+\mathcal{T}. \end{cases} \tag{67}$$

Hence, the **$\mathcal{T}$-programmed problem for spot riskless rate** estimation has the form

min $\quad v_{\mathcal{T}}$ subject to

$$\frac{e^{\beta(t+\mathcal{T})}-e^{\beta t}}{\mathcal{T}\beta} r_0 + \left(\frac{e^{\beta(t+\mathcal{T})}-e^{\beta t}}{\mathcal{T}\beta^2} - \frac{1}{\beta} \right)\alpha + \sum_{k=1}^{i+\mathcal{T}} a_k^i(\beta,t|\mathcal{T}) W_k - v_{\mathcal{T}} \le \hat{R}(t,t+\mathcal{T}),$$

$$\hat{R}(t,t+\mathcal{T}) \le \frac{e^{\beta(t+\mathcal{T})}-e^{\beta t}}{\mathcal{T}\beta} r_0 + \left(\frac{e^{\beta(t+\mathcal{T})}-e^{\beta t}}{\mathcal{T}\beta^2} - \frac{1}{\beta} \right)\alpha \tag{68}$$

$$+ \sum_{k=1}^{i+\mathcal{T}} a_k^i(\beta,t|\mathcal{T}) W_k + v_{\mathcal{T}}, \quad t \in T_i, i = \overline{1,M};$$

$$\alpha_* \le \alpha \le \alpha^*, \quad \beta_* \le \beta \le \beta^*, \quad r_* \le r_0 \le r^*, \quad v_{\mathcal{T}} \ge 0;$$

$$w_* \le w_i \le w^*, \quad i = \overline{1,N}.$$

Table 2. Analog Vasicek model: point impulse (39) and (68).

NPROB	1	2	3	4
β_0	−0.1	−0.5	−1.0	−1.5
r_0^0	0.550814	0.0991620	0.113943	0.113943
α^0	1.343350	2.4453550	4.196010	4.196010
β^0	−0.261763	−0.4832893	−0.834356	−0.834356
$-\frac{\alpha^0}{\beta^0}$	5.131932	5.059816	5.029040	5.029040
v_T^0	0.045	0.045	0.045	0.045
ITER	19	28	25	25
NFUN	19	28	25	25
NQP	19	28	25	25
NGRAD	19	28	25	25
ACCSP	10^{-5}	10^{-5}	10^{-5}	10^{-5}
NSP	2	2	2	2
NCONSP	125	125	125	125
CPU(sec)	204.8	282.27	254.68	253.71

The functions $A(t, s)$ and $B(t, s)$ for (61) have the form (see Appendix A) $B(t, s) = \frac{1 - e^{\beta(s-t)}}{\beta}$,

$$A(t, s) = \frac{\alpha}{\beta}(s - t + B(t, s)) + \frac{1}{\beta}\sum_{k=i+1}^{j}(1 - e^{\beta(s - t_{k-1})})w_k \, .$$

Example 4.2. Consider the data again from Example 4.1. The following initial point

$$\alpha_0 = 0, \quad \beta_0, \quad r_0 = 0, \quad v_0^T = 0, \quad w_k = 0, \quad k = \overline{1, N}$$

was taken for problem (68). The numerical results for different initial points are given in Table 2.

Clearly, problem (68) has optimal solutions.

Figure 5 illustrates the obtained estimate $p(\cdot|\cdot)$ of yield, the observed yield $R(\cdot, \cdot)$ and the obtained estimate of the spot rate $r(\cdot)$ for NPROB = 1. Figure 6 illustrates the obtained estimates of the spot rate $r(\cdot)$ for different initial points β_0. Figure 7 the obtained estimate $p(\cdot|\cdot)$ of yield curve for different intial points β_0. Figure 8 illustrates the obtained estimates of the spot rate $r(\cdot)$ from Examples 4.1 and 5.1. Figure 9 illustrates the obtained estimate $p(\cdot|\cdot)$ of yield from Examples 4.1 and 5.1.

4.2.3. *Analog of Dothan model with impulse perturbations*

Assume that the spot rate follows another linear dynamic system *LDEU* of *Dothan type* with unknown parameters and nonstochastic uncertainty:

$$\dot{r} = rw(t), \quad r(t_0) = r_0; \quad t \in T^N \, . \tag{69}$$

Assume that *a priori* information about unknown parameters of the *LDEU* (69) has the form (53) while the perturbations, $w(\cdot)$, are of impulse type (54). It follows from (33)–(34) that for the model (69), (54) we have

$$\omega = (r_0, w_i, \ i = \overline{1,N}); \quad \omega_* = (r_*, w_{*i}, \ i = \overline{1,N}), \quad \omega^* = (r^*, w_i^*, \ i = \overline{1,N});$$

$$\Omega = \{\omega \in \mathbf{R}^{N+1} : \omega_* \le \omega \le \omega^*\}.$$

By means of the Cauchy formula the solution of the differential Eq. (69) has the form

$$r(t|\omega) = r_0 e^{\sum_{k=1}^{i-1} w_k(t_k - t_{k-1}) + w_i(t - t_{i-1})}, \quad t \in T_i, \quad i = \overline{1,N}. \tag{70}$$

The function $h(\cdot,\cdot)$ is specified by (57) via (70) (analogous to the procedure (56)–(57) for the analog of the Vasicek model), and $p(t,\omega|\mathcal{T})$, is specified next by formula (41), namely,

$$p(t,\omega|\mathcal{T}) = \frac{1}{\mathcal{T}} \int_t^{t+\mathcal{T}} h(\tau,\omega)\, d\tau$$

$$= \frac{1}{\mathcal{T}} \left(\int_t^{t_i} h(\tau,\omega)\, d\tau + \sum_{k=i+1}^{i+\mathcal{T}-1} \int_{t_{k-1}}^{t_k} h(\tau,\omega)\, d\tau + \int_{t_i+\mathcal{T}-1}^{t+\mathcal{T}} h(\tau,\omega)\, d\tau \right)$$

$$= \frac{r_0}{\mathcal{T}} \left(e^{\sum_{k=1}^{i-1} w_k(t_k-t_{k-1})} \int_t^{t_i} e^{w_i(\tau-t_{i-1})}\, d\tau \right.$$

$$+ \sum_{k=i+1}^{i+\mathcal{T}-1} e^{\sum_{j=1}^{k-1} w_k(t_j-t_{j-1})} \int_{t_{k-1}}^{t_k} e^{w_k(\tau-t_{k-1})}\, d\tau$$

$$\left. + e^{\sum_{k=1}^{i+\mathcal{T}-1} w_k(t_k-t_{k-1})} \int_{t_i+\mathcal{T}-1}^{t+\mathcal{T}} e^{w_{i+\mathcal{T}-1}(\tau-t_{i+\mathcal{T}-1})}\, d\tau \right)$$

$$= \frac{r_0}{\mathcal{T}} \left(e^{\sum_{k=1}^{i-1} w_k(t_k-t_{k-1})} \frac{e^{w_i(t_i-t_{i-1})} - e^{w_i(t-t_{i-1})}}{w_i} \right.$$

$$+ \sum_{k=i+1}^{i+\mathcal{T}-1} e^{\sum_{j=1}^{k-1} w_k(t_j-t_{j-1})} \frac{e^{w_k(t_k-t_{k-1})} - 1}{w_k}$$

$$\left. + e^{\sum_{k=1}^{i+\mathcal{T}-1} w_k(t_k-t_{k-1})} \frac{e^{w_{i+\mathcal{T}-1}(t+\mathcal{T}-t_{i+\mathcal{T}-1})} - 1}{w_{i+\mathcal{T}}} \right); \ t \in T_i, \quad i = \overline{1,N}. \tag{71}$$

By substituting (71) into (44) we obtain the *NSIP* problem for estimating the spot rate $r_{\mathcal{T}}(\cdot)$. However, some additional constraints are needed in order to guarantee that $w_i \neq 0$, $i = \overline{1,N}$. The functions $A(t,s)$, $B(t,s)$ for the bond pricing formula (61) have the form developed at the end of Appendix C.

$$B(t,s) = \frac{1}{w_i} - \frac{1}{w_j} e^{\sum_{k=i}^{j+1} w_k(t_k - t_{k-1}) + w_j(s - t_{j-1}) - w_i(t - t_{i-1})},$$

and
$$A(t,s) = -\frac{r_0}{w_i} e^{\sum_{k=1}^{i} w_k(t_k - t_{k-1})}$$

$$- r_0 \sum_{k=i+1}^{j-1} \frac{e^{\sum_{l=1}^{k} w_l(t_l - t_{l-1})} - e^{\sum_{l=1}^{k-1} w_l(t_l - t_{l-1})}}{dw_k}$$

$$+ \frac{r_0}{w_j} e^{\sum_{k=1}^{j-1} w_k(t_k - t_{k-1})}.$$

4.2.4. *Analog of Courtadon model with impulse perturbations*

Assume that the spot rate follows linear dynamic system *LDEU* of *Courtadon type* with unknown parameters and nonstochastic uncertainty:

$$\dot{r} = \alpha + \beta r + w(t)r, \quad r(t_0) = r_0; \quad t \in T^N. \tag{72}$$

In addition, we assume that the *a priori* information about unknown parameters of the *LDEU* (69) has the form (53) and the perturbations $w(\cdot)$ are still of impulse form, (54). Then ω and Ω, are defined by (55). Using the Cauchy formula again the solution of the differential Eq. (72) has the form

$$r(t|\omega) = -\frac{\alpha}{\beta + w_i} + \left(\frac{\alpha}{\beta + w_i} + r(t_{i-1})\right) e^{(\beta + w_i)(t - t_{i-1})}, \ t \in T_i, \ i = \overline{1,N}, \tag{73}$$

$$r(t|\omega) = -\frac{\alpha}{\beta + w_i} + \gamma_i(\omega) e^{(\beta + w_i)(t - t_{i-1})}, \quad t \in T_i, \quad i = \overline{1,N}, \tag{74}$$

where

$$\gamma_1(\omega) = \frac{\alpha}{\beta + w_1} + r_0,$$

$$\gamma_2(\omega) = \frac{\alpha}{\beta + w_2} - \frac{\alpha}{\beta + w_1} + \left(\frac{\alpha}{\beta + w_1} + r_0\right) e^{(\beta + w_1)(t_1 - t_0)},$$

$$\gamma_i(\omega) = \left(\frac{\alpha}{\beta + w_i} - \frac{\alpha}{\beta + w_{i-1}}\right) + \left(\frac{\alpha}{\beta + w_1} + r_0\right) \cdot e^{\sum_{k=1}^{i-1}(\beta + w_k)(t_k - t_{k-1})}$$

$$+ \sum_{k=2}^{i-1} \left(\frac{\alpha}{\beta + w_k} - \frac{\alpha}{\beta + w_{k-1}}\right) e^{\sum_{j=k}^{i-1}(\beta + w_j)(t_j - t_{j-1})}, \quad i = \overline{3,N}.$$

Define the function $h(\cdot|\cdot)$ by (57), and find the function $p(t,\omega|\mathcal{T})$, $t \in T_i$, $i = \overline{1,M}$. From (74), (41) it follows that

$$p(t,\omega|\mathcal{T}) = \frac{1}{\mathrm{cal}T} \int_t^{t+\mathcal{T}} h(\tau,\omega)\,d\tau$$

$$= \frac{1}{\mathcal{T}} \left(\int_t^{t_i} h(\tau,\omega)\,d\tau + \sum_{k=i+1}^{i+\mathcal{T}-1} \int_{t_{k-1}}^{t_k} h(\tau,\omega)\,d\tau + \int_{t_i+\mathcal{T}-1}^{t+\mathcal{T}} h(\tau,\omega)\,d\tau \right)$$

$$= \frac{1}{\mathcal{T}} \left(-\frac{\alpha}{\beta+w_i}(t_i-t) + \frac{\gamma_i(\omega)}{\beta+w_i}(e^{(\beta+w_i)(t_i-t_{i-1})} - e^{(\beta+w_i)(t_i-t_{i-1})}) \right.$$

$$+ \sum_{k=i+1}^{i+\mathcal{T}-1} \left(-\frac{\alpha}{\beta+w_k}(t_k-t_{k-1}) + \frac{\gamma_k(\omega)}{\beta+w_k}(e^{(\beta+w_k)(t_k-t_{k-1})} - 1) \right)$$

$$-\frac{\alpha}{\beta+w_{i+\mathcal{T}}}(t+\mathcal{T}-t_{t_i+\mathcal{T}-1})$$

$$\left. + \frac{\gamma_{i+\mathcal{T}}(\omega)}{w_{i+\mathcal{T}}+\beta}(e^{(\beta+w_{i+\mathcal{T}})(t+\mathcal{T}-t_{i+\mathcal{T}-1})} - 1) \right). \tag{75}$$

By substituting (75) into (44) we obtain the *NSIP* problem for estimating the spot rate $r_{\mathcal{T}}(\cdot)$. However, some additional constraints are needed in order to guarantee that $\beta + w_i \neq 0$, $i = \overline{1,N}$.

4.2.5. *A deterministic model studied by Boyle [4]*

Assume that the spot rate follows the linear dynamic system (*LDEU*) with unknown parameters [27], [4, (5)]:

$$\dot{r} = \alpha(\gamma - r), \quad r(t_0) = r_0; \quad t \in T^N. \tag{76}$$

The *a priori* information about unknown parameters of the *LDEU* (76) shall be of the following form:

$$\omega = (r_0, \alpha, \gamma); \quad \omega_* = (r_*, \alpha_*, \gamma_*); \quad \omega^* = (r^*, \alpha^*, \gamma^*);$$

$$\Omega = \{\omega \in \mathbf{R}^3 : \omega_* \leq \omega \leq \omega^*\}.$$

Using the Cauchy formula the solution of the differential Eq. (76) is

$$r(t|\omega) = \gamma + (r_0 - \gamma)e^{-\alpha(t-t_0)}, \quad t \in T^N, i = \overline{1,N}. \tag{77}$$

Define the function $f(\cdot,\cdot)$ by (57), and derive the function $p(t,\omega|T)$, $t \in T_i$, $i = \overline{1,M}$ according to (77) and (41). It follows that

$$p(t, \omega | T) = \frac{1}{T} \int_t^{t+T} f(\tau, \omega) \, d\tau$$

$$= \frac{1}{T} \int_t^{t+T} \left(\gamma + (r_0 - \gamma) e^{-\alpha(\tau - t_0)} \right) d\tau$$

$$= \gamma + \frac{r_0 - \gamma}{\alpha T} \left(e^{-\alpha(t-t_0)} - e^{-\alpha(t+T-t_0)} \right). \tag{78}$$

By substituting (78) into (44) we obtain the following NSIP problem for estimating the unknown parameters ω of the *LDEU* (76):

$$\min \quad v_T \quad \text{subject to}$$

$$\gamma + \frac{r_0 - \gamma}{\alpha T} (e^{-\alpha(t-t_0)} - e^{-\alpha(t+T-t_0)}) - v_T \le \hat{R}(t, t+T),$$

$$\hat{R}(t, t+T) \le \gamma + \frac{r_0 - \gamma}{\alpha T} (e^{-\alpha(t-t_0)} - e^{-\alpha(t+T-t_0)}) + v_T, \quad t \in T^M; \tag{79}$$

$$\alpha_* \le \alpha \le \alpha^*, \quad \gamma_* \le \gamma \le \gamma^*, \quad r_* \le r_0 \le r^*, \quad v_T \ge 0.$$

Example 4.3. Let us consider the observed values of Treasury yield rates for maturity term 3 months from 1 Oct 1995 to 31 Dec 1995. Let us choose the model (39) for modelling values of observed yield. We started from initial points α_0, r_0, γ_0. The numerical results for different intial points are given in the Table 3.

Figure 14 illustrates the obtained estimates $p(\cdot|\cdot)$ of yield and the observed yield $R(\cdot, \cdot)$ for NPROB $= 1, 2$. Figure 15 illustrates the obtained estimates of the spot rate $r(\cdot)$ for NPROB $= 1, 2, 3$.

Table 3. Deterministic model (76) under (39) with optimization (79).

NPROB	1	2	3
α_0	1	5	10
γ_0	1	5	10
r_0	1	5	10
r_0^0	1	5	10
α^0	1	5	10
γ^0	5.285	5.285	5.285
v^{T0}	0.305	0.305	0.305
ITER	4	2	2
NFUN	4	2	2
NQP	4	2	2
NGRAD	4	2	2
ACCSP	10^{-5}	10^{-5}	10^{-5}
NCONSP	2024	2024	2024
CPU(sec)	< 1	< 1	< 1

In Example 4.3, we see that the estimate of γ is equal to 5.285 in all cases. The maximum value of observed Treasury yield rate is equal to 5.59, and the minimum value is equal to 4.98. So, the estimate of γ is equal to median value of the observed Treasury yield rate and the error of estimation is equal to half of difference between maximum value of observed Treasury yield rate and minimum value of observed Treasury yield rate on *whole interval of observation* (0.305). For our previously considered models with disturbances, the estimation error is equal to *half of maximum jump in the observed Treasury yield rate* (0.065 for these data). In Sec. 4.3 we describe a model which takes into account the jumps occurring in the observed Treasury yield rates.

4.2.6. *A nondifferential equation, continuous-time model*

The models from previous sections were constructed by means of linear differential equations with uncertainty. In this section we consider another way of modelling the problem of estimating spot rates.

Assume that the spot rate $r(\cdot)$ belongs to the class of impulse functions of the form

$$r(t) = r_i, \quad r_{*i} \le r_i \le r_i^*; \quad t \in T_i, \quad i = \overline{1, N}. \tag{80}$$

In this case we obtain

$$\omega = (r_i, \, i = \overline{1, N}); \quad \omega_* = (r_{*i}, \, i = \overline{1, N}); \quad \omega^* = (r_i^*, \, i = \overline{1, N});$$

$$\Omega = \{\omega \in \mathbf{R}^N : \omega_* \le \omega \le \omega^*\}, \text{ with corresponding yield:} \tag{81}$$

$$p(t, \omega | \mathcal{T}) = \frac{1}{\mathcal{T}} \left((t_i - t) w_i + \sum_{k=i+1}^{i+\mathcal{T}-1} (t_k - t_{k-1}) w_k + (t + \mathcal{T} - t_{i+\mathcal{T}-1}) w_{i+\mathcal{T}} \right),$$

$$t \in T_i, \, i = \overline{1, M}, \quad \omega \in \Omega. \tag{82}$$

Denote

$$a_{(i)}(t|\mathcal{T}) = \begin{vmatrix} a_{(i),k}(t) = \begin{cases} 0 & \text{if } k < i, \\[1.5ex] \dfrac{t_i - t}{\mathcal{T}} & \text{if } k = i, \\[2ex] \dfrac{t_k - t_{k-1}}{\mathcal{T}} & \text{if } i < k \le i + \mathcal{T} - 1, \\[2ex] \dfrac{t + \mathcal{T} - t_{i+\mathcal{T}-1}}{\mathcal{T}} & \text{if } k = i + \mathcal{T}, \\[1.5ex] 0 & \text{if } i + \mathcal{T} < k \le N \end{cases} \end{vmatrix},$$

$$t \in T_i, i = \overline{1, M}. \tag{83}$$

Hence the $\mathcal{T}$-**programmed problem of spot riskless rate estimation** is

$$\min \quad v_{\mathcal{T}} \quad \text{subject to}$$

$$\langle a_{(i)}(t), \omega \rangle - v_{\mathcal{T}} \leq \hat{R}(t, t+\mathcal{T}), \quad \hat{R}(t, t+\mathcal{T}) \leq \langle a_{(i)}(t), \omega \rangle + v_{\mathcal{T}}, \quad t \in T_i, \quad i = \overline{1, M};$$

$$\omega \in \Omega, \quad v \geq 0. \tag{84}$$

Problem (84) is a linear semi-infinite programming (LSIP) problem.

4.3. *Programmed estimation of the spot riskless rate: a bilevel perturbation model*

In our previous models for the term structure of interest rates we constructed continuous estimates of observed yield $\hat{R}(t, t+\mathcal{T})$, $t \in T^M$. These forms arose through integrations. For this special case of the model (39) the value of the trust region for the yield cannot be less then the maximum value in the jumps of observed yield (see, for example, Fig. 2). By *trust region* we mean the tube having minimal width that covers the observed yield curve.

In this section we reformulate the model of the term structure of interest rates so that jumps in the observed yield can be taken into account. This is done by introducing perturbations on <u>both</u> the spot rate <u>and</u> the yield.

In this case Assumption 4.2 does not hold because now the yield (price) of the bond depends not only on the short rate process but, in addition, on the second process that acts directly on the observed data.

Assume that the spot riskless rate follows the Vasicek model (52)–(53) either under (a) impulse perturbations (54)–(55) or (b) the point-impulse perturbations (62)–(64).

The definition of the *bilevel model* is completed when we introduce perturbations on the structural from of the yield $R(t, t+\mathcal{T})$, $t \in T^M$ too. The admissible class of these particular perturbations $\{\bar{w}(t), \ t \in T^M\}$ are defined as follows, recalling (22).

(1)
$$\bar{w}(t) = 0, \quad t \in T^M \setminus \tilde{T}^M, \tag{85}$$

(2) the trajectory $R(t, t+\mathcal{T})$ of the yield prevailing at time $t \in T^M$ is described by the Eqs. (35), (41), (supressing the ω notation in R) by

$$R(t, t+\mathcal{T}) = \frac{1}{\mathcal{T}} \int_t^{t+\mathcal{T}} r(\tau|\omega) \, d\tau, \tag{86}$$

on the intervals $]t_{i-1}, t_i[$, $i = \overline{1, M}$ and
(3) the trajectory $R(t, t+\mathcal{T})$ has jumps defined by

$$R(t_i + 0, t_i + 0 + \mathcal{T}) = R(t_i - 0, t_i - 0 + \mathcal{T}) + \mathcal{T} e^{\gamma(t-t_i)} \bar{w}(t_i),$$

$$\bar{w}_* \leq w(t_i) \leq \bar{w}^*, \tag{87}$$

at the points $t = t_i \in \tilde{T}^M$. Introduce additional perturbations into the structural equation for the yield function from the standard model as follows. Denote $\bar{w}_i = \bar{w}(t_i)$, $t_i \in \tilde{T}^M$. Then from (35), (41) and (85)–(87) it follows that

$$R(t, t+T) = \frac{1}{T} \int_t^{t+T} r(\tau|\omega)\, d\tau + T \sum_{k=1}^{i-1} e^{\gamma(t-t_k)} \bar{w}_k, \quad t \in T_i, \ i = \overline{1, M}.$$

(88)

Hence the $\mathcal{T}$-**programmed problem of spot riskless rate estimation** is

$$\min \quad v_T \quad \text{subject to}$$

$$\frac{e^{\beta(t+T)} - e^{\beta t}}{T\beta} r_0 + \left(\frac{e^{\beta(t+T)} - e^{\beta t}}{T\beta^2} - \frac{1}{\beta} \right) \alpha + \sum_{k=1}^{i+T} a_k^i(\beta, t|T) w_k$$

$$+ T \sum_{k=1}^{i-1} e^{\gamma(t-t_k)} \bar{w}_k - v_T \le \hat{R}(t, t+T), \text{ and } \hat{R}(t, t+T)$$

(89)

$$\le \frac{e^{\beta(t+T)} - e^{\beta t}}{T\beta} r_0 + \left(\frac{e^{\beta(t+T)} - e^{\beta t}}{T\beta^2} - \frac{1}{\beta} \right) \alpha + \sum_{k=1}^{i+T} a_k^i(\beta, t|T) w_k$$

$$+ T \sum_{k=1}^{i-1} e^{\gamma(t-t_k)} \bar{w}_k + v_T$$

where $t \in T_i$, $\bar{w}_* \le \bar{w}_i \le \bar{w}^*$, $i = \overline{1, M}$, and

$$\alpha_* \le \alpha \le \alpha^*, \quad \beta_* \le \beta \le \beta^*, \quad r_* \le r_0 \le r^*, \quad v_T \ge 0;$$
$$w_* \le w_i \le w^*, \quad i = \overline{1, N};$$

where $a_k^i(\beta, t|T), k = \overline{1, i+T}, i = \overline{1, M}$, is defined by (59) or (67) depending on the type of perturbation for spot rate we use.

Remark 4.3. The parameter γ influences the longer term limit of the yield. If $\gamma = 0$, then the yield perturbations can be interpreted as "long run factors" since in this case, the perturbations uniformly shift the corresponding yields to maturity. If $\gamma \ne 0$, then the yield perturbations affect the corresponding yields for shorter periods of time and so can be interpreted as "short run factors".

Remark 4.4. It is possible to introduce into (87) actually two forms of the perturbations, namely simultaneously long run factors and short run factors.

Remark 4.5. The recommended value of γ should be obtained from solving the optimal observation problem. Alternatively, it is possible to employ special optimization problems for the selection of γ, where objective functions could stem from improving the estimation error and the quality of forecasting where observed trends in observed data could be of influence.

Remark 4.6. Clearly (89) and (50) can be combined to obtain a bilevel model extension for bonds with differing maturities. While we have not formally done this, we have done experiments on the extension and report on them later.

Remark 4.7. From (88) we have for $T \geq t$, where $t + \mathcal{T} = T$,

$$R(t,T) = \frac{1}{T-t} \int_t^T r(\tau)\, d\tau + (T-t) \sum_{k=1}^{i-1} e^{\gamma(t-t_k)} \bar{w}_k, \quad t \in T_i, \ i = \overline{1,M}. \tag{90}$$

From the yield to maturity definition (24) we have,

$$-\log P(t,T) = \int_t^T r(\tau)\, d\tau + (T-t)^2 \sum_{k=1}^{i-1} e^{\gamma(t-t_k)} \bar{w}_k, \quad t \in T_i, \ i = \overline{1,M}. \tag{91}$$

By definition of the instantaneous forward rate (27), it follows that

$$f(t,T) = r(T) + 2(T-t) \sum_{k=1}^{i-1} e^{\gamma(t-t_k)} \bar{w}_k, \quad t \in T_i, \ i = \overline{1,M}. \tag{92}$$

Note that the linear function $T - t$ could be replaced with a sufficiently regular function $s(t,T)$ satisfying $s(t,t) = 0$.

In general (92) doesn't guarantee a non-negative forward rate for some future time because the forward rate depends on the values of the perturbations appearing in (92). However, it is straightforward to include additional constraints to problem (89) in order guarantee non-negativity, namely, $f(t_M, s) \geq 0$, $s \in [t_M, t_M + T]$.

In modern stochastic theory it is shown that the expected path of the short rate should lie above the forward rate curve, see for example, [5, 20, 19]. This follows from Jensen's inequality applied to the expectation of the exponential function of the stochastic spot rate function. In our approach we use another norm (in contrast to expectation) for uncertainty, so this property is not a feature of our model. But adjoining constraints such as $r(s) \geq f(t_M, s)$, $s \in [t_M, t_M + T]$ to problem (89) is sufficient.

Example 4.4. We use the observed yield data from Example 4.1. Assume that the spot rate follows the *LDEU* (52)–(54). We set $\gamma = 0$ in all numerical experiments.

We solve problem (89) on a grid with 186 points with the following initial point:

$$\alpha_0 = 0, \quad \beta_0, \quad r_0 = 0, \quad v_0^T = 0; \quad w_k = 0, \quad k = \overline{1,N}; \quad \bar{w}_k = 0, \quad k = \overline{1,M},$$

where the choice of β is given in Table 4.

The numerical results for different initial points are given in the Table 4.

The numerical results for different *stopping tolerances ACCSP* for solving the support problem and fixed initial point

$$\alpha_0 = 0, \quad \beta_0 = -0.1, \quad r_0 = 0, \quad v_0^T = 0; \quad w_k = 0, \quad k = \overline{1,N}; \quad \bar{w}_k = 0, \quad k = \overline{1,M},$$

are given in the Table 5.

Table 4. Bilevel perturbations model under (39) with optimization (89) and different starting points.

NPROB	1	2	3
β_0	-0.1	-0.5	-1.0
r_0^0	3.1792402	2.9672991	3.3461018
α^0	4.7066660	3.4655109	4.4081008
β^0	-0.9216965	-0.6798958	-0.8657656
$-\frac{\alpha^0}{\beta^0}$	5.106524	5.097120	5.091563
v_T^0	10^{-8}	10^{-8}	10^{-8}
ITER	238	220	229
NFUN	238	220	229
NQP	238	220	229
NGRAD	238	220	229
NCONSP	372	372	372
ACCSP	10^{-7}	10^{-7}	10^{-7}
CPU(sec)	14700	13810	14004

Table 5. Bilevel perturbations model under (39) with optimization (89) and different accuracy same starting point.

NPROB	4	5	6	7
r_0^0	0.5714122	1.5869700	3.1792462	3.1792444
α^0	2.6360272	5.0857674	4.7066660	4.7066660
β^0	-0.518674	-1.0	-0.9216965	-0.9216965
$-\frac{\alpha^0}{\beta^0}$	5.082242	5.085767	5.106524	5.106524
v_T^0	0.0286725	0.0043419	0.0000125	0.00000001
ITER	33	87	221	238
NFUN	33	87	221	238
NQP	33	87	221	238
NGRAD	33	87	221	238
ACCSP	10^{-4}	10^{-5}	10^{-6}	10^{-7}
CPU(sec)	758.52	4035	12600	14700

Figure 10 illustrates the estimates $p(\cdot|\cdot)$ of the yield we obtained and the observed yield $R(\cdot,\cdot)$ for NPROB = 4, 5. Figure 11 illustrates the estimation error of yield for NPROB = 6. Figure 12 illustrates the trust region for the yield for NPROB = 5. Figure 13 illustrates the obtained estimates of the spot rate $r(\cdot)$ for NPROB = 4, 5, 6.

4.4. *Real-time positional (conditional-adaptive) estimation of STRR*

In the previous models the estimates ω^0 of unknown parameters ω were calculated the conclusion of the observation process. In real situations real-time estimates are needed and these are calculated and based on current observations of $\hat{R}(t, t+\mathcal{T})$.

Our methodology is developed from the point of view of realistic operations of a market. Assume that we have observed values of the yield to maturity $\hat{R}(t, t + \mathcal{T})$ for some fixed time-to-maturity $\mathcal{T}$ during τ days. By a *position of the observed system* we shall mean the pair $y = \{\tau; \hat{R}(t, t + \mathcal{T}), t \in T^\tau\}$, where $T^\tau = \bigcup_{i=1}^\tau T_i$. Let $N(\tau) = \tau + \mathcal{T}$.

Introduce

$$\omega(\tau) = (\omega_i(\tau), \; i = \overline{1, N(\tau)}); \quad \omega_*(\tau) = (\omega_{*i}, \; i = \overline{1, N(\tau)});$$

$$\omega^*(\tau) = (\omega_i^*, \; i = \overline{1, N(\tau)}); \quad \Omega(\tau) = \{\omega(\tau) \in \mathbf{R}^{N(\tau)} : \omega_*(\tau) \le \omega(\tau) \le \omega^*(\tau)\}. \tag{93}$$

Construct the following $NSIP$ problem, which is an extension of (44):

$$\min \quad v_{\mathcal{T}}(\tau) \quad \text{subject to}$$

$$p(t, \omega(\tau) | \mathcal{T}) - v_{\mathcal{T}}(\tau) \le \hat{R}(t, t + \mathcal{T}),$$

$$\hat{R}(t, t + \mathcal{T}) \le p(t, \omega(\tau) | \mathcal{T}) + v_{\mathcal{T}}(\tau), \quad t \in [t_0, \; t_\tau]; \tag{94}$$

$$\omega(\tau) \in \Omega(\tau); \quad v_T(\tau) \ge 0.$$

We shall call problem (94) the **τ-positional problem of estimation** of **T-STRR**; compare with the earlier problems (44) and (50). Assume that $(\omega^0(\tau), v_{\mathcal{T}}^0(\tau))$ is an optimal solution of problem (94). Denote by $M^{\mathcal{P}}$ the operating horizon.

Remark 4.8. Problem (94) has the same formal structure as problem (44) except for the appearance of time τ and $t \in [t_0, \; t_\tau]$ instead of $t \in T^M$. Furthermore, the set $\Omega(\tau)$ contains vectors $\omega(\tau)$ whose components are the same as ω in (55) except they now depend on τ. We are particularly interested in how $\alpha(\tau)$ and $\beta(\tau)$ vary as τ varies.

Definition 4.4. By an **adaptive estimator** we shall mean a system for which it is possible to calculate the estimates $\omega^0(\tau)$ for each $\tau \in [0, \; M^{\mathcal{P}}]$, in real-time.

The basic principles for constructing estimators have appeared in [8]. We describe the main idea for synthesizing the estimator.

Suppose we have observed a signal on the interval $[0, \; \tau]$ and have found a solution $(\omega^0(\tau), v_{\mathcal{T}}^0(\tau))$ of (94). When we obtain the new observation of the signal at time $\bar{\tau} = \tau + h$, where $h > 0$ small enough, then it is clear that the solution $(\omega^0(\bar{\tau}), v_{\mathcal{T}}^0(\bar{\tau}))$ of problem (94) at moment $\bar{\tau}$ will not differ much from $(\omega^0(\tau), v_{\mathcal{T}}^0(\tau))$ (at moment τ). Hence, it is not necessary to solve (94) at $\bar{\tau}$ again. It is sufficient to resolve (94) with the advanced start $(\omega^0(\tau), v_{\mathcal{T}}^0(\tau))$, namely, the optimal solution of problem (94) at the previous position. Special procedures for solving this problem with advanced starts are given in [24].

Following (49), (94) we construct the τ-positional estimation problem for the STRR and the adaptive estimator problem for the STRR. In model (94) the number of constraints is increased during the observation process.

Assume that for estimating STRR only S of the last observations of $\hat{R}(t, t + \mathcal{T})$ are used. Consider the following $NSIP$:

$$
\begin{aligned}
&\min \quad v_{\mathcal{T}}(\tau) \\
&p(t, \omega(\tau)|\mathcal{T}) - v_{\mathcal{T}}(\tau) \leq \hat{R}(t, t + \mathcal{T}), \\
&\hat{R}(t, t + \mathcal{T}) \leq p(t, \omega(\tau)|\mathcal{T}) + v_{\mathcal{T}}(\tau), \quad \forall\, t \in T_{\tau,s}; \qquad (95) \\
&\omega(\tau) \in \Omega(\tau); \quad v_{\mathcal{T}}(\tau) \geq 0, \\
&\text{where } T_{\tau} = [t_{\tau-s},\ t_{\tau}] = \bigcup_{i=\tau-S+1}^{\tau} T_i\,.
\end{aligned}
$$

Assume that $(\omega^0(\tau), v_{\mathcal{T}}^0(\tau))$ is the optimal solution of problem (95).

Definition 4.5. A system for which estimates $\omega^0(\tau)$ can be computed for each $\tau \in [S, M^{\mathcal{P}}]$, in real-time shall be termed an **adaptive estimator with fixed observation period**.

We have performed numerical experiments using the analog of the Vasicek model with impulse perturbations and report on some of them in Sec. 5.

4.5. *A nonarbitrage condition for LDEU*

A natural question about the $LDEU$ approach we are taking is whether the classical non-arbitrage condition is preserved. To address this question, consider an investor who at time t^* issues an amount W_1 of bonds having maturity T_1, and simultaneously buys an amount W_2 of a bond maturing at time T_2. Without an arbitrage opportunity the following equation must hold:

$$
W_1 e^{-\int_{t^*}^{T_1} h(\tau,\omega)\, d\tau} = W_2 e^{-\int_{t^*}^{T_2} h(\tau,\omega)\, d\tau}\,. \qquad (96)
$$

Assume

$$
h(\tau, \omega) \geq 0, \quad \text{for all } \tau \in \overline{T}, \quad \omega \in \Omega\,. \qquad (97)
$$

We may suppose without loss of generality from (96) and (97) that $T_1 < T_2$. Rewrite (96) in the form,

$$
W_1 \left(e^{-\int_{t^*}^{T_1} h(\tau,\omega)\, d\tau} - \frac{W_2}{W_1} e^{-\int_{t^*}^{T_2} h(\tau,\omega)\, d\tau} \right)
$$

$$
= W_1 e^{-\int_{t^*}^{T_1} h(\tau,\omega)\, d\tau} \left(1 - \frac{W_2}{W_1} e^{-\int_{T_1}^{T_2} h(\tau,\omega)\, d\tau} \right) = 0\,.
$$

Since $W_1 \neq 0$ and $e^{-\int_{t^*}^{T_1} h(\tau,\omega)\, d\tau} \neq 0$, it follows from the last equation that

$$
e^{-\int_{T_1}^{T_2} h(\tau,\omega)\, d\tau} = \frac{W_1}{W_2}\,,
$$

or

$$-\int_{T_1}^{T_2} h(\tau, \omega)\, d\tau = \log \frac{W_1}{W_2}\, .$$

Let $W = W(T_1, T_2) = |\log \frac{W_1}{W_2}|$. Since $W_1 < W_2$ we have $\log \frac{W_1}{W_2} < 0$, and hence we obtain

$$\int_{T_1}^{T_2} h(\tau, \omega)\, d\tau = W\, .$$

Since the above equation is valid for any $T_1 < T_2$ and $W > 0$, it follows that we can state the following *nonarbitrage* condition.

Nonarbitrage necessary condition

Let the function $h(\cdot, \cdot)$ satisfy (33), (34), and (97). For any arbitrary T_1, T_2, and W satisfying $T_1 < T_2$, $W > 0$, there exists $\hat{\omega} = \hat{\omega}(T_1, T_2) \in \Omega$ such that the following statement is true:

$$\int_{T_1}^{T_2} h(\tau, \hat{\omega})\, d\tau = W\, . \tag{98}$$

Hence, we can guarantee the existence of non-arbitrage if we include the non-negativity of the spot rate and the Nonarbitrage Necessary Condition into Definition 4.1, (33) and Assumption 4.1, (34) for the class of admissible spot-rate functions.

At first it may seem that condition (98) significantly complicates the application of model (33), (34), (97), (98) for estimating the spot rate.

But now we show that this assertion is not true. For example, consider the analog of the Vasicek model.

Assume that $\hat{\omega}_i = w_i = 0$, $i = \overline{1, N}$, i.e., all perturbations are set to zero. From (56) we obtain

$$h(t, \omega|\hat{\omega}) = e^{\beta t} r_0 + \frac{\alpha}{\beta}(e^{\beta t} - 1)\, .$$

From (98), given arbitrary $T_1, T_2, T_1 < T_2$ and $W > 0$ it follows that

$$\int_{T_1}^{T_2} h(\tau, \omega)|\hat{\omega} = 0)\, d\tau = \int_{T_1}^{T_2} \left(e^{\beta t} r_0 + \frac{\alpha}{\beta}(e^{\beta t} - 1)\right) d\tau$$

$$= r_0 \frac{e^{\beta T_2} - e^{\beta T_1}}{\beta} + \alpha \frac{e^{\beta T_2} - e^{\beta T_1}}{\beta^2} - \frac{\alpha}{\beta}(T_2 - T_1) = W\, .$$

Choose $\beta = \hat{\beta} < 0$. It follows that

$$r_0 = \left| \frac{\hat{\beta}}{e^{\hat{\beta} T_2} - e^{\hat{\beta} T_1}} \right| \left(\left(W - \alpha \frac{T_2 - T_1}{|\hat{\beta}|} + \alpha \left| \frac{e^{\hat{\beta} T_2} - e^{\hat{\beta} T_1}}{\beta^{*2}} \right| \right) \right)$$

$$= \overline{W} + \alpha \left(\frac{1}{|\hat{\beta}|} - \frac{T_2 - T_1}{|e^{\hat{\beta} T_2} - e^{\hat{\beta} T_1}|} \right)\, ,$$

where

$$\overline{W} = W \frac{|\hat{\beta}|}{|e^{\beta T_2} - e^{\beta T_1}|} > 0.$$

Choose $\alpha = \hat{\alpha} \geq 0$ so that

$$\overline{W} + \hat{\alpha} \left(\frac{1}{|\hat{\beta}|} - \frac{T_2 - T_1}{|e^{\beta T_2} - e^{\beta T_1}|} \right) \geq 0. \tag{99}$$

which always holds since $\overline{W} > 0$. Hence, for any arbitrary $W > 0$, T_1, T_2, $T_1 < T_2$, we can choose $\hat{\alpha} \geq 0$ such that $\hat{r}_0 \geq 0$.

The Nonarbitrage Necessary Condition therefore holds, if, for example, we define the admissible set of parameters Ω of Definition 4.1 as follows:

(i) $r_0 \geq 0$, i.e., $r_* = 0$, $r^* = +\infty$
(ii) $\alpha \geq 0$, i.e., $\alpha_* = 0$, $\alpha^* = +\infty$, $\beta < 0$, e.g., $\beta_* = -\infty$, $\beta^* = -\epsilon$, where $\epsilon > 0$ is sufficiently small, and
(iii) $w_{*i} < 0$, $w_i^* > 0$, $i = \overline{1, N}$.

Conditions (i)–(iii) are usually used for describing the unknown parameters. Indeed, constraint (i) serves to guarantee the non-negativity of the initial spot rate, (ii) guarantees the mean-reversion property for the spot rate, where we use the following definition.

Definition 4.6. We say that the spot rate has mean reversion if $\lim_{t \to \infty} r(t)$ exists and is finite.

Finally (iii) permits positive or negative perturbations on the spot rate.

In order to satisfy the Nonarbitrage Necessary Condition (98) only 2 parameters in the *LDEU* model are needed: α and β. This has some meaning. Indeed, the perturbations realized in an actual market have influence only on the value of W. Note that the choice of the values of r_0, α, and β serves the purpose of calibrating the non-arbitrage condition (98). Moreover, as (99) is valid for any initial time t_0, the parameter β plays the role of the speed of discounting of the bond, and parameter α is a step size selection for fitting the non-negativity condition of the spot rate.

Nonarbitrage sufficient condition

Let the function $h(\cdot, \cdot)$ satisfy (33), (4.1), and (97). For all arbitrary T_1, T_2, $W = W(T_1, T_2)$ there exists $\hat{\omega}$ such that (98) holds.

Of course this condition is a strong one. There is also an issue about the existence of $\hat{\omega}$. However, we recommend the following procedure for discovering existence of arbitrage.

STEP (1) Employ the Nonarbitrage Necessary Condition. Construct the set Ω for a particular model of the spot rate leading to the addition of constraints with parameter vector ω in the optimal observation problem, *OOP*.

STEP (2) Solve *OOP* obtaining the objective function value ϵ.

STEP (3) Use ϵ as a measure of the violation of the Nonarbitrage Sufficient Condition. Hence, in general in our model arbitrage exists, but in this case we can quantitatively measure the amount of its violation.

4.6. *Estimating the LDEU market price of risk*

The *market price of risk* is a well-known concept in stochastic differential equations models of the spot riskless rate. Usually it is considered to be a constant with respect to time.

Consider an investor who at time $\hat{t}$ issues an amount W_1 of a bond with maturity T_1 and simultaneously buys an amount W_2 of a bond maturing at time T_2. The basic problem is to determine the profit or loss of this operation at time $s > \hat{t}$ in a market functioning under uncertainty.

Following our approach, we assume that the set Ω, of the *a priori* distribution of unknown parameters, (33) and the family of functions, $h(t, \omega)$, in Assumption 4.1 are defined. Introduce

$$z(\omega, s) = W_1 \exp\left\{ -\int_s^{T_1} h(\tau, \omega)\, d\tau \right\} - W_2 \exp\left\{ -\int_s^{T_2} h(\tau, \omega)\, d\tau \right\}, \quad \omega \in \Omega. \quad (100)$$

By the nonarbitrary property, we have $z(\omega, \hat{t}) = 0$.

Definition 4.7. The function $z(\omega, s)$ defined in (100) is termed the *LDEU-market price of risk*.

Of course the *LDEU*-market price of risk also depends on W_1, W_2, T_1, and T_2, but we assume that these values are given.

A *priori* estimation of the market price of risk

We seek an upper bound, $z^*(s)$ and a lower bound, $z_*(s)$ of $z(\cdot, \cdot)$. A simple approach is to solve two nonlinear problems

$$z^*(s) = z^*_{AP}(s) := \max_{\omega \in \Omega} z(\omega, s), \quad z_*(s) = z_{*AP}(s) = \min_{\omega \in \Omega} z(\omega, s). \quad (101)$$

Note that estimates (101) are obtained only from *a priori* information about the unknown parameters, and therefore we shall term them the *a priori bounds on the LDEU-market price of risk*.

More precise estimates can be obtained by using observations of the yield curve. For simplicity, consider model (44).

Assume observations have been made by observation model (39) or (40) for some given time-to-maturity $\mathcal{T}$ during M days. Clearly, not all members, ω of Ω can generate observed values of the yield. Hence, a subset, $\hat{\Omega}$ of Ω consisting of members ω that can generate points on the yield curve, exists and shall be termed the *a posteriori distribution of unknown parameters* ω. We describe how to obtain extremal values, $z^*(s)$, $z_*(s)$, on certain subsets of $\hat{\Omega}$ itself.

Optimal estimation of the market price of risk

Let $(\omega^0, v_\mathcal{T}^0)$ be a solution of problem (44). We shall estimate the z-extremal values on the subset of $\hat{\Omega}$ of members which can guarantee an optimal estimation error, $v_\mathcal{T}^0$ of the observed yield. For this purpose, define

$$\Omega_{OPT} = \{\omega \in \hat{\Omega} \mid |p(t, \omega|\mathcal{T}) - \hat{R}(t, t + \mathcal{T})| \leq v_\mathcal{T}^0, \quad \forall t \in T^M\}.$$

We term the set Ω_{OPT}, the *optimal a posteriori distribution of unknown parameters* ω. Analogous to (101) we obtain the following *NSIP* problems:

$$z^*(s) = z^*_{OPT}(s) := \max_{\omega \in \Omega_{OPT}} z(\omega, s), \quad z_*(s) = z_{*OPT}(s) = \min_{\omega \in \Omega_{OPT}} z(\omega, s). \quad (102)$$

The values $z^*_{OPT}(s)$ and $z^*_{OPT}(s)$ are termed the *optimal observation bounds for the market price of risk*.

Estimating the market price of risk under fixed observed present structure

Let $(\omega^0, v_\mathcal{T}^0)$ be a solution of problem (44). Denote $\overline{k} = k - \mathcal{T}$. The goal here is to fix estimates of unknown parameters corresponding to the current moment of time. Define a subset of $\hat{\Omega}$ as follows:

$$\Omega_{FIX} = \{\omega \in \hat{\Omega} \mid \omega_i = w_i^0, \; i = \overline{1, \overline{k}}\}.$$

Analogous to (101) and (102) we obtain the *NSIP* problems:

$$z^*(s) = z^*_{FIX}(s) := \max_{\omega \in \Omega_{FIX}} z(\omega, s), \quad z_*(s) = z_{*FIX}(s) = \min_{\omega \in \Omega_{FIX}} z(\omega, s). \quad (103)$$

In all cases the function $z^*(s)$ and $z_*(s)$ can be generated by the adaptive estimators (94) or (95) applied to problems (101)–(103). However, problems (101)–(103) are nonlinear nonconvex so that global optimization methods must be used to compute bounds for the market price of risk.

5. Numerical Experiments for Periods of Extended Observations

In this section we describe numerical experiment based on models stated above. We choose

$$w_* = \beta_* = -10, \; w^* = r^* = \alpha^* = 10, \; \alpha_* = r_* = 0,$$

$$\beta^* = -10^{-5}; \quad r_0 = 5, \; \alpha_0 = 0.0, \; \beta_0 = -0.1;$$

$ACCSP = 10^{-5}$ for all examples. We solve generated $NSIP$ on the grid with stepsize equal to 0.5. In some cases we use stepsize 0.1 (when period of observation equals 3 months).

Tables 6, 7, 8, 9, and 10 describe the results of numerical experiments for different periods of observation of Treasury yield rates for maturity term 3 months and different models of the spot rate. We use the model (39) for modelling values of

Table 6. Analog of Vasicek model with impulse perturbations over four horizons.

TT	1Oct–1Dec95	1Jul–31Dec95	1Apr–31Dec95	3Jan–31Dec95
r_0^0	4.882779	5.028930	5.204785	4.997817
α^0	0.300465	0.413059	0.255036	0.222343
β^0	−0.055194	−0.076130	−0.045490	−0.039272
$-\frac{\alpha^0}{\beta^0}$	5.443798	5.425706	5.606419	5.661616
v_T^0	0.065	0.075	0.075	0.085
MAX	5.59	5.68	5.94	6.07
MIN	4.98	4.98	4.98	4.98
NCONSP	1104	1104	1650	2178
NVAR	188	280	371	459
ITER	21	18	44	45
NFUN	21	18	45	46
NQP	21	18	44	45
NGRAD	21	18	44	45
CPU(h:mm:ss)	0:03:21	0:06:27	0:41:31	1:25:13

Table 7. Analog of Vasicek model with impulse perturbations over four additional horizons.

TT	3Jan–31Mar95	3Jan–30Jun95	3Jan–30Sep95	3Jan–31Dec95
r_0^0	4.898917	4.976673	5.035318	4.997817
α^0	0.748241	0.701506	0.637142	0.222343
β^0	−0.126938	−0.120722	−0.112358	−0.039272
$-\frac{\alpha^0}{\beta^0}$	5.894539	5.816921	5.670642	5.661616
v_T^0	0.085	0.085	0.085	0.085
MAX	5.59	5.68	5.94	6.07
MIN	4.98	4.98	4.98	4.98
NCONSP	528	1074	1626	2178
NVAR	182	275	367	459
ITER	18	35	38	45
NFUN	18	35	38	46
NQP	18	35	38	45
NGRAD	18	35	38	45
CPU(h:mm:ss)	0:01:54	0:14:08	0:33:30	1:25:13

Table 8. Analog of Vasicek model with point impulse perturbations with four horizons of Table 6.

TT	1Oct–31Dec95	1Jul–31Dec95	1Apr–31Dec95	3Jan–31Dec95
r_0^0	4.909963	5.031434	5.204212	5.025502
α^0	0.304368	0.413545	0.247247	0.206829
β^0	−0.056144	−0.076373	−0.044135	−0.036717
$-\frac{\alpha^0}{\beta^0}$	5.421203	5.414806	5.602062	5.633058
v_T^0	0.065	0.075	0.075	0.085
MAX	5.59	5.68	5.94	6.07
MIN	4.98	4.98	4.98	4.98
NCONSP	1104	1104	1650	2178
NVAR	188	280	371	459
ITER	20	17	38	42
NFUN	20	17	39	43
NQP	20	17	38	42
NGRAD	20	17	38	42
CPU(h:mm:ss)	0:02:43	0:06:04	0:32:07	1:27:47

Table 9. Analog of Dothan model with four horizons of Table 6.

TT	1Oct–31Dec95	1Jul–31Dec95	1Apr–31Dec95	3Jan–31Dec95
r_0^0	5.001765	5.002267	5.003071	5.003747
v_T^0	0.065	0.075	0.075	0.085
NCONSP	1104	1104	1650	2178
NVAR	188	280	371	459
ITER	8	9	15	13
NFUN	10	12	21	15
NQP	8	9	15	13
NGRAD	8	9	15	13
CPU(h:mm:ss)	0:02:47	0:05:28	0:19:46	2:14:45

observed yield. The common additional legend for all the tables to follow is:

- TT denotes the period of observation.
- NCONSP denotes number of constraints of NLP problem obtained from *NSIP* problem.
- NVAR denotes number of variables in the corresponding NLP problem.
- MAX denotes maximum value of observed Treasury yield rate.
- MIN denotes minimum value of observed Treasury yield rate.

Figure 16 illustrates the estimates of spot rate for different models of spot rate received by observation of yield to maturity from 3 January 95 to 31 December 95. Figure 17 illustrates the observed yield to maturity and the estimates of yield

Table 10. Impulse model with four horizons of Table 6.

TT	1Oct–31Dec95	1Jul–31Dec95	1Apr–31Dec95	3Jan–31Dec95
v_T^0	0.065	0.075	0.075	0.085
NCONSP	1104	1104	1650	2178
NVAR	185	277	368	456
ITER	30	41	71	76
NFUN	30	41	71	76
NQP	30	41	71	76
NGRAD	30	41	71	76
CPU(h:mm:ss)	0:02:59	0:07:25	0:35:35	1:50:34

Table 11. Piecewise-linear model of observations with spot rate models of the analogs of the Vasicek and Dothan models.

NPROB	1	2	3
r_0^0	3.550961	3.938536	4.979456
α^0	0.651080	0.615566	
β^0	−0.120340	−0.113778	
$-\frac{\alpha^0}{\beta^0}$	5.4103373	5.4102374	
v_T^0	0.018687	0.003889	0.017648
ITER	127	182	53
NFUN	133	194	94
NQP	127	182	53
NGRAD	127	182	53
CPU(h:mm:ss)	0:18:26	0:51:28	0:16:01

to maturity for different models of spot rate received by observation of yield to maturity from 3 January 95 to 31 December 95. Figure 18 illustrates the estimates of spot rate for different period of observation of yield to maturity received by using the analog of Vasicek model with impulse perturbations. Figure 19 illustrates the estimate of term structure of interest rates received by means of the analog of Vasicek model with piecewise-constant perturbations for observation of yield to maturity from 1 October 95 to 31 December 95.

Table 11 presents the results of numerical experiments for the piecewise-linear model (40) of observation and different models of spot rates by observation yield to maturity from 1 October 95 to 31 December 95.

Here NPROB = 1 corresponds to analog of Vasicek model with impulse perturbations, NPROB = 2 corresponds to analog of Vasicek model with point impulse perturbations and NPROB = 3 corresponds to the analog of Dothan model.

Figure 20 illustrates the observed yield to maturity and the estimates of yield to maturity for different models of spot rate and piecewise-linear model (40) of observation. Figure 21 illustrates the estimates of spot rate for piecewise-constant (39) and piecewise-linear (40) models of observation received by means of analog of Vasicek model with impulse perturbations.

Figure 22 gives the estimated yields and observed yields for model (50) (bonds with differing maturities) and model (56) for the spot rate. The observations are on yields to maturity terms of 3 months and of 6 months during the observation period 3 January to 30 June 1995.

Figure 23 shows the estimated yields and observed yields for the extended bilevel model with differing maturities, see Remark 4.6, and spot rate model (56). The observation period is same as above (3 January to 30 June 1995), and the observations are on yields to maturity of 3 months and 6 months.

With respect to Remark 4.8 we sought to test how $\alpha(\tau)$ and $\beta(\tau)$ vary in model (94)–(95) with τ. Figures 24 and 26 give the computed functions as τ varies in half-day increments from 31 March 95 to 31 December 95. We chose $S = 3$ months for the estimator in (95). Figures 25 and 27 present the number of iterations required for re-solving problems (94)–(95) in each position τ.

Figure 28 presents the forecasted yield curve and estimated yield curves for different periods of observations with current time 1 April 95 obtained with the analog of the Vasicek model with impulse perturbations.

6. Characteristics of Moments in Linear Dynamical Systems under Uncertainty with Perturbations

From the theory of stochastic processes applied to the law of motion of a risky asset over time, there exist unique characteristics such as drift and volatility for each stochastic processes model. An issue with our approach to modelling asset prices with linear dynamical systems under non-stochastic uncertainty and perturbations arises about the existence of companion or analogous concepts of "drift" and "volatility". We address this issue in this section.

6.1. *Mean path solutions to LDSU (1)*

Drift in a stochastic process describes the mean (first moment) behavior of the stochastic process. By analogy, the mean function of the estimate of uncertainty embedded in our approach should have the following three properties:

(i) the mean function should be a solution to LDSU (1) for the spot rate, and it should approximate the observed signal well. In our application, the integral of an optimal solution should approximate the observed yields,

(ii) the influence or force of uncertainty on the mean function estimate should be absent or at least minimal, and

(iii) a mean function should be uniquely determined.

Consider the analog of the Vasicek model (52), with perturbations (54) and (62)–(64).

We assume now that β is fixed, say $\beta = \hat{\beta} < 0$, and that the accuracy obtained in approximating the observed signal within ϵ. Let

$$\Omega^\epsilon = \{\omega \,|\, |\hat{R}(t, t + T) - p(t, \omega | T| \leq \epsilon, \, \forall\, t \in T^N\}. \tag{104}$$

Note that the set Ω^ϵ consists of only those members of Ω for which an estimated trajectory exists within an ϵ approximation of the observed data, in the sense of the L_1 norm.

Motivated by Properties (i)–(iii) we construct the following optimization problem:

$$V(\hat{\beta}, \epsilon) = \min_{\omega \in \Omega^\epsilon} \frac{1}{T^N} \int_{T^N} w^2(\tau)\, d\tau u,$$
$$r_{0*} \leq r_0 \leq r_0^*, \quad \alpha \geq 0, \quad w_i \in \mathbf{R}, \, i = \overline{1, N}. \tag{105}$$

Remark 6.1. We have considered two classes of perturbation functions, $\{w(t), t \in T^N\}$, specified in (54) and (62)–(64). Such functions are among the unknown parameters in Ω, as part of any $\omega \in \Omega$. For both of these cases the perturbation functions can be characterized by real-valued constants, $\{w_i\}_{i=\overline{1,N}}$ namely those appearing in (105). Note that in (105) there are no constraints on the perturbations. The objective function is strictly convex quadratic in the variables, w_i, and the constraints are linear semi-infinite.

Assume that ϵ is chosen so that (105) has an optimal solution, $w^0(\epsilon, \hat{\beta})$ which necessarily will be unique. The unique solution corresponds to the estimate of the spot rate having minimal possible values of the perturbations in sum of squares sense, for the fixed parameters β and ϵ.

Definition 6.1. A feasible solution $\omega \in \Omega$ to (105) is termed an ϵ-approximation of the observed data. If $w^0(\epsilon, \hat{\beta})$ is the optimal solution, then the function $\mu_T(t) = r(t|\hat{\beta}, \epsilon) = h(t, w^0(\epsilon, \hat{\beta}))$, $t \in T^N$ is termed the mean path of the spot rate.

The *mean path* corresponds to the unique estimate of the spot rate that comes from an ϵ-approximation with minimal possible influence of the perturbations. This estimate describes the mean tendency with maximal possible stable behavior.

Remark 6.2. In [21] it was shown that within the framework of affine stochastic models for the term structure of interest rates with constant parameters, observations of the yield rates process do not in general uniquely determine the parameters of the model. In our approach we obtain a unique estimate of the spot rate for some fixed specification of the parameter β and preassigned error of the estimation. Naturally, the question arises, how does the the estimated trajectory of the spot rate depend on the choices of these two parameters, namely, β and the maximum admissible error?

The following experiment shows the dependence of the spot rate trajectory on different values of β with a fixed estimation error of $\epsilon = 0 : 001$.

We observe the 3M Treasury Bill Rate Auction Average taken from the St. Louis Federal Reserve Bank's file wtb3mo for the period of observation defined by 2 July 1997 through 2 June 1998.

The computations were for

$$\beta \in \{-0.001, \ -0.01, \ -1, \ -5, \ -10, \ -100\}.$$

All estimates of the spot rate trajectories are close to each other. The following figure provides an illustration of estimated spot rate trajectories for the two extreme cases, namely, $\beta = -0.001$ and $\beta = -100$.

Hence, the parameter β determines the type of spot rate function that is generated. For large negative value of β we obtain a spot rate trajectory that is approximately an impulse function. When values of β are chosen close to zero, then the spot rate trajectory is approximately piecewise linear.

6.2. *Minimax amplitude*

We examine the *LDEU* for the purpose of estimating the amplitude of possible values around some characteristic of the estimate of the spot rate. It is natural to choose the mean reversion limit of the estimated spot rate for one such characteristic.

By our conservative *minimax* approach we automatically generate lower, guaranteed bounds for this type of amplitude. Upper bounds are more difficult because these bounds will depend on realized values (e.g., implied bounds) of the perturbations.

We begin by freeing the parameter β while keeping the accuracy of the approximation fixed at ϵ. We seek an estimate of the amplitude according to the following optimization problem:

$$\sigma_\epsilon^0 = \min_{\omega \in \Omega} \max_{t \in T^N} \left| r(t|\omega) - \left| \frac{\alpha}{\beta} \right| \right|. \tag{106}$$

Problem (106) is equivalent to:

$$\begin{aligned}
&\min_{\omega \in \Omega} \sigma_\epsilon, \\
&|\hat{R}(t, t + \mathcal{T}) - p(t, \omega \,|\, \mathcal{T})| \leq \epsilon, \quad \forall \, t \in T^N \\
&\left| r(t|\omega) + \frac{\alpha}{\beta} \right| \leq \sigma_\epsilon, \quad \forall \, t \in T^N \\
&r_{0*} \leq r_0 \leq r_0^*, \quad \alpha \geq 0, \quad \beta \leq -\gamma, \quad w_i \in \mathbf{R}, \ i = \overline{1, N},
\end{aligned} \tag{107}$$

where γ is sufficiently small.

Definition 6.2. We term the optimal value, σ_ϵ^0, the minimax amplitude of the estimate of the spot rate.

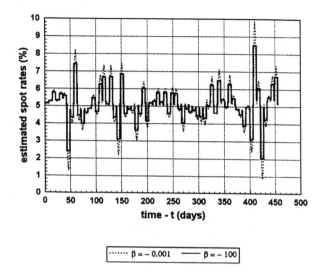

Fig. 6.2. Comparison of estimated spot rates for different values of β.

Table 12. Comparison of SDE-derived properties with dynamical systems under nonstochastic uncertainy with perturbations.

Item	Stochastic	Dynamic
Type of Uncertainty	stochastic process	unknown function of class of perturbation
model for spot rate	Stochastic Differential Equation	DE with nonstochastic uncertainty
norm of uncertainty	Probabilistic Expectation	minimax L_∞: sum squares L_2
nonarbitrage condition	risk free measure	constraints in Observation Problem and nonarbitrage nec. condition
moments of uncertainty	drift and volatility	mean path max/min Realized Perturbations, minimax amplitude
Other Features	Computing Non-Negative Spot and Forward rates not always possible	additional constraints alleviates problem

The value σ_ϵ^0 has the following interpretation. For any estimate $r(\cdot|\cdot)$ of the spot rate $r(\cdot)$ that stems from an ϵ-approximation of the observed data, the maximum deviation of the values of the spot rate coming from an ϵ-approximation around the mean reversion limit on the whole interval cannot be less than σ_ϵ^0. It is natural to expect that different types of perturbations for the spot rate will give different values of σ_ϵ^0. This means that we can use the value of σ_ϵ^0 as a criterion for the choice of perturbations to use in a given situation. We should choose a perturbation of the spot rate that gives the minimal value of σ_ϵ^0 among all available classes of perturbations.

7. Concluding Comments and Discussion

In our approach a broad class of admissible functions $h(t,\omega)$, say $\mathcal{C}$ is defined to which our postulated short rate function $r(t|\omega)$ belongs. Members of $\mathcal{C}$ share similar properties to those of probability distributuons of the mixed-type, namely, where continuous with discrete distributions are mixed. We are focusing on the distribution-type functions rather than the probability measure defined through integration or expectation.

Typically these properties are specified (a) **marginally**, $[h(t.\cdot), t$ varying and $h(\cdot,\omega), \omega$ varying] and (b) **jointly** $[h(\cdot,\cdot), t, \omega$ varying together.

The selection of a member $h(t,\omega) = r(t,|\omega)$ is achieved according to a minimax optimal programmed problem of short rate estimation, where real observed rates data are input.

In contrast, the expectation operation requires a sampling procedure or an estimate of the probability distribution of the short rate process. Apparently, real observed data must be used to estimate relevant statistics of the stochastic short rate process.

We summarize the differences between (1), the stochastic differential approach and (2), and our dynamical systems approach to nonstochastic uncertainty via perturbations under minimax optimization in the following table.

We conclude with a list of comments which refer to the figures in Sec. 8. The comments include some financial interpretations stemming from the dynamical systems approach.

(1) Table 1 and Figs. 6 and 7 show that that problem (44) coupled with the observation model (41) unfortunately does not have a unique solution. Additional criteria could be used for identifying one solution from all the alternate optima, for example, choosing the estimate of the spot rate that minimizes a pre-selected norm; see Sec. 6.

(2) In general we cannot guarantee a global extremum of problem (44), but we can analyze the computed solution, for example, by the following procedure. If the optimal solution yields a cost value equal to $\frac{1}{2}$ of the maximal jump of the observed yield to maturity, then the computed solution is globally optimal for problem (44).

(3) The proposed models are sensitive to rapid changes in the observed yield. For example, in Fig. 1 the values of the observed yield to maturity decrease from 5.16 to 5.08 at time = 20. As is well known, the spot rate is the instantaneous rate of increase of the bond price. If the yield to maturity rapidly decreases, then the price of the bond increases rapidly. Hence, the spot rate should be rapidly increasing at this particular moment too. In Fig. 6 we see that at time 20 the trajectory of the spot has a very high peak (see also Figs. 16–17).

On the other hand, the spot rate represents the instantaneous yield to maturity (see (28), so if the bond price increases rapidly, then the instantaneous yield to maturity should sharply increase too. Moreover, the instantaneous yield to maturity should sharply decrease at the moment of time equalling the current time plus the term of maturity. We do observe the bottom valley of the spot rate at time, 20 + 3 mo. = 111.

(4) Figs. 8 and 9 show a comparison between two types of perturbations tested within one particular model of the spot rate, namely the impulse and point-impulse, respectively. The figures show that the spot rates for both cases are almost identical. It supports the hypothesis that the type of perturbation used is not important. Moreover, from Figs. 16 and 17 we can see that the type of model for generating the spot rate (differential equation or some unknown function) is also not significant. On the other hand from Fig. 21 we see that the trajectories of the estimated spot rate differ for different models of observation of the yield to maturity. We can conclude that when we use a one-factor model for the spot rate the task of significance is how to describe the observed values of the yield to maturity, but not what model is selected for the spot rate or what type of perturbation is chosen.

(5) Figure 12 shows that when we introduce bilevel perturbations into the model, then with increasing accuracy of the approximation, the role of the yield perturbations is increased, and trajectories of the estimated spot rate become more "smooth". The interpretation is that the behavior of the spot rate has an average character, and the estimates of the yield perturbations play the main role for providing higher accuracy in the estimations. The task is to find a balance between the accuracy of estimation and the influence of the spot rate on the observed yield to maturity.

(6) Additional constraints on the spot rate may be readily adjoined to problem (44). For example,

$$z_* \le h(t,\omega) \le z^*, \quad \forall\, t \in T^N, \quad \omega \in \Omega;$$

see (34). "Box" constraints of this type may be used to guarantee non-negativity of the spot rate, i.e., $h(t,\omega) \ge 0, \forall\, t \in T^N, \omega \in \Omega$, see (97).

(7) We remark on the computational results on estimating the spot rate under observation of yield curves by means of

(a) the deterministic model (76)
(b) one-factor model such as (52), and
(c) bilevel perturbation model (89).

We draw the following conclusions.

(i) From the deterministic model we see that the error of the estimation is equal to *half the difference between the maximum value of the observed yield and the minimum value of the observed yield for the entire interval of observation.*

(ii) Using the one-factor model under nonstochastic uncertainty yields an error of the estimation which equals *half of the maximum jump of the observed yield.*

(iii) Using the bilevel perturbation model yields an error of estimation that *doesn't depend on the structure of the observed yields*, but depends on the stopping criterion used for solving problem (89).

(8) In this paper we have developed observation models for the case of common maturities (as one observed curve, see (44)), or for observations of yields having different maturities (as a family of observed curves, see (50))). Based on observed yield curves, we can assume that there are two independent arguments of time, namely, *current* time, t, and *maturity* time, T. Then we construct an observation surface-model by interpolating the values of the observed yield curves according to the argument T, as follows.
Let

$$\tilde{T}_j = [T(j = -1), T(j)], \quad j = \overline{2, L} \text{ and } \tilde{T}^L = \cup_{j=1}^{L} \tilde{T}_j.$$

Then the interpolation is the following one:

$$\hat{R}(t, \tau) = R_{i-1, j-1} + \frac{t - t_{i-1}}{t_i - t_{i-1}} (R_{i, j-1} - R_{i-1, j-1})$$

$$+ \frac{\tau - T(j - 1)}{T(j) - T(j - 1)} (R_{i-1, j} - R_{i-1, j-1})$$

$$+ \frac{(t - t_{i-1})(\tau - T(j - 1))}{(t_i - t_{i-1})(T(j) - T(j - 1))} (R_{i, j} - R_{i, j-1} - R_{i-1, j} - R_{i-1, j-1}),$$

$$t \in T_i, \quad i = \overline{1, M}, \quad \tau \geq t; \quad \tau \in \tilde{T}_j^L, \quad j = \overline{2, L}.$$

For this case the form of the Ω-based yield is

$$p(t, \tau, \omega) = \frac{1}{\tau - t} \int_t^\tau h(s, \omega) \, ds, \quad \omega \in \Omega, \ t \in T^N, \ \tau \in [t, t + T(L)].$$

Hence, we obtain problem $NSIP$ as problem (44) but with two-dimensional infinite index set for t, T.

(9) The models proposed in this paper are *mean-reverting*; see Definition 4.6. For example, in the analysis of the analog of the Vasicek model we have

$$\lim_{t \to \infty} r(t) = -\frac{\alpha}{\beta} \, .$$

For the Dothan model

$$r(t) = r_0 \exp\left(\sum_{i=1}^{N} w_i (t_i - t_{i-1})\right), \quad \text{for all } t > t^N \, .$$

(10) By using our estimator we obtain estimates $w^0(\tau)$ which depend on the current moment of time τ. Hence we may assert that in this case the unknown parameters of any *LDEU* modelling of the spot rate also depend on τ. Therefore, by analogy with the analog of the Vasicek model we really have the parametric dependence, (see Figs. 24 and 26).

$$\alpha = \alpha(\tau), \ \beta = \beta(\tau), \ w(t) = w(t, \tau), \ \tau \in T^{M^P} \, .$$

(11) Let τ be the current time of observation and $(v^0(\tau), w^0(\tau))$ be a solution to problem (94). Then by (38) we have,

$$f(\tau, t) = r(t|w^0(\tau)) \text{ for any } t \in [\tau, t_{N(\tau)}] \, .$$

Hence, we can calculate the forward rate corresponding to the current time of observation. By using the estimator it is possible to construct values for the forward rates, namely,

$$f(\tau, t), \ \tau \in [0, t_{M^P}], \ t \in [\tau, t_{N(\tau)}], \ \text{for each current moment of time } \tau \, .$$

(12) From Figs. 22 and 23 it is clear for the one-factor model for the spot rate that good estimation accuracy cannot be obtained for the case of estimating yields to maturity from observations of yields having differing terms of maturity.

We believe the stochastic equations models for the term structure of interest rates and the approach taken in this paper complement each other. Our dynamic systems approach to modelling uncertainty of asset prices provides a new computational methodology for obtaining numerical solutions to some important applications which have resisted numerical treatment or whose underlying assumptions have not been entirely realistic. For mathematical finance there will now be another alternative to current uncertainty modelling that relies heavily on the underlying stochastic process being of Brownian motion type. The dynamical system approach we propose incorporates measurement signals conditionally and adaptively into the underlying mathematical model.

8. Twenty-Eight Geometrical Figures

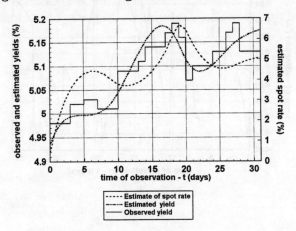

Fig. 1

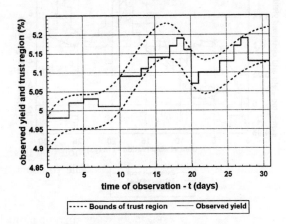

Fig. 2

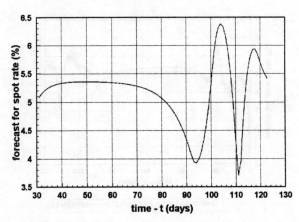

Fig. 3

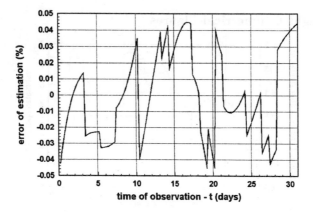

Fig. 4

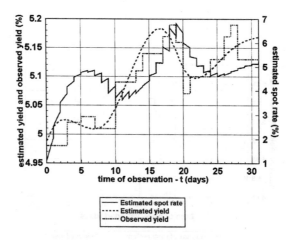

Fig. 5

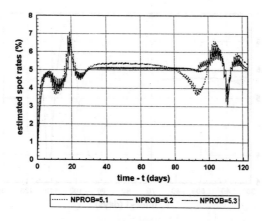

Fig. 6

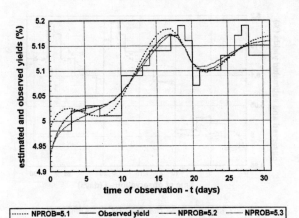

Fig. 7

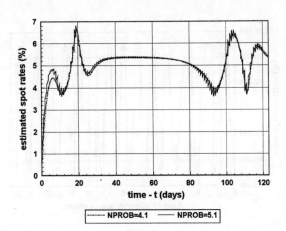

Fig. 8

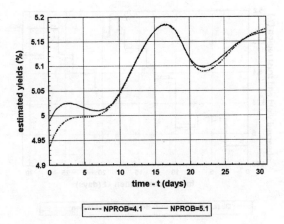

Fig. 9

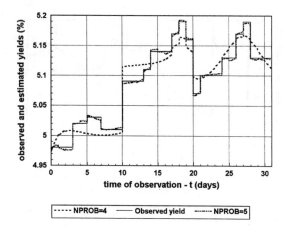

Fig. 10

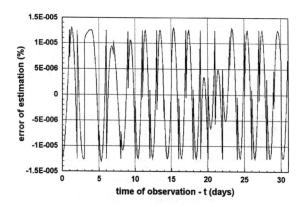

Fig. 11

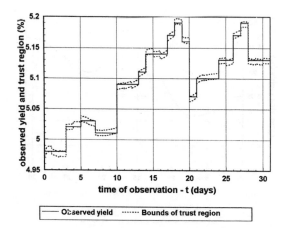

Fig. 12

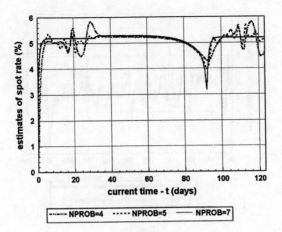

Fig. 13

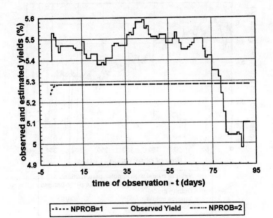

Fig. 14

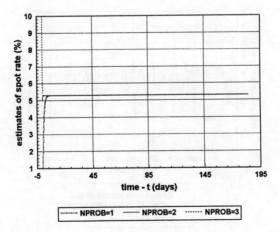

Fig. 15

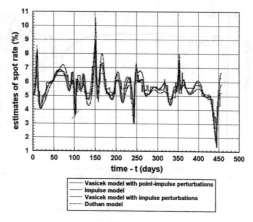

Fig. 16

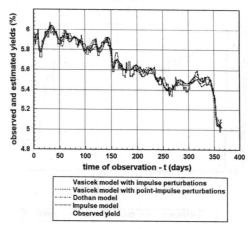

Fig. 17

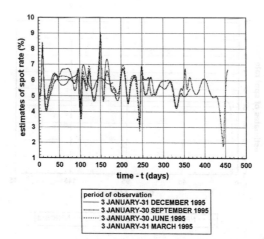

Fig. 18

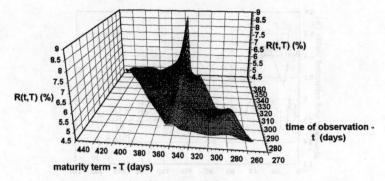

Fig. 19

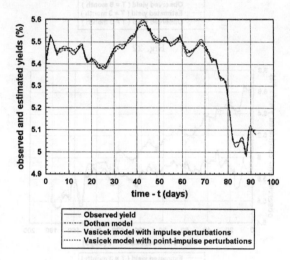

Fig. 20

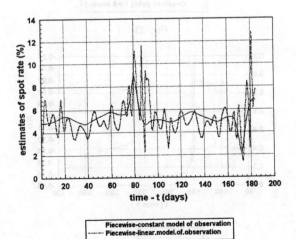

Fig. 21

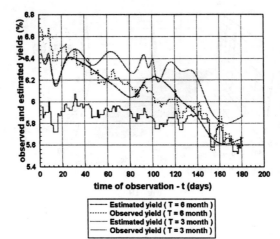

Fig. 22

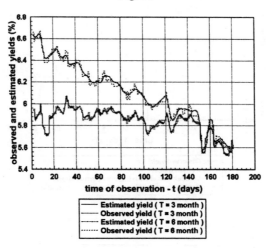

Fig. 23

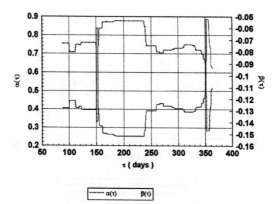

Fig. 24

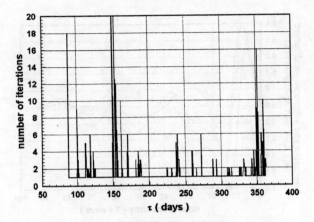

Fig. 25

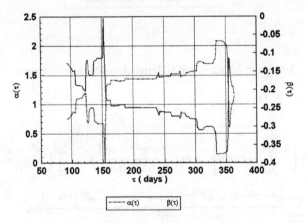

Fig. 26

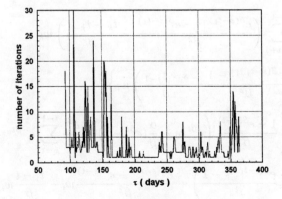

Fig. 27

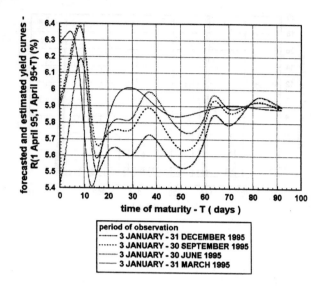

Fig. 28

Appendix A. Vasicek Model with Impulse Perturbations

Case 1. $t \in T_i$, $s \in T_j$, $i \leq j$.

Then

$$P(t,s) = \exp\left\{-\int_t^s r(\tau)\,d\tau\right\} = \exp\left\{-\frac{e^{\beta s} - e^{\beta t}}{\beta}r_0 - \frac{\alpha}{\beta}\left(\frac{e^{\beta s} - e^{\beta t}}{\beta} - (s-t)\right)\right.$$

$$-\sum_{k=1}^{i-1}\frac{e^{\beta(s-t_{k-1})} - e^{\beta(s-t_k)} + e^{\beta(t-t_k)} - e^{\beta(t-t_{k-1})}}{\beta^2}w_k$$

$$-\left(\frac{e^{\beta(s-t_{i-1})} - e^{\beta(s-t_i)} - e^{\beta(t-t_{i-1})} + 1}{\beta^2} - \frac{t_i - t}{\beta}\right)w_k$$

$$-\sum_{k=i+1}^{j-1}\left(\frac{e^{\beta(s-t_{k-1})} - e^{\beta(s-t_k)}}{\beta^2} - \frac{t_k - t_{k-1}}{\beta}\right)w_k$$

$$\left.-\left(\frac{e^{\beta(s-t_{j-1})} - 1}{\beta^2} - \frac{s - t_{j-1}}{\beta}\right)w_i\right\}$$

$$= \exp\left\{\frac{1 - e^{\beta(s-t)}}{\beta}\left(e^{\beta t}r_0 + \frac{\alpha}{\beta}(e^{\beta t} - 1) + \frac{e^{\beta t}}{\beta}\sum_{k=1}^{i-1}(e^{-\beta t_{k-1}} - e^{-\beta t_k})w_k\right.\right.$$

$$\left.\left.+ w_i\frac{e^{\beta(t-t_{i-1})} - 1}{\beta}\right) + \frac{e^{\beta(s-t_i)} - e^{\beta(s-t)}}{\beta^2}w_i + \frac{t_i - t}{\beta}w_i\right.$$

$$-\sum_{k=i+1}^{j-1}\left(\frac{e^{\beta(s-t_{k-1})}-e^{\beta(s-t_k)}}{\beta^2}-\frac{t_k-t_{k-1}}{\beta}\right)w_k$$

$$-\left(\frac{e^{\beta(s-t_{j-1})}-1}{\beta^2}-\frac{s-t_{j-1}}{\beta}\right)w_j+\frac{\alpha}{\beta}(s-t)$$

$$=\exp\left\{\frac{1-e^{\beta(s-t)}}{\beta}r(t)+\frac{\alpha}{\beta}(s-t)+\frac{\alpha}{\beta}\frac{1-e^{\beta(s-t)}}{\beta}\right.$$

$$+\left(\frac{e^{\beta s}}{\beta^2}(e^{\beta t_i}-e^{-\beta t})+\frac{t_i-t}{\beta}\right)w_i$$

$$-\sum_{k=i+1}^{j-1}\left(\frac{e^{\beta s}}{\beta}\left(\frac{e^{-\beta t_{k-1}}-e^{-\beta t_k}}{\beta}\right)-(t_k-t_{k-1})\right)\frac{w_k}{\beta}$$

$$\left.-\left(\frac{e^{\beta(s-t_{j-1})}-1}{\beta}-(s-t_{j-1})\right)\frac{w_j}{\beta}\right\}.$$

Denote

$$B(t,s)=\frac{1-e^{\beta(s-t)}}{\beta},\quad A(t,s)=\frac{\alpha}{\beta}(s-t+B(t,s))$$

$$+\left(\frac{e^{\beta s}}{\beta}(e^{\beta t_i}-e^{-\beta t})+t_i-t\right)\frac{w_i}{\beta}$$

$$-\sum_{k=i+1}^{j-1}\left(\frac{e^{\beta s}}{\beta}(e^{-\beta t_{k-1}}-e^{-\beta t_k})-(t_k-t_{k-1})\right)\frac{w_k}{\beta}$$

$$-\left(\frac{e^{\beta(s-t_{j-1})}-1}{\beta}-(s-t_{j-1})\right)\frac{w_j}{\beta}.$$

Then we have (61).

Case 2. $t\in T_i$, $s\in T_j$, $i=j\leq N$.

Then

$$P(t,s)=\exp\left\{-\int_t^s r(\tau)\,d\tau\right\}$$

$$=\exp\left\{-\frac{e^{\beta s}-e^{\beta t}}{\beta}r_0-\frac{\alpha}{\beta}\left(\frac{e^{\beta s}-e^{\beta t}}{\beta}-(s-t)\right)\right.$$

$$-\sum_{k=1}^{i-1}\frac{e^{\beta(s-t_{k-1})}-e^{\beta(s-t_k)}+e^{\beta(t-t_k)}-e^{\beta(t-t_{k-1})}}{\beta^2}w_k$$

$$- \left(-\frac{e^{\beta(t-t_{i-1})} - e^{\beta(s-t_{i-1})}}{\beta^2} - \frac{s-t}{\beta} \right) w_i \Big\}$$

$$= \exp \left\{ \frac{1 - e^{\beta(s-t)}}{\beta} \left(e^{\beta t} r_0 + \frac{\alpha}{\beta}(e^{\beta t} - 1) \right) \right.$$

$$+ \frac{1}{\beta} \sum_{k=1}^{i-1} (e^{\beta(t-t_{k-1})} - e^{\beta(t-t_k)}) w_k + \frac{w_i}{\beta}(e^{\beta(t-t_{i-1})} - 1) \Big)$$

$$+ \frac{\alpha}{\beta}(s-t) + \frac{\alpha}{\beta} \cdot \frac{1 - e^{\beta(s-t)}}{\beta} + \frac{s-t}{\beta} w_i + \frac{w_i}{\beta} \cdot \frac{1 - e^{\beta(s-t)}}{\beta} \Big\}$$

$$= \exp \left\{ \frac{1 - e^{\beta(s-t)}}{\beta} r(t) + \left(\frac{1 - e^{\beta(s-t)}}{\beta} + s - t \right) \left(\frac{\alpha}{\beta} + \frac{w_i}{\beta} \right) \right\}.$$

Let $B(t,s) = \dfrac{1 - e^{\beta(s-t)}}{\beta}$. Then $A(s,t) = (B(t,s) + s - t) \left(\dfrac{\alpha}{\beta} + \dfrac{w_i}{\beta} \right)$.

Case 3. $t \in T_i$, $s \in T_j$, $i > N$.

Then

$$P(t,s) = \exp \left\{ -\int_t^s r(\tau)\, d\tau \right\}$$

$$= -\frac{e^{\beta s} - e^{\beta t}}{\beta} r_0 - \frac{\alpha}{\beta} \left(\frac{e^{\beta s} - e^{\beta t}}{\beta} - (s-t) \right)$$

$$- \sum_{k=1}^N \frac{e^{\beta(s-t_{k-1})} - e^{\beta(s-t_k)} - e^{\beta(t-t_k)} - e^{\beta(t-t_{k-1})}}{\beta^2} w_k$$

$$= \exp \left\{ \frac{1 - e^{\beta(s-t)}}{\beta} \left(e^{\beta t} r_0 + \frac{\alpha}{\beta}(e^{\beta t} - 1) \right) \right.$$

$$+ \sum_{k=1}^N (e^{\beta(t-t_{k-1})} - e^{\beta(t-t_k)}) w_k \Big)$$

$$+ \frac{\alpha}{\beta}(s-t) + \frac{\alpha}{\beta} \cdot \frac{1 - e^{\beta(s-t)}}{\beta} \Big\}$$

$$= \exp \left\{ \frac{1 - e^{\beta(s-t)}}{\beta} r(t) + \frac{\alpha}{\beta} \left(\frac{1 - e^{\beta(s-t)}}{\beta} + s - t \right) \right\}.$$

Let $B(t,s) = \dfrac{1 - e^{\beta(s-t)}}{\beta}$. Then $A(s,t) = \dfrac{\alpha}{\beta}(s - t + B(t,s))$.

Appendix B. Vasicek Model with Point-Impulse Perturbations

Let $t \in T_i$, $s \in T_j$, $i \leq j$. Then,

$$P(t,s) = \exp\left\{-\int_t^s r(\tau)\,d\tau\right\}$$

$$= \exp\left\{-\frac{e^{\beta s} - e^{\beta t}}{\beta}r_0 - \alpha\left(\frac{e^{\beta s} - e^{\beta t}}{\beta} - \frac{s-t}{\beta}\right)\right.$$

$$\left. -\sum_{k=1}^i \frac{e^{\beta(s-t_{k-1})} - e^{\beta(t-t_{k-1})}}{\beta}w_k - \sum_{k=i+1}^j \frac{e^{\beta(s-t_{k-1})} - 1}{\beta}w_k\right\}$$

$$= \exp\left\{\frac{1 - e^{\beta(s-t)}}{\beta}\left(e^{\beta t}r_0 + \frac{\alpha}{\beta}(e^{\beta t} - 1) + \sum_{k=1}^i e^{\beta(t-t_{k-1})}\right)\right.$$

$$\left. +\frac{1}{\beta}\sum_{k=i+1}^j (1 - e^{\beta(s-t_{k-1})})w_k + \frac{\alpha}{\beta}\frac{1 - e^{\beta(s-t)}}{\beta} + \frac{\alpha}{\beta}(s-t)\right\}$$

$$= \exp\left\{\frac{1 - e^{\beta(s-t)}}{\beta}r(t) + \frac{\alpha}{\beta}\left(\frac{1 - e^{\beta(s-t)}}{\beta} + (s-t)\right)\right.$$

$$\left. +\frac{1}{\beta}\sum_{k=i+1}^j (1 - e^{\beta(s-t_{k-1})})w_k\right\}.$$

Denote $B(t,s) = \dfrac{1 - e^{\beta(s-t)}}{\beta}$, $A(s,t) = \dfrac{\alpha}{\beta}(s-t+B(t,s)) + \dfrac{1}{\beta}\sum_{k=i+1}^j (1 - e^{\beta(s-t_{k-1})})w_k$.

Then we also obtain formula (61).

Appendix C. Dothan Model with Impulse Perturbations

Let $t \in T_i$, $s \in T_j$, $i \leq j$. Then,

$$P(t,s) = \exp\left\{-\int_t^s r(\tau)\,d\tau\right\}$$

$$= \exp\left\{-r_0\left(e^{\sum_{k=1}^{i-1} w_k(t_k - t_{k-1})} \cdot \frac{e^{w_i(t_i - t_{i-1})} - e^{w_i(t_i - t_{i-1})}}{w_i}\right.\right.$$

$$\left. +\sum_{k=i+1}^{j-1} e^{\sum_{l=1}^{k-1} w_l(t_l - t_{l-1})} \cdot \frac{e^{w_k(t_k - t_{k-1})} - 1}{w_k}\right.$$

$$\left.\left. + e^{\sum_{k=1}^{j-1} w_k(t_k - t_{k-1})} \cdot \frac{e^{w_j(s - t_{j-1})} - 1}{w_j}\right)\right\}$$

$$= \exp\left\{ \frac{r_0}{w_i} e^{\sum_{k=1}^{i-1} w_k(t_k-t_{k-1})+w_i(t-t_{i-1})} \right.$$

$$- \frac{r_0}{w_j} e^{\sum_{k=1}^{j-1} w_k(t_k-t_{k-1})+w_j(s-t_{i-1})+w_i(t-t_{k-1})-w_i(t-t_{k-1})}$$

$$- \frac{r_0}{w_i} e^{\sum_{k=1}^{i} w_k(t_k-t_{k-1})} - r_0 \sum_{k=i+1}^{j-1} \frac{e^{\sum_{l=1}^{k} w_l(t_l-t_{l-1})} - e^{\sum_{l=1}^{k-1} w_l(t_l-t_{l-1})}}{w_k}$$

$$\left. + \frac{r_0}{w_j} e^{\sum_{k=1}^{j-1} w_k(t_k-t_{k-1})} \right\}$$

$$= \exp\left\{ -r(t)\left(-\frac{1}{w_i} + \frac{1}{w_j} e^{\sum_{k=i}^{j+1} w_k(t_k-t_{k-1})+w_j(s-t_{j-1})-w_i(t-t_{i-1})} \right) \right.$$

$$\left. + A(t,s) \right\} = \exp\{A(t,s) + r(t)B(t,s)\}, \text{ which is (61)},$$

where $B(t,s) = \dfrac{1}{w_i} - \dfrac{1}{w_j} e^{\sum_{k=i}^{j+1} w_k(t_k-t_{k-1})+w_j(s-t_{j-1})-w_i(t-t_{i-1})}$, and

$$A((t,s)) = -\frac{r_0}{w_i} e^{\sum_{k=1}^{i} w_k(t_k-t_{k-1})} - r_0 \sum_{k=i+1}^{j-1} \frac{e^{\sum_{l=1}^{k} w_l(t_l-t_{l-1})} - e^{\sum_{l=1}^{k-1} w_l(t_l-t_{l-1})}}{w_k}$$

$$+ \frac{r_0}{w_j} e^{\sum_{k=1}^{j-1} w_k(t_k-t_{k-1})}.$$

References

[1] Nicola Anderson, Francis Breedon, Mark Deacon, Andrew Derry, and Gareth Murphy, *Estimating and Interpreting the Yield Curve*, Financial Economics and Quantitatice Analysis, John Wiley & Sons, Chichester, New York, Brisbane, Toronto, Singapore (1996).

[2] Kerry Back, *Yield curve models: a mathematical review*, in *Option Embedded Bonds*, eds. J. Ledermann, R. Klein, and I. Nelkin, Chapter One, Irwin Publishing, New York (1996), 1–35.

[3] Martin Baxter and Andrew Rennie, *Financial Calculus, An Introduction to Derivative Pricing*, Cambridge Univ. Press, Cambridge CB2 1RP (1997).

[4] P. P. Boyle, *Recent models of the term structure of interest rates with actuarial applications*, Transactions of the 21st International Congress 4 (1980) 95–104.

[5] Roger H. Brown and Stephen M. Schaefer, *Interest rate volatility and the shape of the term structure*, Phil. Trans. R. Soc. London A 347 (1994) 563–576.

[6] K. C. Chan, G. Andrew Karolyi, Francis A. Longstaff, and Anthony B. Sanders, *An empirical comparison of alternative models of the short-term interest rate*, J. Finance 47 (1992) 1209–1227.

[7] F. L. Chernousko, *The Estimating of State for Dynamic Systems*, Nauka, Moscow, Russia (1989) in Russian.

[8] R. Gabasov, F. M. Kirillova, and S. Prischepova, *Optimal Feedback Control*, Number 207 in Lecture Notes in Economica and Information Systems, Springer-Verlag, Berlin–Heidelberg–New York–Tokyo (1987).

[9] Hans U. Gerber and Elias S. W. Shiu, *Option pricing by Esscher transforms*, Transactions of the Society of Actuaries **46** (1994) 99–140.

[10] D. Heath, R. Jarrow, and A. Morton, *Bond pricing and the term structure of interest rates, a new methodology*, Econometrica **60** (1992) 77–105.

[11] R. Hettich and K. O. Kortanek, *Semi-infinite programming: theory, methods, and applications*, SIAM Review **35** (1993) 380–429.

[12] John C. Hull, *Options, Futures, and Other Derivative Securities*, Prentice Hall, Englewood Cliffs, New Jersey, 3 edition (1993).

[13] R. E. Kalman, *A new approach to linear filtering and prediction problems*, J. Basic Engineering **821** (1960) 34–45.

[14] V. Kolmanovskii and A. Myshkis, *Applied Theory of Functional Differential Equations*, Mathematics and Its Application, Soviet Series, Kluwer Academic Publishers, Dordrecht–Boston–London, 1992.

[15] K. O. Kortanek and V. G. Medvedev, *Models for estimating the structure of interest rates from observations of yield curves*, Technical report, The University of Iowa, College of Business Administration, Iowa City, Iowa 52242, June 1998. Revision of September 1997 Technical Report: web site http://kwel.biz.uiowa.edu/ announced in *Risks and Rewards*, Vol. 30 (March 1998).

[16] K. O. Kortanek and H. No, *A central cutting plane algorithm for convex semi-infinite programming problems*, SIAM J. Optimization **3** (1993) 901–918.

[17] N. N. Krasovsky, *Theory of Control with Movement*, Nauka, Moscow, Russia (1976) in Russian.

[18] A. B. Kurzansky, *Control and Observation in Indefiniteness Conditions*, Nauka, Moscow, Russia (1977) in Russian.

[19] Alexander Levin, *Linear systems theory in stochastic pricing models*, ed. Ronald J. Ryan, pp. 123–146, Glenlake Publ. Co., Ltd., Chicago–London–New Delhi (1997).

[20] Alexander Levin and D. James Daras, *A Methodology for Market Rate Analysis and Forecasting*, ed. Ronald J. Ryan, pp. 57–73, Glenlake Publ. Co., Ltd., Chicago–London–New Delhi, (1997).

[21] Gennady Medvedev and Samuel H. Cox, *The market price of risk for affine interest rate term structures* in Proceedings of the 6th International AFIR-Colloquium Nuremberg Vol. 1, pp. 913–924, VVW Karsruhe, Germany (1996), Aktuarielle Ansätze für Finanz–Risken AFIR (1996).

[22] V. G. Medvedev, *Algorithm for solving the optimal observation problem with nonconvex a priori distribution of initial state*, Automatic Remote Control **10** (1993) 121–130.

[23] V. G. Medvedev, *Optimal observations of initial state and input disturbances for dynamic systems*, SAMS **14** (1994) 275–288.

[24] V. G. Medvedev, *Positional algorithm for optimal observations of linear dynamical systems*, SAMS **16** (1994) 93–111.

[25] V. G. Medvedev, *The method of construction of approximate solutions of linear optimal control problems with state constraints*, Technical report, Department of Optimal Control Methods, Byelorussian State Univ., F. Skorina 4, 220050 Minsk, Belarus (1995) appeared in Proc. XII International Conference on Proclaw, Poland (September 1995) 12–15.

[26] Riccardo Rebonato, *Interest-Rate Option Models*, Financial Engineering, John Wiley & Sons, Chichester, New York, Brisbane, Toronto, Singapore (1998) 2 edition.

[27] Scott F. Richard, *An arbitrage model of the term structure of interest rates*, J. Financial Economics **6** (1978) 33–57.

[28] Oldrich Vasicek, *An equilibrium characterization of the term structure*, J. Financial Economics **5** (1977) 177–188.

[29] Kenneth R. Vetzal, *A survey of stochastic continuous time models of the term structure of interest rates*, Insurance: Mathematics and Economics **14** (1994) 139–161.

[30] Norbert Wiener, *The extrapolation, interpolation, and smoothing of stationary time series*, J. Wiley, New York (1949).

[31] Yong Yao, *One-factor continuous-time models of the term structure of interest rates*, Technical report, Department of Statistics & Actuarial Science, The University of Iowa, Iowa City IA 52242 (1997).

Reprinted from Appl. Math. Finance
4(1) (1987) 37–64
© Routledge

CALIBRATING VOLATILITY SURFACES VIA
RELATIVE-ENTROPY MINIMIZATION[*]

MARCO AVELLANEDA, CRAIG FRIEDMAN,
RICHARD HOLMES and DOMINICK SAMPERI

*Courant Institute of Mathematical Sciences, New York University,
251 Mercer Street, New York, NY 10012, USA*

We present a framework for calibrating a pricing model to a prescribed set of option prices
quoted in the market. Our algorithm yields an arbitrage-free diffusion process that minimizes the relative entropy distance to a prior diffusion. We solve a constrained (minimax)
optimal control problem using a finite-difference scheme for a Bellman parabolic equation combined with a gradient-based optimization routine. The number of unknowns in
the optimization step is equal to the number of option prices that need to be matched,
and is independent of the mesh-size used for the scheme. This results in an efficient,
non-parametric calibration method that can match an arbitrary number of option prices
to any desired degree of accuracy. The algorithm can be used to interpolate, both in
strike and expiration date, between implied volatilities of traded options and to price
exotics. The stability and qualitative properties of the computed volatility surface are
discussed, including the effect of the Bayesian prior on the shape of the surface and
on the implied volatility smile/skew. The method is illustrated by calibrating to market prices of Dollar-Deutschemark over-the-counter options and computing interpolated
implied-volatility curves.

Keywords: Option pricing, implied volatility surface, calibration, relative entropy,
stochastic control, volatility smile and skew.

1. Introduction

1.1. *Deriving a diffusion model from option prices*

It is well known that the constant-volatility assumption made in the Black–
Scholes framework for option pricing is not valid in real markets. For example,
S&P 500 index options are such that out-of-the money puts have higher implied
volatilities than out-of-the money calls. In the currency options markets, implied
volatilities exhibit a "smile" and a "skew" (in both maturity and strike) whereby
at-the-money options trade at lower volatilities than other strikes, and a premium
for puts in one of the two currencies is manifest in the price of "risk-reversals".[a] To
model the strike- and maturity-dependence of implied volatility, researchers have

---

[*]This research was supported by the National Science Foundation.
[a]A risk-reversal is a position consisting in being long a call and short a put with symmetric strikes.

proposed using arbitrage-free diffusion models for the underlying index in which the spot volatility coefficient is a function of the index level and time. The problem is then to determine what this volatility "surface" should be, given the observed option prices.

This paper present a simple, rigorous, method for constructing such an arbitrage-free diffusion process. The basic idea is to assume an initial Bayesian prior distribution for the evolution of the index and to modify it to produce a calibrated model such that the corresponding probability is as close as possible to the prior. For this, we use the concept of Kullback–Leibler information distance, or relative entropy.

The basic approach is as follows. Let

$$\frac{dS_t}{S_t} = \sigma_t dZ_t + \mu \, dt \tag{1.1}$$

represent the process that we wish to determine. Here σ_t is a random process adapted to the standard information flow and μ is the risk-neutral drift, which we assume is known.[b] The calibration condition for M traded options can be written as

$$\mathbf{E}^\sigma[e^{-rT_i}G_i(S_{T_i})] = C_i, \ i = 1, 2, \ldots, M, \tag{1.2}$$

where r is the interest rate, $\mathbf{E}^\sigma[\cdot]$ denotes the expectation with respect to the measure corresponding to (1.1) and $G_i(S_{T_i}), C_i, \ C_i, \ i = 1, 2, \ldots, M$ represent, respectively, the payoffs and prices of the M options that we wish to match.

We will show that minimizing relative entropy is essentially equivalent to minimizing the functional

$$\mathbf{E}^\sigma \left[\int_0^T \eta(\sigma_s^2) ds \right], \tag{1.3}$$

where $\eta(\sigma_s^2)$ is a strictly convex function which vanishes at the volatility of the prior distribution.

This constrained stochastic control problem is equivalent to a Lagrange multiplier problem in which we maximize the augmented objective function

$$\mathbf{E}^\sigma \left[-\int_0^T \eta(\sigma_s^2) ds + \sum_{i=1}^M \lambda_i \, e^{-rT_i} G_i(S_{T_i}) \right] - \sum_{i=1}^M \lambda_i \, C_i, \tag{1.4}$$

over all adapted volatility processes σ_t and then minimize the result over $(\lambda_1, \ldots, \lambda_M)$.[c]

---

[b]μ is the interest rate differential (carry) in foreign exchange and the interest rate minus the dividend yield for equity indices. We assume therefore a "risk-adjusted" drift.
[c]In practice, we shall restrict our search to volatility processes that satisfy uniform bounds $0 < \sigma_{min} \leq \sigma_s \leq \sigma_{max}$. This constraint will typically not be binding except in a neighborhood of points in the (S, t)-plane corresponding to each strike/expiration date.

We show that in the absence of arbitrage opportunities the value function $V(\lambda_1, \ldots, \lambda_M)$ corresponding to (1.4) is smooth and strictly convex in λ. In particular, it has a unique minimum. The first-order condition at the minimum,

$$\frac{\partial V}{\partial \lambda_i} = \mathbf{E}^{\sigma^*} \left[e^{-rT_i} G_i(S_{T_i}) \right] - C_i = 0, \ i = 1, 2, \ldots, M,$$

ensures that the model is calibrated to market prices. Hence, in this approach, *calibrating the model to the M option prices is equivalent to finding the minimum of a convex function of M variables.*

The algorithm for computing $V(\lambda_1, \ldots, \lambda_M)$ for a given set of Lagrange multipliers consists in solving the Bellman partial differential equation corresponding to (1.4), *viz.*,

$$V_t + \frac{1}{2} e^{rt} \Phi \left(\frac{e^{-rt}}{2} S^2 V_{SS} \right) + \mu S V_S - rV$$

$$= - \sum_{t < T_i}^{M} \lambda_i (G_i(S_{T_i}) - e^{rT_i} C_i) \, \delta(t - T_i),$$

where Φ is the Legendre dual of η. This is done numerically for successive choices of $(\lambda_1, \ldots, \lambda_M)$ until the minimum of

$$V(\lambda_1, \ldots, \lambda_M) = V(S, 0; \ \lambda_1, \ldots, \lambda_M)$$

is reached. The optimal volatility surface is identified as

$$\sigma_t^* = \sigma(S, t) = \sqrt{\Phi' \left(\frac{e^{-rt} S_2 V_{SS}(S, t)}{2} \right)}. \tag{1.5}$$

The volatility function thus obtained is what is traditionally called a "conditional volatility surface".

A few remarks are in order. First, this approach permits the user to impose his or her preference ordering via the specification of a Bayesian prior: the diffusion selected by the model matches market prices and is also as close as possible to the prior. The specification of prior distribution is a key feature of the procedure.[d] Minimizing the relative entropy with respect to the prior stabilizes the far-tails of the probability distribution for the underlying index and implies smoothness of the volatility surface (1.5).[e] The procedure leads to a simple and numerically stable

---

[d]The prior volatility need not be a constant. It can be, for instance, a function of time and/or price.

[e]As we shall see, a unique prescription of the volatility surface far way from traded strikes cannot be obtained precisely from option prices. The introduction of the Bayesian prior serves as an "extrapolation mechanism" for characterizing the volatility in regions where the price information is weak, e.g. for strikes which are deeply away-from-the-money, as well as a mechanism for smoothing the volatility surface.

method for calibrating a pricing model. The small number of input parameters that need to be adjusted makes it tractable in practice. This is in contrast to the proposals where ad-hoc adjustments are required to achieve a stable algorithm.

Figures 1 and 2 display a calibrated spot volatility surface $\sigma(S, t)$ corresponding to a dataset corresponding to Dollar-Deutchemark over-the-counter options for the date of August 23, 1995, provided to us by a market-maker. It consisted of 25 option prices, corresponding to 20- and 25-delta puts and calls and 50-delta calls for maturities of 30, 60, 90, 180 and 270 days.[f] The model was calibrated to mid-market quotes to an accuracy of 10^{-4} (in relative terms). The complete dataset is included in Appendix B.

Generically, the volatility surface corresponding to calibrating to a finite number of option prices converges to the prior volatility surface for (S, t) far away from strikes/expiration dates. Significant variations of the volatility surface occur near strikes/expiration dates. These distortions are sharp near the strikes/expiration dates and diffuse smoothly away from these points. The peaks near strikes/maturities are caused by the infinite Gamma of option payoffs near expiration.[g] As we shall see, these "peaks" in the volatility surface do not affect the continuous dependence

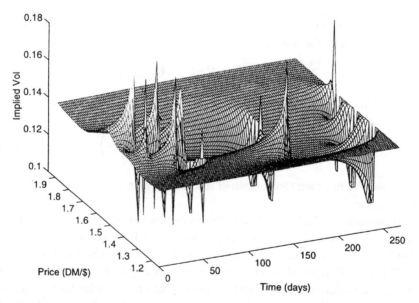

Fig. 1. Calibrated volatility surface for a set of 25 options on Dollar-Deutchemark (dataset in Appendix B). The prior in this calculation is $\sigma_0 = 0.141$. The surface consists of "humps" and "troughs" originating near each strike/expiration date which are smooth away from these points. At the strike/expiration points, the volatility peaks are $\sigma_{\min} = 0.10$ or $\sigma_{\max} = 0.20$. Notice that the surface converges to the prior volatility away from the input strikes.

[f]A 25-delta put is a put with a Black–Scholes delta of -0.25, etc. This is standard terminology for over-the-counter currency options.
[g]The volatility surface obtained using our method is smooth in the S-variable if the option payoffs are regularized prior to implementing the algorithm.

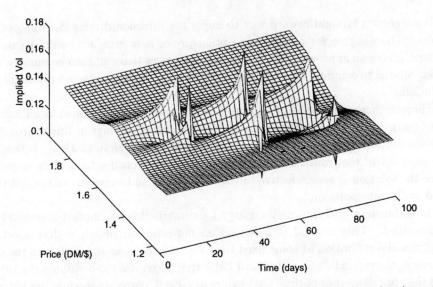

Fig. 2. Detail of the volatility surface of Fig. 1 corresponding to the first 100 days after the trading date. Notice that the price information corresponding to maturities after 90 days affects the earlier values of the surface at earlier dates, as a trough inherited from the later maturities appears in the period from 90 to 100 days.

of the model values on the input prices: the model value of any contingent claim with a payoff which is continuous except on a set of Wiener measure zero is Lipschitz-continuous with respect to the parameters $C_1, \ldots, C_M$.[h]

1.2. *Previous approaches to the "implied tree" problem*

To our knowledge, the first solution of the implied diffusion problem was proposed by Breeden and Litzenberger (1978), and applied to capital budgeting problems in Banz and Miller (1978). Recently, there have been important contributions by Dupire (1994), Shimko (1993), Rubinstein (1994), Derman and Kani (1994), Barle and Kakici (1995) and Chriss (1996), among others. This "smooth-and-differentiate" approach is based on the observation that a call option price can be written as

$$C(K,T) = \int e^{-rT} \max(S_T - K, 0) p(S_T|S_0) dS_T, \qquad (1.6)$$

where $p(S_T|S_0)$ is the conditional probability corresponding to the pricing measure Q associated with the diffusion driving S_t. Differentiating this equation twice with respect to K, we obtain

$$p(K, S_0) = e^{rT} \frac{\partial^2 C(K,T)}{\partial K^2}.$$

[h]In practical applications, filtering techniques can be used to generate smoother conditional volatility surfaces, in conjunction wiht this algorithm.

This suggests a straightforward way to imply the diffusion driving S_t from option prices. The discrete set of observed option prices is interpolated onto a smooth surface, giving an approximating complete set of prices that can then be numerically differentiated to compute the conditional distribution corresponding to the unknown diffusion.

However, since the price of an option is not uniquely determined in an incomplete market (there is more than one pricing measure), implicit in this approach is the assumption that we can find an "approximating complete market", *before the computation of the transition probabilities*. These approaches tend to be unstable since the solution is very sensitive to the smoothness and convexity of the function used in the interpolation.[i]

In Rubinstein (1994), a methodology for constructing an implied binomial tree is described. This method is based on an optimization principle that selects a conditional distribution at some fixed time T that is as close as possible to the distribution corresponding to a standard CRR tree (Cox, Ross and Rubinstein 1979), and that prices a set of options that expire at time T correctly modulo the bid/ask spread.

Rubinstein's approch is revisited in Jackwerth and Rubinstein (1995), where empirical results are discussed and a penalty approach is introduced to smooth the estimated conditional probability function. The fact that Rubinstein's original approach uses only one expiration date has recently been addressed (Jackwerth 1996a; Jackwerth 1996b). The proposed methodology involves the solution of a large scale optimization problem, with number of variables roughly equal to the number of nodes in the tree.

We mention also the recent paper of Bodhurta and Jermakian (1996) who propose to compute a volatility surface in the form of a perturbation series, where each term in the series is computed by solving a partial differential equation containing source terms determined by the previous term. The coefficients in the partial differential equations are computed as they are required by solving a least-squares problem. This approach effectively solves a series of linear partial differential equations to compute approximate prices and an approximate volatility surface, with the approximation improving as more terms are computed.

In Rubinstein (1994) a least-squares criterion is used to measure the distance between two distributions, but the possible benefits of using other measures, including the relative entropy distance, are discussed. Recent work in the one-period setting has suggested that the relative entropy may be a good choice for such a measure. For example, it is shown in Stutzer (1995) that if we select a distribution that

---

[i]This well-known instability is a consequence of the fact that the problem that we are trying to solve is ill-posed. This is obvious when we compare it to the problem of numerically differentiating a function when we only have discrete noisy observations. At a more fundamental level we note that if we fix T and let K vary in (1.6), we obtain a Volterra integral equation for the transition probabilities. Such equations are known to be ill-posed, and specialized techniques such as regularization, smoothing, filtering, etc., are typically required to solve them (Tikhonov and Arsenin 1977; Banks and Kunish 1989; Banks and Lamm 1985).

minimizes the relative entropy to a prior subject to pricing constraints, the resulting distribution is maximally unbiased and absolutely continuous with respect to the prior. Relative entropy minimization is also studied in the one-period context in Buchen and Kelly (1996) and Gulko (1995, 1996). The present paper can be seen as an extension of these ideas to the multi-period setting.

1.3. *Relationship to the uncertain volatility model*

In Avellaneda, Levy and Parás (1995) the Uncertain Volatility Model (UVM) was introduced for hedging a position in a portfolio of derivative securities by selecting the worst possible volatility path with respect to this portfolio. This model was combined with a Lagrange multiplier approach in Avellaneda and Parás (1996) in order to minimize the risk of the worst-case hedge by using options as part of the hedge.

There exists a duality between the problem of finding the worst-case volatility path and the problem of implementing a one-sided hedge (that is, one that perfectly protects either a short or a long position). This duality and its game-theoretical implications were studied Samperi (1995), where it was shown that the duality applies even when the derivative claim to be hedged is path-dependent.

The entropy-based approach introduced in this paper can be viewed as an application of the aforementioned framework to a path-dependent "volatility option". Specifically, consider a contingent claim that pays $\int_0^T \eta(\sigma_s^2)ds$ at time T, i.e., pays $\eta(\sigma_s^2)$ for each "day" that the spot volatility is different from the prior.[j] The solution of the stochastic control problem can then be interpreted as the maximum income that an investor with a long position in this claim can earn by hedging his position with the M options. It is worthwhile to point out that this approach can be used to modify the problem by adding other contingent claims to the portfolio to be hedged, thus combining the entropy-minimization idea with the Lagrangian Uncertain Volatility (Avellaneda and Parás 1995).

1.4. *Outline*

In Sec. 2 we study the notion of Kullback–Leibler relative entropy in the context of diffusions which are mutually singular. This section has the purpose of motivating the constrained stochastic control problem mentioned above.

In Sec. 3, we present a solution to the stochastic control problem using the Bellman dynamic programming principle, and characterize the calibrated volatility surface in terms of partial differential equations.

In Sec. 4 we present the basic numerical algorithm, which involves solving simultaneously a system of $M+1$ partial differential equations for the value-function and its gradient with respect to $(\lambda_1,\ldots,\lambda_M)$.

---

[j]This assumes, however, that the spot volatility is observable, which is not the case in practice. Notice also that the payoff is not discounted by the time-value of money, due to the way the pseudo-entropy is derived from the Kullback–Leibler entropy distance (see Sec. 2). We could also choose to discount the "volatility payoff" at some rate with qualitatively the same results.

In Sec. 5, we discuss the qualitative properties of the volatility surface, on the one hand, and present the calculation of "volatility smiles", which consist in interpolation of the implied volatility data at different maturities. We also analyze the effect of varying the prior, and how this affects the shape of the smile.

In Sec. 6 we discuss the stability of the method with respect to perturbations in the option prices.

The conclusions are presented in Sec. 7.

Mathematical proofs which are overly technical or otherwise standard are presented in an Appendix.

2. Minimizing the Relative Entropy of Pricing Measures and the Constrained Stochastic Control Problem

2.1. *Relative entropy of measures in path-space*

Given two probability measures P and Q on a common probability space $\{\Omega, \Sigma\}$, the relative entropy, or Kullback–Leibler distance, of Q with respect to P is defined as

$$\mathcal{E}(Q; P) = \int_\Omega \ln\left(\frac{dQ}{dP}\right) dQ,$$ (2.1)

where dQ/dP is the Radon–Nikodym derivative of Q with respect to P. $\mathcal{E}(Q; P)$ provides a measure of the relative "information distance" of Q compared to P, where P represents a Bayesian prior distribution. It is well known that

  (i) $\mathcal{E}(Q; P) \geq 0$,
  (ii) $\mathcal{E}(Q; P) = 0 \Longleftrightarrow Q = P$,
  (iii) $\mathcal{E}(Q; P) = \infty$ if Q is not absolutely continuous with respect to P.

Large values of $\mathcal{E}$ correspond to a large information distance (so that Q is very different from the prior P) and $\mathcal{E} \approx 0$ corresponds to low information distance, i.e. proximity to the Bayesian prior P.[k]

We shall study the relative entropy of no-arbitrage pricing measures for derivative securities depending on a single underlying index. Accordingly, consider a pair of probability measures P and Q defined on the set of continuous paths $\Omega = \{S_\theta, \ 0 \leq \theta \leq T\}$ such that

$$\frac{dS_t}{S_t} = \sigma_t^P dZ_t^P + \mu_t^P dt, \quad \text{under } P$$ (2.2a)

and

$$\frac{dS_t}{S_t} = \sigma_t^Q dZ_t^Q + \mu_t^Q dt, \quad \text{under } Q$$ (2.2b)

in the sense of Itô. Here, $\sigma^\bullet$, $\mu^\bullet$ are assumed to be bounded, progressively measurable processes and $Z^\bullet$ are Brownian motions under the respective probabilities.

---

[k]For background on information theory and entropy see Cover and Thomas (1991); Georgescu-Roegen (1971); McLaughlin (1984); Jaynes (1996).

The computation of $\mathcal{E}(Q;\,P)$ is straightforward if $\sigma^P = \sigma^Q \equiv \sigma$ with probability 1 under Q. In this case, dQ/dP can be found explicitly using Girsanov's Theorem and we have

$$\mathcal{E}(Q;\,P) = \frac{1}{2}\mathbf{E}^Q \left\{ \int_0^T \left(\frac{\mu_t^Q - \mu_t^P}{\sigma^t} \right)^2 dt \right\}. \tag{2.3}$$

For applications to the calibration of volatility surfaces we should consider situations where the volatilities of the processes in (2.2) are *not* equal with probability 1. In this case the relative entropy is formally equal to $+\infty$, due to the fact that P and Q are mutually singular. To overcome this problem we shall consider discrete-time approximations to these processes and analyze the behavior of the sequence of entropies as the mesh-size tends to zero.

Consider to this end two probability measures P and Q defined on discrete paths

$$S_0,\, S_1, \ldots, S_N \,,$$

where N is some integer. The P-probability that such a path occurs can be written as

$$\prod_{n=0}^{N-1} \pi_n^P \,,$$

where π_n^P is the conditional probability given the information set at time n that the price S_{n+1} will occur at date $n+1$. An analogous notation will be used for Q. From (2.1) the relative entropy of Q with respect to P is then given by

$$\mathcal{E}(Q;\,P) = \sum_{\text{paths}} \left(\prod_{n=0}^{N-1} \pi_n^Q \right) \cdot \ln \left(\frac{\displaystyle\prod_{n=0}^{N-1} \pi_n^Q}{\displaystyle\prod_{n=0}^{N-1} \pi_n^P} \right)$$

$$= \mathbf{E}^Q \left\{ \ln \left(\frac{\displaystyle\prod_{n=0}^{N-1} \pi_n^Q}{\displaystyle\prod_{n=0}^{N-1} \pi_n^P} \right) \right\}$$

$$= \mathbf{E}^Q \left\{ \sum_{n=0}^{N-1} \ln \left(\frac{\pi_n^Q}{\pi_n^P} \right) \right\}$$

$$= \mathbf{E}^Q \left\{ \sum_{n=0}^{N-1} \left(\mathbf{E}_n^Q \left[\ln \left(\frac{\pi_n^Q}{\pi_n^P} \right) \right] \right) \right\}. \tag{2.4}$$

In (2.4), the symbol $\mathbf{E}_n^Q$ represents the conditional expectation operator given the information set at time n. The last equality states that the relative entropy is obtained by summing the *conditional relative entropies* $\mathbf{E}_n^Q\left[\ln\left(\frac{\pi_n^Q}{\pi_n^P}\right)\right]$ along each path and averaging with respect to the probability Q.

Let us focus on a special class of approximations to the Itô processes in (2.2) for which the entropy can be computed explicitly as $N \to \infty$. These processes are based on trinomial trees and are thus well-suited for numerical computation. We assume, specifically, that

$$S_{n+1} = S_n\, H_{n+1}, \quad n = 0, 1, \ldots$$

where

$$H_{n+1} = \begin{cases} e^{\bar\sigma\sqrt{dt}}, & \text{with probability } P_U, \\ 1, & \text{with probability } P_M, \\ e^{-\bar\sigma\sqrt{dt}}, & \text{with probability } P_D, \end{cases}$$

with transition probabilities given by

$$P_U = \frac{p}{2}\left(1 - \frac{\bar\sigma\sqrt{dt}}{2}\right) + \frac{\mu\sqrt{dt}}{2\bar\sigma},$$

$$P_M = 1 - p,$$

$$P_D = \frac{p}{2}\left(1 + \frac{\bar\sigma\sqrt{dt}}{2}\right) - \frac{\mu\sqrt{dt}}{2\bar\sigma}. \tag{2.5}$$

Here, $dt = T/N$ represents the time-step (measured in years). Notice that the logarithm of S_n follows a random walk on the lattice $\{\nu\bar\sigma\sqrt{dt},\ \nu \text{ integer}\}$. In (2.5), the probabilities have been arranged so that the instantaneous mean and variance of $\ln S_n$ are, respectively, $\mu - (1/2)p\bar\sigma^2$ and $p\bar\sigma^2$, consistently with (2.2). Thus, μ and $\bar\sigma\sqrt{p}$ can be interpreted, respectively, as the carry (interest-rate differential for FX, interest rate minus dividend yield for equities) and the volatility of the index. This model accommodates, by varying the local value of p, processes with variable volatilities in the range $0 < \sigma_t \le \bar\sigma$.[1]

The parameters corresponding to the two probabilities P and Q will be denoted by p_0, μ_0 and p, μ respectively. After some computation, we find that[m]

$$\mathbf{E}_n^Q\left[\ln\left(\frac{\pi_n^Q}{\pi_n^P}\right)\right] = p\ln\left(\frac{p}{p_0}\right) + (1-p)\ln\left(\frac{1-p}{1-p_0}\right)$$

$$+ \frac{p}{2}\left(\frac{\mu}{p\bar\sigma} + \frac{\mu_0}{p_0\bar\sigma}\right)^2 dt + o(dt), \quad dt \ll 1. \tag{2.6}$$

[1]This last statement is true only for dt small enough so that the probabilities in (2.5) are positive. Notice that this setup produces approximations to diffusion processes (in which the local volatility depends on the price and time-to-maturity) as well as more general random-volatility processes. This latter can be obtained by sampling the volatility from a random distribution.

[m]The sum of the conditional relative entropies is finite if and only if $p = p_0$. In this case, the result (2.3) is recovered by replacing the sum of the dt-terms in the right-hand side of (2.6) by an integral. On the other hand, for $p \ne p_0$, the total relative entropy diverges as $dt \to 0$.

In the sequel, we shall assume for simplicity that the two processes have identical, constant drift, i.e., $\mu = \mu_0$, that the Bayesian prior P constant volatility given by

$$\sigma_0^2 = p_0 \overline{\sigma}^2 \,,$$

and that p varies stochastically under Q. Defining the instantaneous volatility for the Q-process at time $t_n = n\,dt$ by

$$\sigma^2(t_n) = \overline{\sigma}^2 p(t_n) \,,$$

we conclude from (2.6) that the conditional relative entropy at time t_n of Q with respect to P is equal to $\eta(\sigma^2(t_n))$ to leading order in dt, where

$$\eta(\sigma^2) \equiv \frac{\sigma^2}{\overline{\sigma}^2} \ln\left(\frac{\sigma^2}{\overline{\sigma}_0^2}\right) + \left(1 - \frac{\sigma^2}{\overline{\sigma}^2}\right) \ln\left(\frac{\overline{\sigma}^2 - \sigma^2}{\overline{\sigma}^2 - \sigma_0^2}\right) . \qquad (2.7)$$

Substituting expression (2.7) into (2.4) and taking into account the estimate of equation (2.6) for the remainder, we conclude that

$$\mathcal{E}(Q;P) = \mathbf{E}^Q \left\{ \sum_{n=0}^{N-1} \left(\eta(\sigma^2(t_n)) + O(dt) \right) \right\}$$

$$= \frac{1}{dt} \mathbf{E}^Q \left\{ \sum_{n=0}^{N-1} \eta(\sigma^2(t_n)) dt \right\} + O(1)$$

$$= \frac{1}{dt} \mathbf{E}^Q \left\{ \int_0^T \eta(\sigma^2(t)) dt \right\} + O(1)$$

$$= N \cdot \frac{1}{T} \mathbf{E}^Q \left\{ \int_0^T \eta(\sigma^2(t)) dt \right\} + O(1)$$

where $T = N\,dt$ and $\mathbf{E}^Q$ represents the expresents the expectation operator with respect to the probability distribution of the *continuous-time* process (2.2b). The relevant information-theoretic quantity for $dt \ll 1$ is thus

$$\frac{1}{T} \mathbf{E}^Q \left\{ \int_0^T \eta(\sigma^2(t)) dt \right\} , \qquad (2.8)$$

which represent the relative entropy per unit time-step of Q with respect to P.

The notion of entropy per unit time-step is not a property of the Ito processes (2.2), but rather of the pair of approximating sequences, (P_N, Q_N). In fact, the function $\eta(\sigma^2) \approx \mathbf{E}_n^Q \left[\ln\left(\frac{\pi_n^Q}{\pi_n^P}\right) \right]$ depends on the discretization used to appropriate the pair (P, Q). To illustrate the non-uniqueness of η, we consider, for example, a discrete-time approximation of (2.2) in which σ_t^P is constant and σ_t^Q is piecewise constant on time-intervals of length dt. In this case, the single-period distributions are conditionally Gaussian and

$$\eta(\sigma^2) = -\frac{1}{2}\left[\ln\left(\frac{\sigma^2}{\sigma_0^2}\right) + 1 - \frac{\sigma^2}{\sigma_0^2}\right].$$ (2.9)

Notice that the function η in (2.7) depends on the lattice constant $\bar{\sigma}$. For large values of $\bar{\sigma}$ in (2.7), we have

$$\eta(\sigma^2) \approx \frac{1}{\bar{\sigma}^2}\left(\sigma^2 \ln\left(\frac{\sigma^2}{\sigma_0^2}\right) - \sigma^2 + \sigma_0^2\right).$$

Thus, we may choose to minimize instead the functional (2.8) with

$$\eta(\sigma^2) = \sigma^2 \ln\left(\frac{\sigma^2}{\sigma_0^2}\right) - \sigma^2 + \sigma_0^2.$$ (2.10)

2.2. *Stochastic control problem*

Due to the non-uniqueness of η, it is mathematically convenient to develop a framework for optimization of the function (2.8) in which $\eta(\sigma^2)$ belongs to a general class of functions which includes (2.7), (2.9) and (2.10) as special cases.

Definition. A pseudo-entropy (PE) function $\eta(\sigma^2)$ with σ_0 is a smooth, real-valued function defined on $O < \sigma^2 < +\infty$, such that

  (i) $0 \le \eta(\sigma^2) < \infty$,
  (ii) $\eta(\sigma^2)$ is strictly convex,
  (iii) $\eta(\sigma^2)$ attains the minimum value of zero at $\sigma^2 = \sigma_0^2$.

The reader can easily check that (2.7), (2.9) and (2.10) are PE functions.[n] The simplest PE function with prior σ_0^2 is the quadratic function[o]

$$\eta(\sigma^2) = \frac{1}{2}(\sigma^2 - \sigma_0^2)^2, \quad \sigma^2 \ge 0.$$ (2.11)

To model the minimization of the Kullback–Leibler distance in the continuous-time setting, we consider the problem:

Given a pseudo-entropy function η,

$$minimize \quad \mathbf{E}^Q\left\{\int_0^T \eta(\sigma^2(s))\,ds\right\}$$

$$subject\ to \quad \mathbf{E}^Q\{e^{-T_i r}G_i(S_{T_i})\} = C_i, \quad i = 1,\ldots,M$$

---

[n]We note that the function in (2.7) is defined only on the interval $0 < \sigma^2 < \bar{\sigma}^2$. To generate a PE function we can extend it arbitrarily as a convex function for $\sigma^2 > \bar{\sigma}^2$.
[o]This function will be used in numerical computations due to its simplicity. As a rule, the choice of the PE function does not effect qualitatively the results that will follow.

amomg all probability distributions Q of Itô processes of the form

$$\frac{dS_t}{S_t} = \sigma_t dZ_t + \mu \, dt \,,$$

such that σ_t is a progressively measurable process satisfying $0 < \sigma_{\min} \leq \sigma_t \leq \sigma_{\max} < +\infty$.

To avoid degeneracies, we assume that there is a unique option per strike/maturity and that there is at least one strike different from zero.[P]

The constraint imposed on σ_t,

$$\sigma_{\min} \leq \sigma_t \leq \sigma_{\max} \,, \quad 0 \leq t \leq T \,, \tag{2.12}$$

where $\sigma_{\min}$ and $\sigma_{\max}$ are positive constants, is made for technical reasons. This assumption guarantees that the class of diffusions considered in the control problem is closed with respect to the topology of weak convergence of measures on continuous paths (Bilingsley 1968). It is equivalent to the uniform parabolicity of the associated Hamilton–Jacobi–Bellman equation, a desirable feature for achieving stability of standard finite-difference schemes. Specifying *a-priori* bounds on volatility could also be useful in order to incorporate beliefs about extreme volatilities.

We view the optimization problem as a means to achieving a balance between "subjective beliefs", represented by the diffusion

$$\frac{dS_t}{S_t} = \sigma_0 \, dZ_t + \mu \, dt \,,$$

and the objective information provided by the market prices C_i. Minimization of the relative entropy implies that the pricing measure deviates as little as possible from the prior, while incorporating the observed price information. Thus, entropy minimization corresponds, roughly speaking, to a "minimal" modification of the prior which leads to an arbitrage-free model.[q]

As mentioned in the introduction, the prior plays a significant role in the algorithm. The prior probability determines the behavior of the transition probabilities far away from the mean position (where the information contributed by option prices is "weak" because the options have low Gamma). In practice, σ_0 should be chosen so that (a) it is near the implied volatilities corresponding to $C_1, \ldots, C_M$, e.g. their geometric or arithmetic mean and (b) it corresponds to the user's expectations about the implied volatility of very low or very high strikes. For instance, to adjust the prior to a market with many expiration dates, one can assume a time-dependent initial prior, $\sigma_0 = \sigma_0(t)$, taking into account the forward-forward

---

[P]In particular, we do not consider puts and calls with same strike and maturity, since their prices should be exactly related by put-call parity in the absence of arbitrage.
[q]While this interpretation is motivated by the calculation of the previous section, it is valid only in an "asymptotic" sense. Here and in sequel, we refer to a solution of the stochastic control problem as a "minimum-entropy measure" irrespective of the choice of the PE function.

volatilities derived from the volatility term-structure. Finally, to incorporate beliefs about the implied volatility at extreme strikes one could consider a prior of the form $\sigma_0 = \sigma_0(S,t)$, with a prescribed behavior for $S \ll 1$ or $S \gg 1$.

3. Solution via Dynamic Programming

We start with an elementary result from convex duality (Rockafellar, 1970). Let η be a PE function with prior σ_0. For $0 \le \sigma_{\min} < \sigma_0 < \sigma_{\max} \le +\infty$. Define

$$\Phi(X) = \sup_{\sigma^2_{\min} < \sigma^2 < \sigma^2_{\max}} \left[\sigma^2 X - \eta(\sigma^2) \right] . \tag{3.1}$$

Lemma 1. *A. If $\sigma_{\max} < +\infty$, then*

(i) $\Phi(X)$ *is convex in* X,

(ii) $\Phi(0) = 0$,

(iii) $\Phi'(0) = \sigma_0^2$,

(iv) $\dfrac{\Phi(X)}{X} \to \sigma^2_{\max}$ *as* $X \to +\infty$,

(v) $\dfrac{\Phi(X)}{X} \to \sigma^2_{\min}$ *as* $X \to -\infty$,

(vi) $\Phi'(X) = \sigma^2_{\max}$ *for* $X \ge \eta'(\sigma^2_{\max})$,

(vii) $\Phi'(X) = \sigma^2_{\min}$ *for* $X \le \eta'(\sigma^2_{\min})$.

B. If $\sigma_{\max} = +\infty$, and

$$\lim_{\sigma^2 \to +\infty} \frac{\eta(\sigma^2)}{\sigma^2} = +\infty,$$

then $\Phi(X)$ is convex and differentiable for all X and (i)–(vii) hold with $\sigma_{\max} = +\infty$.

We shall refer to Φ as the *flux function* associated with the pseudo-entropy η and the bounds $\sigma_{\min}$, $\sigma_{\max}$. In the rest of this section, we assume the η, $\sigma_{\min}$ and $\sigma_{\max}$ are fixed and that $0 < \sigma_{\max} < \infty$.

There exists a one-to-one correspondence between PE functions and flux functions, in the sense that every flux function satisfying assumptions (i) through (vii) of Lemma 1 corresponds to a PE function. In particular, any monotone-increasing function which interpolates between the values $\sigma^2_{\min}$ and $\sigma^2_{\max}$ and takes the intermediate value σ_0^2 at $X = 0$ can be regarded as the derivative $\Phi'(X)$ of a flux function of a PE function η with prior volatility σ_0.

Proposition 2. *Given a vector of real numbers $(\lambda_1, \lambda_2, \ldots, \lambda_M)$, let $W(S,t) = W(S,t; \lambda_1, \lambda_2, \ldots, \lambda_M)$ be the solution of the final-value problem*

$$W_t + e^{rt}\Phi \left(\frac{e^{-rt}}{2} S^2 W_{SS} \right) + \mu S W_S - rW$$

$$= - \sum_{t < T_i \le T} \lambda_i \delta(t - T_i) G_i(S), \quad S > 0, t \le T, \tag{3.2}$$

with final condition $W(S, T + 0) = 0$.[r] Let $\mathcal{P}$ represent the class of probability distributions of admissible Itô processes satisfying (2.12). Then,

$$W(S, t) = \sup_{Q \in \mathcal{P}} \mathbf{E}_t^Q \left[-e^{rt} \int_t^T \eta(\sigma_s^2) \, ds + \sum_{t < T_i \leq T} \lambda_i e^{-(T_i - t)} G_i \right], \quad (3.3)$$

where $\mathbf{E}_t^Q$ is the conditional expectation operator with respect to the information set at time t and $S = S_t$. Moreover, the supremum in (3.3) is realized by the diffusion process

$$\frac{dS_t}{S_t} = \sigma(S, t) \, dZ_t + \mu \, dt \, ,$$

with

$$\sigma^2(S, t) = \Phi' \left(\frac{e^{-rt}}{2} S^2 W_{SS}(S, t) \right) .$$

The final-value problem (3.2) is well-posed because the partial differential equation is uniformly parabolic. This follows from the properties of Φ listed in Lemma 1. The proof of this Proposition follows the standard procedure for "verification theorems" in Control Theory (Krylov (1980); Fleming and Soner (1992)) It is given in Appendix A.

Proposition 3. *The function* $W(S, t; \lambda_1, \lambda_2, \ldots, \lambda_M)$ *defined in Proposition 2 is continuously differentiable and strictly convex in* $(\lambda_1, \lambda_2, \ldots, \lambda_M)$.

For a proof, see also Appendix A. Differentiating Eq. (3.2) with respect to λ we see that the gradient of $W(S, t; \lambda_1, \lambda_2, \ldots, \lambda_M)$ with respect to the λ variables,

$$W_i = \frac{\partial W}{\partial \lambda_i} \, ,$$

satisfies the partial differential equation

$$W_{it} + \frac{1}{2} \Phi' \left(\frac{e^{-rt}}{2} S^2 W_{SS} \right) S^2 W_{iSS} + \mu S W_{iS} - r W_i = -\delta(t - T_i) G_i \, , \quad (3.4)$$

for $1 \leq i \leq M$, with $W_i(S, T+0) = 0$. These equations can be interpreted as pricing equations for the M input options using the diffusion with volatility $\sigma^2(S, t) = \Phi' \left(\frac{e^{-rt} S^2 W_{SS}(S, t)}{2} \right)$. In particular, the model will be calibrated if

$$W_i(S, 0) = C_i \, , \quad \text{or} \quad \frac{\partial W(S, 0; \lambda_1, \lambda_2, \ldots, \lambda_M)}{\partial \lambda_i} = C_i \, .$$

[r]Subscripts indicate partial derivatives; e.g. $W_t = \partial W / \partial t$, etc. $W(S, T + 0)$ represents the value of W for t infinitesimally larger than T. This notation is used to be consistent with the way in which the final conditions corresponding to different option maturities are expressed in (3.2).

This shows that calibration is equivalent to minimizing the function $W(S, 0; \lambda_1, \lambda_2, \ldots, \lambda_M) - \sum \lambda_i C_i$. The next proposition formalizes this and shows that this choice of volatility solves the stochastic control problem.

Proposition 4. *Define*

$$V(S, t; \lambda_1, \lambda_2, \ldots, \lambda_M) = W(S, t; \lambda_1, \lambda_2, \ldots, \lambda_M) - \sum_{i=0}^{M} \lambda_i C_i.$$

Suppose that, for fixed S, $V(S, 0; \lambda_1, \lambda_2, \ldots, \lambda_M)$ attains a global minimum at the point $(\lambda_1^*, \lambda_2^*, \ldots, \lambda_M^*)$ in $\mathbf{R}^M$. Then, the class of probability measures satisfying the price constraints and the volatility bounds (2.12) is non-empty and the stochastic control problem problem admits a unique solution. The solution corresponds to the diffusion process with volatility

$$\sigma(S, t) = \sqrt{\Phi'\left(\frac{e^{-rt}}{2} S^2 W_{SS}\right)}, \quad 0 \le t \le T,$$

where W is the solution of the final-value problem (3.2) with $\lambda_i = \lambda_i^*$, $i = 1, \ldots, M$.

Proof. To establish (i), observe that

$$V(S, 0; \lambda_1, \lambda_2, \ldots, \lambda_M) = \sup_{Q \in \mathcal{P}} \left(a(Q) + \sum_{i=1}^{M} \lambda_i \, b_1(Q)\right), \qquad (3.5)$$

where

$$a(Q) = \mathbf{E}^Q \left\{ -\int_0^T \eta(\sigma_s^2) \, ds \right\}$$

and

$$b_i(Q) = \mathbf{E}^Q = \mathbf{E}^Q \{e^{-rT_i} G_i(S_{T_i})\} - C_i, \quad 1 = 1, 2, \ldots, M.$$

Suppose the function attains a global minimum at some M-tuple $(\lambda_1^*, \ldots, \lambda_M^*)$ and let Q^* denote the unique measure that realizes the sup in (3.5) for these λ-values. (The measure Q^* is unique, by Proposition 2.) The linear function

$$a(Q^*) + \sum_{i=1}^{M} \lambda_i \, b_i(Q^*)$$

can be viewed as the graph of a supporting hyper-plane to the graph of V passing through the minimum. In particular, the smoothness of V (a consequence of Proposition 2) implies that this hyper-plane is tangent to the graph of V and thus that

$$\left(\frac{\partial V}{\partial \lambda_i}\right)_{\lambda=\lambda^*} = b_i(Q^*) = \mathbf{E}^{Q^*}\{e^{-rT_i}G_i(S_{T_i})\} - C_i = 0$$

for $i = 1, \ldots, M$. The subset of measures of type $\mathcal{P}$ which satisfy the price constraints is therefore non-empty: it contains at least the element Q^*.

Suppose now that Q' is another measure in the class $\mathcal{P}$ such that

$$\mathbf{E}^{Q'}\{e^{-rT_i}G_i(S_{T_i})\} = C_i, \quad i = 1, \ldots, M.$$

Then, $b_i(Q') = 0$, so

$$a(Q') = a(Q') + \sum_{i=1}^{M} \lambda_i^* b_i(Q')$$

$$\leq \sup_{Q \in \mathcal{P}} \left(a(Q) + \sum_{i=1}^{M} \lambda_i^* b_i(Q)\right),$$

$$= a(Q^*).$$

This establishes that Q^* has the smallest relative entropy among all measures of type $\mathcal{P}$ satisfying the price constraints.

4. Numerical Implementation

The numerical solution consists in computing the function $V(S, 0; \lambda_1, \ldots, \lambda_M)$ and searching for its minimum in λ-space. For this purpose, we consider a system of PDEs for the evaluation of this function and its derivatives,

$$V_i(S, 0; \lambda_1, \ldots, \lambda_M) = W_i(S, 0; \lambda_1, \ldots, \lambda_M) - C_i, \quad i \leq i \leq M,$$

namely,

$$V_t + e^{rt}\Phi\left(\frac{e^{-rt}}{2}S^2 V_{SS}\right) + \mu S V_S - rV$$

$$= -\sum_{t<T_i}^{M} \lambda_i\left(G_i(S_{T_i}) - e^{rT_i}C_i\right)\delta(t - T_i), \tag{4.1}$$

$$V_{it} + \frac{1}{2}\Phi'\left(\frac{e^{-rt}}{2}S^2 V_{SS}\right)S^2 V_{iSS} + \mu S V_{iS} - rV_i$$

$$= -\sum_{t<T_i}^{M}\left(G_i(S_{T_i}) - e^{rT_i}C_i\right)\delta(t - T_i), \tag{4.2}$$

for $1 \leq i \leq M$.

Concretely, the algorithm for finding the minimum of $V(S, 0; \lambda_1, \ldots, \lambda_M)$ consists in

- rolling back the values of the vector $(V, V_1, \ldots, V_M)$ to the date $t = 0$,
- updating the estimate of $(\lambda_1, \ldots, \lambda_M)$ using the computed value of the gradient with a gradient-based optimization subroutine,
- repeating the above steps until the minimum is found.

Our numerical method for solving (4.1)–(4.2), uses a finite-difference scheme (trinomial tree) presented in Sec. 2.1, with the risk-neutral probabilities in (2.5). We implemented, for simplicity, the quadratic pseudo-entropy function in (2.11). The corresponding flux function is

$$\Phi(X) = \begin{cases} \frac{1}{2}, X^2 + \sigma_0^2 X\,, & \sigma_{\min}^2 - \sigma_0^2 < X < \sigma_{\max}^2 - \sigma_0^2\,, \\[2mm] \sigma_{\min}^2 X - \frac{1}{2}(\sigma_{\min}^2 - \sigma_0^2)^2\,, & X \le \sigma_{\min}^2 - \sigma_0^2\,, \\[2mm] \sigma_{\max}^2 X - \frac{1}{2}(\sigma_{\max}^2 - \sigma_0^2)^2 & X \ge \sigma_{\max}^2 - \sigma_0^2\,, \end{cases}$$

The derivative of Φ varies linearly between $\sigma_{\min}^2$ and $\sigma_{\max}^2$. It is given by

$$\Phi'(X) = \begin{cases} X + \sigma_0^2\,, & \sigma_{\min}^2 - \sigma_0^2 \le X \le \sigma_{\max}^2 - \sigma_0^2\,, \\[2mm] \sigma_{\min}^2\,, & X \le \sigma_{\min}^2 - \sigma_0^2\,, \\[2mm] \sigma_{\max}^2\,, & X \ge \sigma_{\max}^2 - \sigma_0^2\,. \end{cases}$$

As a numerical approximation for the "dollar Gamma" $\frac{1}{2}S^2 V_{SS}$ in the lattice, we take

$$\left(\frac{1}{2}S^2 V_{SS}\right)_n^j \longleftrightarrow \frac{1}{\bar{\sigma}^2\,dt}\left[\left(1 - \frac{\bar{\sigma}\sqrt{dt}}{2}\right)\cdot V_n^{j+1} + \left(1 + \frac{\bar{\sigma}\sqrt{dt}}{2}\right)\cdot V_n^{j-1} - 2V_n^j\right] \quad (4.3)$$

The partial differential equations are approximated by local "roll-backs" using the probabilities (2.5) with the appropriate choice for the parameter p at each node, dictated by the value of (4.3). The "local volatility" in the trinomial tree is

$$(\sigma_n^j)^2 = p_n^j\bar{\sigma}^2\,,$$

so we take

$$p_n^j = \frac{1}{\bar{\sigma}^2}\left[\frac{\Phi\left(e^{-rt}(\frac{1}{2}S^2 V_{SS})_n^j\right)}{e^{-rt}(\frac{1}{2}S^2 V_{SS})_n^j}\right]$$

in Eq. (4.1) and

$$p_n^j = \frac{1}{\bar{\sigma}^2}\Phi'\left(e^{-rt}\left(\frac{1}{2}S^2 V_{i,SS}\right)_n^j\right)$$

for Eq. (4.2).

The scheme implemented for this study was explicit Euler with trimming of the tails after 3.5 standard deviations.[s] For the numerical optimization, we used the BFGS algorithm(Byrd *et al.* (1994); Byrd *et al.* (1996); Zhu *et al.* (1994)).

5. The Volatility Surface

5.1. *Spot volatility*

We study in more detail the spot volatility surface computed by this algorithm. To simplify the analysis, we perform a change of variables that eliminates μ and r from the right-hand side of the PDE (38), namely:

$$\tilde{V} = e^{-rt}V, \quad \tilde{S} = e^{-\mu t}S.$$

With these new variables,[t] Eq. (4.1) becomes

$$\tilde{V}_t + \frac{1}{2}\Phi\left(\frac{\tilde{S}^2\tilde{V}_{SS}}{2}\right) = -\sum_{T_i<t\leq T}^{M}\lambda_i\left[e^{-rT_i}G_i(\tilde{S}e^{\mu T_i}) - C_i\right]\delta(t - T_i), \qquad (5.1)$$

with $\tilde{V}(S, T + 0) = 0$.

Differentiating this equation twice with respect to $\tilde{S}$ and multiplying both sides by $\frac{\tilde{S}^2}{2}$ we obtain an evolution equation for the "dollar-Gamma" of the value function

$$\tilde{\Gamma} = \frac{\tilde{S}^2\tilde{V}_{\tilde{S}\tilde{S}}}{2} = \frac{e^{-rt}S^2V_{SS}}{2}.$$

Dropping the tildes to simplify notation, the equation thus obtained is

$$\Gamma_t + \frac{S^2}{2}(\Phi(\Gamma))_{SS} = -\sum_{T_i<t\leq T}^{M}\lambda_i e^{-(r-\mu)T_i}\delta(S - e^{-\mu T_i}K_i)\,\delta(t - T_i), \qquad (5.2)$$

or

$$\Gamma_t + \frac{S^2}{2}(\Phi'(\Gamma)\Gamma_S)_S = -\sum_{T_i<t\leq T}^{M}\lambda_i e^{-(r-\mu)T_i}\delta(S - e^{-\mu T_i}K_i)\,\delta(t - T_i). \qquad (5.3)$$

The latter equation clarifies the nature of the volatility surface

$$\sigma^2 = \Phi'(\Gamma). \qquad (5.4)$$

For instance, if the option prices C_i are exactly the Black–Scholes prices with volatility σ_0, the solution of the stochastic control problem has $\lambda_i^* = 0$ for all i and $\Gamma = 0$, consistently with the fact that $\sigma^2(S, t) = \sigma_0^2$ is the minimum-entropy solution. (In this case no information is added by considering option prices.) On the other hand,

---

[s]See Parás (1995) for a proof of consistency fo the scheme.
[t]The new variables correspond to the value of assets measured in dollars at time $t = 0$.

if one or more option prices are inconsistent with the prior, the Lagrange multipliers are not all zero. Each non-zero λ_i^*, gives rise to a Dirac source in (5.2)–(5.3). The resulting Γ profile is initially singular (it is similar to the Gamma of an option portfolio) and diffuses progressively into the (S, t)-plane as a smooth function. Instantaneous smoothing of Γ is guaranteed by the bounds on the volatility Φ' which follow from (2.12) (cf. Lemma 1). Using Eq. (5.4), we find that, immediately before time T_i and near the strike, σ^2 is equal to $\sigma_{\min}$ or $\sigma_{\max}$, according to the sign of λ_i^*. As $T_i - t$ increases, the surface becomes smoother and the constraint $\sigma_{\min} \leq \sigma_t \leq \sigma_{\max}$ is non-binding. Generically, each point (K_i, T_i) gives rise to a disturbance of the volatility surface, which looks like a "ridge" ($\lambda_i^* > 0$) or a "trough" ($\lambda_i^* < 0$). To complete the picture, note that the disturbances "interact" with each other due to the nonlinearity of the equation. The overall topography of the surface is determined by the relative strengths of the Lagrange multipliers $\lambda_1^*, \ldots, \lambda_M^*$.[u]

5.2. *Implied volatility: interpolating between traded strikes*

The main application of the implied volatility surface is to calculate the fair values of derivative securities which are not among the M input options. An interesting diagnostic for our algorithm consists in analyzing the implied volatility profiles that can be generated after calibrating the model to a finite number of option prices. There are two features of interest here: the shape of the curve between strikes (interpolation) and the shape of the curve for strikes which are smaller or larger than the ones used for calibration (extrapolation).

A first set of numerical experiments was done using the Dollar/Mark dataset of Appendix B; cf. Figs. 1 and 2. At each of the standard maturities, ranging from 30 days to 270 days, we have 5 traded strikes. After calibrating to the mid-market prices of these options using the parabolic PE function with prior $\sigma_0 = 0.141$ (a rough average of the implied volatilities of traded options), we computed option prices for a sequence of strikes at each expiration date using a fine mesh. We then computed the corresponding implied volatilities and generated an "implied smile" for each standard maturity.

The curves are shown in Fig. 3.

Notice that the shapes are influenced by the relation between the implied volatilities and the prior. This market corresponds to an "inverted" volatility term-structure, with near-term options trading at more than 14% or higher and 270-day options trading at approximately 13% volatility.

Given our choice of prior (arbitrarily chosen), σ_0 is lower than the volatilities of traded options with short-maturities and higher than the implied volatilities of

---

[u]In numerical calculations, point sources corresponding to small values of λ^* may not always be observable, due to the discrete approximation of the Delta functions. Thus, weak point sources may become "masked" by the Γ produced by other options with larger λ^*.

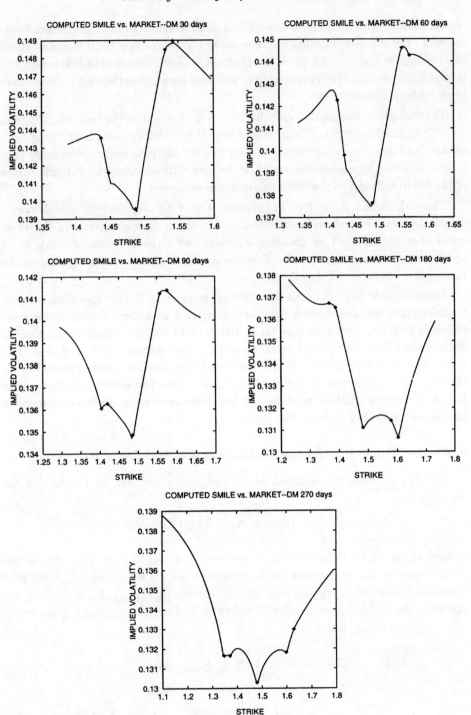

Fig. 3. Implied volatility curves at different expiration dates computed using a constant prior of 0.141. The data is given in Appendix B.

traded options with long maturities. The minimum relative-entropy criterion tends to "pull" the implied volatility curve towards the prior. The "pull-to-prior" effect can be seen in the way the curve interpolates between strikes. For low priors, the interpolation tends to be a convex curve while for high priors the interpolated curve tends to be concave.

The "wings" of the implied volatility curves are lower than the (extreme) 20-delta volatilities for short-term options, higher than the 20-delta volatilities for long-term options and are practically horizontal for the 90-day puts and 60-day calls, that have volatilities approximately equal to the prior. In all cases, the extreme values of the volatility tend to the prior volatility, as we expect.

This calculation show that, in practice, it may be necessary to consider prior volatilities that depend on both S and t. A more conventional form of the smile could then be achieved by choosing σ_0 using the term-structure of volatility of at-the-money-forward options for S between traded strikes and a higher prior to extrapolate beyond traded strikes.

To investigate in more detail the effect of the prior on the interpolation between traded strikes, we considered a hypothetical market with three traded options, expiring in 30 days, with strikes equal to 100, 95 and 105 percent of the spot price. We assumed that the implied volatilities of the options were 14%, 15% and 16%, respectively and that $\mu = r = 0$. We calibrated four different volatility surfaces for this dataset, using priors of 11%, 13%, 14% and 17%. The results are displayed in Fig. 4. These calculations confirms our previous conclusions on the sensitivity of the implied volatility curve to the prior.

6. Convex Duality, Lagrange Multipliers and Stability Analysis

We may visualize the solution of the optimality problem by considering the function

$$W(S, 0; \lambda_1, \ldots, \lambda_M),$$

defined in Eq. (3.2). This function depends on r, μ, K_i, T_i, $i = 1, \ldots, M$, on the pseudo-entropy function η and on the volatility bounds σ_{min} and σ_{max}. We have established that W is smooth and strictly convex in $(\lambda_1, \ldots, \lambda_M)$. Solving the optimization problem corresponds therefore to finding, for a given a price vector $(C_1, \ldots, C_M)$, the quantity

$$U(C_1, \ldots, C_M) \equiv \inf_{\lambda_1, \ldots, \lambda_M} V(S, 0; \lambda_1, \ldots, \lambda_M)$$

$$= \inf_{\lambda_1, \ldots, \lambda_M} \left[W(S, 0; \lambda_1, \ldots, \lambda_M) - \sum_{i=1}^{M} \lambda_i C_i \right] \tag{6.1}$$

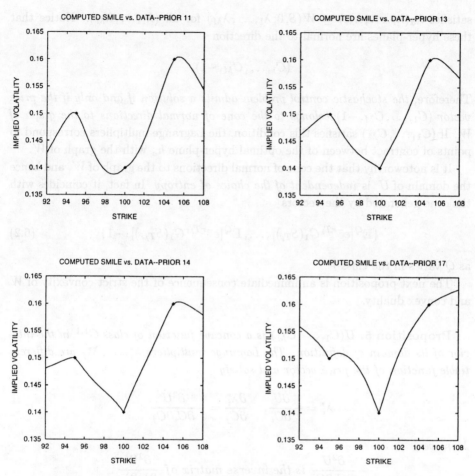

Fig. 4. Effect of varying the prior volatility on the interpolated implied volatility curve (smile). The data consists of 3 options with maturity 30 day and volatilities 14%(strike=100), 15%(strike=95) and 16%(strike=105). Interest rates were taken to be zero.

and the Lagrange multipliers. The function $U(C_1, \ldots, C_M)$ represents the "maximum entropy per lattice site" of measures in the class $\mathcal{P}$ (i.e. Ito processes with drift μ and volatility satisfying the *a priori* volatility bounds) which match market prices. It is the dual of W, in the sense of convex duality.[v]

Geometrically, $U(C_1, \ldots, C_M)$ corresponds to the largest value of a for which the hyper-plane in $\mathbf{R}^{M+1}$

$$h_a(\lambda_1, \ldots, \lambda_M) = \sum_{I=1}^{M} \lambda_i C_i + a$$

[v]Strictly speaking, $-U$ is the Legendre dual of W. The functions η and Φ are in a similar correspondence, if we redefine η to be $+\infty$ for σ^2 outside the interval $[\sigma_{\min}, \sigma_{\max}]$.

satisfies $h_a(\lambda_1, \ldots, \lambda_M) \leq W(S, 0; \lambda_1, \ldots, \lambda_M)$ for all $(\lambda_1, \ldots, \lambda_M)$. Notice that these hyper-planes are normal to the direction

$$(C_1, \ldots, C_M, -1).$$

Therefore, *the stochastic control problem admits a solution if and only if the price vector* $(C_1, \ldots, C_M, -1)$ *belongs to the cone of normal directions to the graph of* W. If $(C_1, \ldots, C_M)$ satisfies this condition, the Lagrange multipliers correspond to points of contract between of the optimal hyper-plane h_a with the graph of W.

It is noteworthy that the cone of normal directions to the graph of W, and hence the domain of U, is *independent of the choice of entropy*. In fact, it coincides with the cone generated by the vectors

$$\{\mathbf{E}^Q[e^{-rT_1} G_1(S_{T_1})], \ldots, \mathbf{E}^Q[e^{-rT_M} G_1(S_{T_M})], -1\} \tag{6.2}$$

as Q varies in the class $\mathcal{P}$.[w]

The next proposition is an immediate consequence of the strict convexity of W and convex duality.

Proposition 5. $U(C_1, \ldots, C_M)$ *is a concave function of class* $C^{1,1}$ *in the interior of its domain of definition. The Lagrange multipliers* $\lambda_1^*, \ldots, \lambda_M^*$ *are differentiable function of the price vector and satisfy*

$$\lambda_i^* = -\frac{\partial U}{\partial C_i}, \quad \frac{\partial \lambda_i^*}{\partial C_j} = -\frac{\partial^2 U}{\partial C_i \partial C_j}.$$

Moreover,

$$-\frac{\partial^2 U}{\partial C_i \partial C_j} \text{ is the inverse matrix of } \frac{\partial^2 W}{\partial \lambda_i \partial \lambda_j}.$$

Thus, if $(C_1, \ldots, C_M)$ varies in a compact subset of the domain of U, the sensitivities $\frac{\partial \lambda_i^*}{\partial C_j}$ remain uniformly bounded. We conclude that the functions $W(S, t)$ and $W_S(S, t)$ are Lipschitz continuous functions of the C_i, uniformly in (S, t). The same is true for the second derivative W_{SS} in any closed region of the (S, t)-plane which excludes the points $(K_i, T_i), i = 1, \ldots, M$. At these points, the second derivative of W is singular, because $G_{i,SS} = \delta(S - K_i)$ and hence $W_{SS}(S, T_i) = \lambda_i^* \delta(S - K_i)$. A discontinuity of $W_{SS}(K_i, T_i)$ with respect to $(C_1, \ldots, C_M)$ will occur when the Lagrange multiplier λ_i^* crosses zero and W_{SS} changes sign.[x] In particular, the volatility surface

$$\sigma(S, t) = \sqrt{\Phi'\left(\frac{e^{-rt} S^2 W_{SS}(S, t)}{2}\right)}$$

---

[w]The reason for this is that the latter cone is the tangent cone at infinity to the cone of normals to W.

[x]We note, however, that W_{SS} is Lipshitz continuous in C_i as a signed measure.

is uniformly Lipschitz-continuous as a function of $(C_1, \ldots, C_M)$ for (S, t) bounded away from the points (K_i, T_i), $i = 1, \ldots, M$. Note, however, that the prices of contingent claims obtained with this model are continuous in $(C_1, \ldots, C_M)$, since W depends smoothly on the Lagrange multipliers and hence on the price vector. The algorithm is therefore stable with respect to perturbations of the price vector.

The stability of the algorithm deteriorates, however, as the price vector approaches the boundary of the domain of definition of U, due to the fact that the Lagrange multipliers increase indefinitely and U tends to $-\infty$ as $(C_1, \ldots, C_M, -1)$ approaches the boundary of the cone (6.2). To increase the stability of numerical computations in these cases, the volatility band should be widened until the Lagrange multipliers are of order 1.

7. Conclusions

The calibration of a diffusion model to a set of option prices can be cast as a minimax problem which corresponds to the minimization of the relative entropy distance between the surface that we wish to find and a Bayesian prior distribution.

The minimax problem can be solved by dynamic programming combined with the minimization of a function of M variables, where M is the number of prices that we seek to match. The evaluation of the function that we wish to minimize and of its gradient is done by solving a system of $M + 1$ partial differential equations on a trinomial tree.

The resulting volatility surfaces are essentially the minimal perturbations of the Bayesian prior that match all option prices. Accordingly, the method allows for constructing a surface that takes into account not only option prices but also the user's expectations about volatility (via the prior). Qualitatively, the surface consists of ridges or troughs superimposed on the prior surface, which are sharp near the strike/expiration points (K_i, T_i) and diffuse smoothly away from these points. Roughly speaking, the shapes of the distortions are close to the shape of the Gamma-surface of an option.

We have shown that the prices of contingent claims generated by the model vary continuously with the input option prices $(C_1, \ldots, C_M)$. The stability and height of the volatility surface at the strike/maturity points is controlled by the bounds $\sigma_{\min}$ and $\sigma_{\max}$.

Numerical calculations show that the algorithm can be used to interpolate between the implied volatilities of traded options. The curves obtained in this fashion depend, however, on the choice of prior distribution. In particular, for extrapolation beyond traded strikes, prior volatilities that take into account subjective expectations about volatilities conditional upon extreme market moves should be used. These and other qualitative features of the algorithm will be studied in future publications.

Appendix A

A.1. *Proof of Proposition 2*

Consider the Itô process

$$\frac{dS_t}{S_t} = \mu\,dt + \sigma_t dZ_t\,,$$

where σ_t is an adapted random process such that

$$\sigma_{\min}^2 \leq \sigma^2 \leq \sigma_{\max}^2\,.$$

By Itô's Lemma, we have

$$d(e^{-rt}W(S_t,t))$$

$$= e^{-rt} \cdot \left\{ W_S(S_t,t)dS_t + \left[W_t + \frac{1}{2}\sigma_t^2 S_t^2 W_{SS}(S_t,t) - rW(S_t,t) \right] dt \right\}. \quad (A.1)$$

From the inequality

$$X\sigma^2 \leq \Phi(X) + \eta(\sigma^2)\,, \quad (A.2)$$

which follows from the definition of Φ, we have

$$\frac{1}{2}S^2 W_{SS} \cdot \sigma^2 = e^{rt} \cdot \left(\frac{e^{-rt}}{2} S^2 W_{SS} \cdot \sigma^2 \right)$$

$$\leq e^{rt} \cdot \Phi\left(\frac{e^{-rt}}{2} S^2 W_{SS} \right) + e^{rt}\eta(\sigma_t^2)\,. \quad (A.3)$$

Substituting this inequality into (A.1) and rearranging terms, we obtain

$$d(e^{-rt}W(S_t,t))$$

$$= e^{-rt}W_S S_t \sigma_t^2 dZ_t + e^{-rt}t\left\{ W_t + \frac{1}{2}\sigma_t^2 S_t^2 W_{SS} + \mu\,S_t - rW \right\} dt$$

$$\leq e^{-rt}W_S S_t \sigma_t^2 dZ_t$$

$$+ e^{-rt}\left\{ W_t + e^{rt}\Phi\left(\frac{e^{-rt}}{2} S_t^2 W_{SS} \right) + \mu\,S_t - rW \right\} dt + \eta(\sigma_t^2)\,dt$$

$$= e^{-rt}W_S S_t \sigma_t^2 dZ_t - \sum_{t<T_i\leq T} \lambda_i e^{-rt}\delta(t-T_i)\,G_i(S_t)\,dt + \eta(\sigma_t^2)\,dt\,, \quad (A.4)$$

where we used Eq. (3.2) to derive the last equality. Integrating with respect to t and taking the conditional expectation at time t, we obtain

$$\mathbf{E}_t^Q\left(e^{-rT}W(S_T, T+0) \right) - e^{rt}W(S,t)$$

$$\leq -\mathbf{E}^Q\left\{ \sum_{t<T_i\leq T} \lambda_i e^{-rT_i}\,G_i(S_{T_i}) - \int_t^T \eta(\sigma_s^2)\,ds \right\}$$

or, since, $W(S_T, T+0) = 0$,

$$\mathbf{E}^Q \left\{ \sum_{t < T_i \le T} \lambda_i e^{-r(T_i - t)} G_i(S_{T_i}) - e^{rt} \int_t^T \eta(\sigma_s^2) \, ds \right\} \le W(S, t) . \tag{A.5}$$

This shows that the function $W(S,t)$ is an upper bound on the possible values taken by the left-hand side of (28) as Q ranges over the family of probabilities $\mathcal{P}$. The calculation also shows that the inequality becomes equality when the volatility of the Itô process is chosen to be precisely

$$\sigma_t = \sqrt{\Phi'\left(\frac{e^{-rt}}{2} S_t^2 W_{SS}(S_t, t)\right)} \tag{A.6}$$

because the inequalities in (A.3) and (A.4) are saturated for this particular σ_t. Consequently, W is the value function for this control problem and (A.6) characterizes the measure where the supremum is attained.

A.2. *Proof of Proposition 3*

We set

$$W_i = \frac{\partial W}{\partial \lambda_i} \quad \text{and} \quad W_{ij} = \frac{\partial^2 W}{\partial \lambda_i \partial \lambda_j} .$$

Differentiating Eq. (3.2) with respect to the variables λ_i, we obtain

$$W_{it} + \frac{1}{2} \Phi'\left(\frac{e^{-rt}}{2} S^2 W_{SS}\right) S^2 W_{iSS} + \mu S W_{iS} - r W_i$$

$$= -\delta(t - T_i) G_i , \quad 1 \le i \le M , \tag{A.7}$$

with $W_i(S, T+0) = 0$, and

$$W_{ijt} + \frac{1}{2} \Phi'\left(\frac{e^{-rt}}{2} S^2 W_{SS}\right) S^2 W_{ijSS} + \frac{e^{-rt}}{4} S^4 \Phi''\left(\frac{e^{-rt}}{2} S^2 W_{SS}\right) W_{iSS} W_{jSS}$$

$$+ \mu S W_{ijS} - r W_{ij} = 0 , \quad 1 \le i, j \le M , \tag{A.8}$$

with $W_{ij}(S, T+0) = 0.$[y]

Equation (A.7) describes the evolution of the gradient of W with respect to λ. It is a Black–Scholes-type equation in which the volatility parameter, depends on S and t.

The second equation has a "source term"

$$\frac{e^{-rt}}{4} S^4 \Phi''\left(\frac{e^{-rt}}{2} S^2 W_{SS}\right) W_{iSS} W_{jSS}$$

---

[y]The smoothness of the function $\Phi(X)$ justifies this formal differentiation procedure.

and no explicit dependence on the G_i's. The show that W is convex, it is sufficient to verify that, for all $\theta \in \mathbf{R}^M$, we have

$$H = \sum_{i=0,j=0}^{M} \theta_i \theta_j W_{ij} \geq 0. \tag{A.9}$$

But it follows from (A.8) that H satisfies

$$H_t + \frac{1}{2} \Phi' \left(\frac{e^{-rt}}{2} S^2 W_{SS} \right) S^2 H_{SS} + \frac{e^{-rt}}{4} S^4 \Phi'' \left(\frac{1}{2} S^2 W_{SS} \right) \cdot \left(\sum_{1}^{M} \theta_i W_{iSS} \right)^2$$

$$+ \mu S H_S - rH = 0 \tag{A.10}$$

with final condition $H(S, T) = 0$. Due to the convexity of Φ, we have

$$\frac{e^{-rt}}{4} S^4 \Phi'' \left(\frac{e^{-rt}}{2} S^2 W_{SS} \right) \cdot \left(\sum_{1}^{M} \theta_i W_{iSS} \right)^2 \geq 0. \tag{A.11}$$

Hence, by the Maximum Principle applied to equation (A.10), we conclude that $H(S, t) \geq 0$ for all S and all $t < T$.

Finally, we show that $H(S, t) > 0$ to establish strict convexity. For this, it is sufficient to show that the left-hand side of (A.11) is positive on a subset of the (S, T) plane of positive Lebesgue measure.[z] Notice that $\Phi''(X) > 0$ in a neighborhood of $X = 0$, which implies

$$\Phi'' \left(\frac{e^{-rt}}{2} S^2 W_{SS} \right) > 0$$

in regions of the (S, t) plane where $|S^2 W_{SS}|$ is sufficiently small. Recalling that the curvature of the payoffs decays as S tends to zero or infinity, we conclude that the latter inequality is satisfied if S is sufficiently far away from all strikes.

In addition, we note that

$$\sum_{1}^{M} \theta_i W_{iSS}$$

cannot vanish on *any* open subset of the (S, t) plane. Indeed, by the Unique Continuation Principle, this would imply that it vanishes identically and thus that

$$\sum_{1}^{M} \theta_i W_i$$

is a linear function. This is impossible unless

$$\sum \theta_i G_i$$

---

[z]This implies the positivity of H because the fundamental solution of Eq. (A.10) is strictly positive. Here we use the fact that $\Phi'(X) \geq \sigma_{\min} > 0$.

is linear at each payoff date, which is ruled out by the assumption that there is a single option per strike and at least one nonzero strike. We conclude that strict inequality holds in (A.11) for (S, t) in some open set. Thus, W is strictly convex in $(\lambda_1, \ldots, \lambda_M)$.

Appendix B. Dataset for the USD/DEM example

Table 1. Dataset used for the calibration example of Figs. 1, 2 and 3. Contemporaneous USD/DEM option prices (based on bid-ask volatilites and risk-reversals) provided to us by a marketmaker on August 23, 1995. The options correspond to 20-delta and 25-delta USD/DEM puts and calls and 50-delta calls. Implied volatility corresponding to mid-market prices for each option fare displayed in the last column. The other market parameters are: spot FX = 1.4885/4890; DEM deposit rate = 4.27%; USD deposit rate = 5.91%.

Maturity	Type	Strike	Bid	Offer	Mid	IVOL
	Call	1.5421	0.0064	0.0076	0.0070	14.9
	Call	1.5310	0.0086	0.0100	0.0093	14.8
30 days	Call	1.4872	0.0230	0.0238	0.0234	14.0
	Put	1.4479	0.0085	0.0098	0.0092	14.2
	Put	1.4371	0.0063	0.0074	0.0069	14.4
	Call	1.5621	0.0086	0.0102	0.0094	14.4
	Call	1.5469	0.0116	0.0135	0.0126	14.5
60 days	Call	1.4866	0.0313	0.0325	0.0319	13.8
	Put	1.4312	0.0118	0.0137	0.0128	14.0
	Put	1.4178	0.0087	0.0113	0.0100	14.2
	Call	1.5764	0.0101	0.0122	0.0112	14.1
	Call	1.5580	0.0137	0.0160	0.0149	14.1
90 days	Call	1.4856	0.0370	0.0385	0.0378	13.5
	Put	1.4197	0.0141	0.0164	0.0153	13.6
	Put	1.4038	0.0104	0.0124	0.0114	13.6
	Call	1.6025	0.0129	0.0152	0.0141	13.1
	Call	1.5779	0.0175	0.0207	0.0191	13.1
180 days	Call	1.4823	0.0494	0.0515	0.0505	13.1
	Put	1.3902	0.0200	0.0232	0.0216	13.7
	Put	1.3682	0.0147	0.0176	0.0162	13.7
	Call	1.6297	0.0156	0.0190	0.0173	13.3
	Call	1.5988	0.0211	0.0250	0.0226	13.2
270 days	Call	1.4793	0.0586	0.0609	0.0598	13.0
	Put	1.3710	0.0234	0.0273	0.0254	13.2
	Put	1.3455	0.0173	0.0206	0.0190	13.2

References

M. Avellaneda, A. Levy and A. Parás, "Pricing and hedging derivative securities in markets with uncertain volatilities", *Applied Mathematical Finance* **2** (1995) 73–88.

M. Avellaneda and A. Parás, "Managing the volatility risk of portfolios of derivative securities: the Lagrangian uncertain volatility model", *Applied Mathematical Finance* **3** (1996) 21–52.

H. Banks and P. D. Lamm, "Estimation of variable coefficients in parabolic distributed systems", *IEEE Transactions on Automatic Control* **30**(4) (1985) 386–398.

H. Banks and K. Kunisch, *Estimation Techniques for Distributed Parameter Systems*, Boston: Birkhäuser.

R. W. Banz and M. H. Miller, "Prices for state-contingent claims: some estimates and applications", *Journal of Business* **51**(4) (1978) 653–672.

S. Barle and N. Cakici, "Growing a smiling tree", *Risk Magazine* **8**(10) (1995, October) 76–81.

P. Billingsley, *Convergence of Probability Measures*, John Wiley & Sons, New York (1968).

J. N. Bodurtha Jr. and M. Jermakyan, "Non-parametric estimation of an implied volatility surface", Georgetown University working paper (1996, October).

D. T. Breeden and R. H. Litzenberger, "Prices of state-contingent claims implicit in option prices", *Journal of Business* **51**(4) (1978) 621–651.

P. W. Buchen and M. Kelly, "The maximum entropy distribution of an asset inferred from option prices", *Journal of Financial and Quantiatative Analysis* **31**(1) 143–159.

R. H. Byrd, P. Lu, J. Nocedal and C. Zhu, "A Limited Memory Algorithm for Bound Constrained Optimization", Northwestern University, Department of Electrical Engineering NAM-08, Evanston Ill. (1994, May).

R. H. Byrd, J. Nocedal and R. B. Schnabel, "Representation of Quasi-Newton Matrices and their use in Limited Memory Models", Northwestern University, Department of Electrical Engineering NAM-03, Evanston Ill. (1996, January).

N. Chriss, "Transatlantic tree", *Risk Magazine* **9**(7) (1996, July).

T. M. Cover and J. A. Thomas, *Elements of Information Theory*, John Wiley & Sons, New York (1991).

J. C. Cox, S. A. Ross and M. Rubinstein, "Option pricing: a simplified approach", *Journal of Financial Economics* **7** (1979) 229–263.

E. Derman and I. Kani, "Riding on a smile", *Risk Magazine* **7**(2) (1994, February).

B. Dupire, "Pricing with a smile", *Risk Manazine* **7**(1) (1994, January).

W. H. Fleming and M. Soner, *Controlled Markov Processes and Viscosity Solutions*, Springer-Verlag, New York (1992).

A. Friedman, *Partial Differential Equations of Parabolic Type*, Prentice-Hall, Englewood Cliffs, N.J. (1964).

N. Georgescu-Roegen, *The Entropy Law and the Economic Process*, Harvard University Press, Cambridge, Massachusetts (1971).

L. Gulko, "The entropy theory of bond option pricing", Yale University Working Paper (1995, October).

L. Gulko, "The entropy theory of stock option pricing", Yale University Working Paper (1996, May).

J. C. Jackwerth, "Generalized binomial trees", Berkeley Working Paper (1996a, August).

J. C. Jackwerth, "Implied binomial trees: Generalizations and empirical tests", Berkeley Working Paper (1996b, August).

J. C. Jackwerth and M. Rubinstein, "Recovering probability distribution from contemporaneous security prices", *Journal of Finance* **51:5** (1996, December) 1611–1631.

E. T. Jaynes, "Probability theory: the logic of science", Unpublished Manuscript, Washington University, St. Louis, MO (1996, March).

D. W. Mclaughlin (ed.), *Inverse Problems*, Providence, Rhode Island. SIAM and AMS: American Mathematical Society. Volume 14 (1984, April).

A. Parás, *Non-linear partial differential equations in finance: A study of volatility risk and transaction costs*, Ph. D. Thesis, New York University (1995).

R. T. Rockafellar, *Convex Analysis*, Princeton University Press, Princeton, New Jersey (1970).

M. Rubinstein, "Implied binomial tree", *The Journal of Finance* **69**(3) (1994, July) 771–818.

D. J. Samperi, "Implied trees in incomplete markets", Courant Institute Research Report, New York University (1995).

D. Shimko, "Bounds on Probability", *Risk Magazine* October (1993) 59–66.

M. Stutzer, "A bayesian approach to diagnosis of asset pricing models", *Journal of Econometrics* **68** (1995) 367–397.

A. N. Tikhonov and V. Y. Arsenin, *Solution of Ill-Posed Problems*, Wiley, New York (1977).

C. Zhu, R. H. Boyd, P. Lu and J. Nocedal, "L-BFGS-B: FORTRAN Subroutines for Large-Scale Bound Constrained Optimization", Northwestern University, Department of Electrical Engineering Evanston Ill. (1994, December).

Reprinted from J. Finance **LIII** (3) (1998) 1165–1190

STATIC HEDGING OF EXOTIC OPTIONS

PETER CARR, KATRINA ELLIS and VISHAL GUPTA[*]

This paper develops static hedges for several exotic options using standard options. The method used relies on a relationship between European puts and calls with different strike prices. The analysis allows for constant volatility or for volatility smiles or frowns.

This paper generalizes a relationship due to Bates (1988) between European puts and calls with different strikes. We term the generalized result *put–call symmetry* (PCS) and use it to develop a method for valuation and static hedging of certain exotic options. We focus on path-dependent options which change characteristics at one or more critical price levels, for example, barrier and lookback options and their extensions. We do not examine American or Asian options.

While these options may be valued and *dynamically* hedged in a lognormal model,[a] we offer valuation and *static* hedging in a slightly more general diffusion setting. As in Bowie and Carr (1994) and in Derman, Ergener, and Kani (1994), we create static portfolios of standard options whose values match the payoffs of the path-dependent option at expiration and along the boundaries. Since the path-dependent options we examine often have high gammas, static hedging using standard options will be considerably easier and cheaper than dynamic hedging. Furthermore, in contrast to dynamic hedging, our static positions in standard options are invariant to the volatility, interest rates, and dividends, bypassing the need to estimate them.[b] Because the path-dependent options we examine are often highly

---

[*]Carr is a VP in Equity Derivatives Research at Morgan Stanley. Ellis is a graduate student at the Johnson Graduate School of Management, Cornell University. Gupta is at Goldman Sachs; this work was completed while he was an MBA student at the Johnson Graduate School of Management, Cornell University. We are grateful to Warren Bailey, Hal Bierman, Sergio Bienstock, Phelim Boyle, Linda Canina, Melanie Cao, Narat Charupat, Neil Chriss, Emanuel Derman, Zhenyu Duanmu, Larry Harris, Eric Jacquier, Robert Jarrow, Iraj Kani, Antoine Kotze, James Kuczmarski, Robert Merton, Alan Shapiro, and especially Jonathan Bowie for their comments. We would similarly like to thank participants of presentations at the Fields Institute and at *Risk* conferences on volatility, equity derivatives, and exotic options. Finally, we thank participants of finance workshops at Cornell University, Goldman Sachs, Harvard University, J. P. Morgan, Morgan Stanley, NationsBank, Salomon Brothers, and the University of Southern California. We also acknowledge the outstanding research assistance of Cem Inal. They are not responsible for any errors.
[a]For example, barrier options are valued in the Black–Scholes (1973) model in Merton (1973).
[b]However, we assume a certain structure on the price process to achieve these invariance results. In particular, we assume that the cost of carrying the underlying is zero, and that its volatility satisfies a symmetry restriction.

sensitive to volatility, the hedging error due to volatility misspecification may be substantial with dynamic hedging.

Our PCS relationship can be viewed as both an extension and a restriction of the widely known put–call parity (PCP) result. The generalization involves allowing the strikes of the put and the call to differ in a certain manner. The restrictions sufficient to achieve this result are essentially that the underlying price process has both zero drift and a symmetric volatility structure, which is described below.

The rest of the paper is organized as follows. Section 1 presents the assumptions and the intuition behind PCS, which is the foundation for our hedging strategy. Section 2 reviews the static replication of single barrier options. Section 3 focuses on exotic options involving multiple barriers, such as double knockouts, roll-down, ratchet, and lookback options. In Sec. 4, we relax the assumption of zero drift and provide tight bounds on the static hedges developed in the earlier sections. Section 5 concludes the paper while the Appendix contains the mathematical details supporting our results.

1. Put–Call Symmetry

Throughout this paper we assume that markets are frictionless and there are no arbitrage opportunities. Let $P(K)$ and $C(K)$ denote the time 0 price of a European put and call, respectively, with both options struck at K and maturing at T. Because maturity is the same for all instruments we consider in any given example, we suppress dependence on the time to maturity to ease notation. Let B denote the time 0 price of a pure discount bond paying one dollar at T. Then put–call parity expressed in terms of the forward price F for time T delivery is

$$C(K) = [F - K]B + P(K). \qquad (1)$$

PCP implies that if the common strike of the put and call is the current forward price, then the options have the same value. Since put values increase with increasing strikes and call values decrease, we can write inequalities for European puts and calls whose strikes are on the same side of the forward. By contrast, PCS is an equality between scaled puts and calls whose strikes are on opposite sides of the forward.

To obtain PCS, certain restrictions are imposed on the stochastic process governing the underlying asset's price. In particular, we assume that the underlying price process is a diffusion, with zero drift under any risk-neutral measure, and where the volatility coefficient satisfies a certain symmetry condition. Thus, we rule out jumps in the price process and assume that the process starts afresh at any stopping time, such as at a first passage time to a barrier.

The assumption of zero risk-neutral drift is innocuous for options written on the forward or futures price of an underlying asset. For options written on the spot price, the assumption implies zero carrying costs.[c] Thus, the no-drift restriction implies that options written on the spot price behave as if they were written on the forward price. We relax the assumption of zero drift in Sec. 4 and obtain tight bounds on the value of options whose payoffs depend on the spot price path.

Throughout this paper, We assume that the volatility of the forward price is a known function $\sigma(F_t, t)$ of the forward price F_t and time t. We also assume the following symmetry condition:

$$\sigma(F_t, t) = \sigma(F^2/F_t, t), \quad \text{for all } F_t \geq 0 \text{ and } t \in [0, T], \tag{2}$$

where F is the current forward price. Thus the volatility at any future date is assumed to be the same for any two levels whose geometric mean is the current forward.

This symmetry condition is satisfied in the Black (1976) model, where volatility is deterministic, i.e., $\sigma(F_t, t) = \sigma(t)$. The symmetry arises when the volatility is graphed as a function of $X_t \equiv \ln(F_t/F)$. Letting $v(X_t, t) \equiv \sigma(F_t, t)$, the equivalent condition is

$$v(x, t) = v(-x, t), \quad \text{for all } x \in \Re \text{ and } t \in [0, T]. \tag{3}$$

Thus, the symmetry condition is also satisfied in models with a symmetric smile[d] in the log of K/F. As a result, a graph of volatility against K/F will be asymmetric, with higher put volatility than call volatility for strikes equidistant from the forward. Finally, the symmetry condition also allows for volatility frowns or even for more complex patterns.

Given the above assumptions, the Appendix proves:[e]

European Put–Call Symmetry: *Given frictionless markets, no arbitrage, zero drift, and the symmetry condition, the following relationship holds:*

$$C(K)K^{-1/2} = P(H)H^{-1/2}, \tag{4}$$

where the geometric mean of the call strike K and the put strike H is the forward price F:

$$(KH)^{1/2} = F. \tag{5}$$

Consider a numerical illustration of the PCS result: When the current forward is \$12, a call struck at \$16 has the same value as $\frac{4}{3}$ puts struck at \$9. The example is depicted in Fig. 1. The reason the call has much greater value, even though it

[c]Thus, for options written on a single stock or on a stock index, the no-drift assumption implies that the dividend yield always equals the risk-free rate. For options on spot FX, the no-drift assumption implies that the foreign interest rate always equals the domestic rate. For options on the spot price of a commodity, the assumption implies that the convenience yield is the riskless rate.
[d]Note that the assumed smile is in the local volatility as opposed to the Black (1976) model implied volatility.
[e] Bates (1988) first proves this result for constant volatility. See Bates (1991) for an excellent exposition of the implications of asymmetry for implying out crash premia.

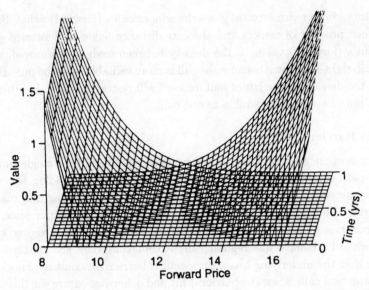

Fig. 1. Illustration of put–call symmetry (PCS). A call with strike 16 is equal to $\frac{4}{3}$ puts with strike 9 when the forward price is 12.

is further out-of-the-money arithmetically, is that our diffusion process has greater *absolute* volatility[f] when prices are high than when prices are low. Because call and put payoffs are determined by the arithmetic distance between terminal price and strike, the higher absolute volatility at higher prices leads to higher call values.

One intuition for PCS arises from generalizing the following intuition for put–call parity. Imagine a graph of the "risk-neutral" density of terminal prices and suppose that the horizontal axis is placed on a wedge with the objective of finding the fulcrum. The fulcrum is found by balancing the product of density and distance from the wedge integrated across terminal prices. In other words, the fulcrum occurs at the expected value under the risk-neutral distribution, which is the current forward price. Summing the product of density and distance from the wedge on the right of the fulcrum gives the (forward) price of a European call struck at the forward. Similarly, summing the product of density and absolute distance from the wedge on the left of the fulcrum gives the forward price of an at-the-money forward European put. Because the options' forward prices coincide, their spot prices also coincide by a simple cost of carry argument.

The PCS result (Eq. (4)) implies that a call struck at twice the current forward has twice the value of a put struck at half the current forward. To extend the above balancing intuition to these "winger" options, we now imagine that the horizontal axis is placed on two wedges, one located at half the current forward price and the other at twice the current forward price. Then the summed product of density

---

[f]Absolute volatility is defined as the standard deviation of price changes, i.e., $\text{Std}(dF)$. In contrast, the usual volatility is defined as the standard deviation of relative price changes, i.e., $\text{Std}(dF/F)$.

and distance above twice forward gives the winger call's (forward) value. Similarly, the summed product of density and absolute distance below half forward gives the winger put's (forward) value. If the density between wedges is removed, then the axis will tip right-side down because the call is more valuable than the put. However, doubling the density to the left of half forward will restore balance. In other words, two such puts have the same value as one call.

2. Single Barrier Options

In this section we represent path-dependent options with a single barrier[g] in terms of path-independent standard options. The key to providing this result is put–call symmetry, which is assumed to hold when the underlying first reaches the barrier price. Thus, the axis of symmetry for volatility is the barrier price.

Without loss of generality, we concentrate on valuing and hedging knockout calls. Such calls behave like regular calls except that they are knocked out the first time that the underlying hits a prespecified barrier. In contrast, knock in calls become standard calls when the barrier is hit and otherwise expire worthless. Given a valuation result and hedging strategy for knockout calls, the corresponding results for knock in calls can be recovered using the following parity relation:[h]

$$OC(K, H) = C(K) - IC(K, H), \tag{6}$$

where $IC(K, H)(OC(K, H))$ is an in-call (out-call) with strike K and barrier H.

2.1. *Down-and-out calls*

By definition, a down-and-out call (DOC) with strike K and barrier $H < K$ becomes worthless if H is hit at any time during its life. If the barrier has not been hit by the expiration date, the terminal payoff is that of a standard call struck at K.

To hedge a down-and-out call we need to match the terminal payoff and the payoff along the barrier. Thus a first step in constructing a hedge is to match the terminal payoff, which is done by purchasing a standard call, $C(K)$. Now let us consider option values along the barrier. When $F = H$, the DOC is worthless, while our current hedge $C(K)$ has positive value. Thus we need to sell off an instrument that has the same value as the European call when the forward price is at the barrier. Using PCS when $F = H$, we obtain[i]

$$C(K) = KH^{-1}P(H^2K^{-1}).$$

Thus, we need to write KH^{-1} European puts struck at H^2K^{-1} to complete the hedge.

---

[g]We focus on call options, leaving analogous results for puts as an exercise to the reader.
[h]This result does not hold for American options. See Chriss (1996) for a lucid discussion.
[i]The required put strike is $K_p = H^2K^{-1}$, from Eq. (5), and substituting this into Eq. (4) gives the result.

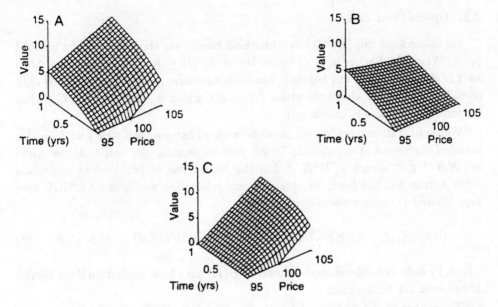

Fig. 2. Static hedge for a down-and-out call ($K = 100$, $H = 95$). Panel A shows the value of a European call struck at 100 for stock prices ranging from 95 to 105 and times to maturity up to a year. The axis labeled Time indicates time to maturity. Panel B shows the value of 1.0526 puts struck at 90.25. Note that the graphs are identical along the barrier of 95. Panel C shows the difference between the first two graphs, indicating that the replicating portfolio value vanishes along the barrier.

Thus the complete replicating portfolio for a DOC is a buy-and-hold strategy in standard options which is purchased at the initiation of the option

$$DOC(K, H) = C(K) - KH^{-1}P(H^2K^{-1}), \quad H < K. \tag{7}$$

If the barrier is hit before expiration, the replicating portfolio should be liquidated with PCS guaranteeing that the proceeds from selling the call are exactly offset by the cost of buying back the puts. If the barrier is not hit before expiration, then the long call gives the desired terminal payoff and the written puts expire worthless, as $H^2K^{-1} < H$ when $H < K$.

Figure 2 illustrates the replication of a down-and-out call with strike $K = \$100$, barrier $H = \$95$, and an initial maturity of one year. Panel A is of a standard call with the same strike and maturity as the down-and-out. Along the barrier $F = \$95$, the call has positive value. Panel B is of $KH^{-1} = 1.0526$ puts struck at $H^2K^{-1} = \$90.25$. Notice that the value of these puts along the barrier $F = \$95$ matches that of the standard call. When Panel B is subtracted from Panel A, the result is Panel C. Panel C shows that the replicating portfolio has zero value along the barrier $F = \$95$ and the payoff of a standard call struck at \$100 at expiration.

2.2. *Up-and-out call*

An up-and-out call (UOC) has a knockout barrier set above the current forward price. When the barrier is at or below the strike $(H \leq K)$, the UOC is worthless as it is always knocked out before it can have a positive payoff. Thus we need only consider barriers set above the strike $(H > K)$, which implies that the UOC has intrinsic value before it knocks out.

We again start our replicating portfolio with a European call struck at K as this matches our payoff at expiration. To get zero value along the barrier H, we could sell KH^{-1} puts struck at H^2K^{-1}, but this would give us problems at expiration if the barrier has not been hit. Instead, our replicating portfolio for a UOC uses Eqs. (6) and (1) with up-and-in securities:

$$UOC(K, H) = C(K) - UIP(K, H) - (H - K)UIB(H), \quad H > K, F, \quad (8)$$

where, by definition, the up-and-in bond $UIB(H)$ pays \$1 at expiration if the barrier H has been hit before then.

To see that the portfolio matches the payoffs of the UOC, consider the payoff of the UOC if the barrier is never touched — the required payoff is that of a standard call struck at K. In the replicating portfolio, the up-and-in put and bonds expire worthless, while the standard call provides the desired payoff. Conversely, at the first passage time to the barrier, the up-and-out call knocks out just as the up-and-in put and bonds knock in. Since the forward price is at H, put–call parity implies that the replicating portfolio can be liquidated at zero cost. The up-and-in put struck at K with barrier H matches the time value of the standard call $C(K)$ at the barrier and the $(H - K)$ up-and-in bonds match its intrinsic value.

The advantage of representing an up-and-out call in terms of up-and-in puts and bonds is that Eq. (8) holds for any continuous process for the underlying's price. The disadvantage is that the up-and-in securities may not trade or may only trade with heavy frictions. We can apply PCS to show that the $UIP(K, H)$ can be replicated with KH^{-1} calls struck at H^2K^{-1}. The Appendix shows that a $UIB(H)$ can be replicated by buying two *binary* calls (BC) struck at H and H^{-1} European calls struck at H:

$$UIB(H) = 2BC(H) + H^{-1}C(H), \quad H > F. \quad (9)$$

By definition, the binary calls pay \$1 at expiry if the underlying finishes above H then. The intuition for the replication of the UIB is that when the forward is at the barrier, each binary call is valued at approximately the probability of finishing in-the-money. If this probability were exactly 0.5, then the two binary calls alone would suffice. The positive skew of the terminal price distribution implies that the probabilities are slightly less than 0.5, entailing a minor correction using calls as in Eq. (9).

Rewriting Eq. (8) in terms of standard and binary calls gives

$$UOC(K, H) = C(K) - KH^{-1}C(H^2K^{-1})$$

$$- (H - K)[2BC(H) + H^{-1}C(H)], \quad H > K, F. \qquad (10)$$

It is well known that binary calls can be synthesized using an infinite number of vertical spreads of standard calls[j]

$$BC(H) = \lim_{n \uparrow \infty} n[C(H) - C(H + n^{-1})]. \qquad (11)$$

As a result, the up-and-out call can be replicated using standard calls alone.

Table 1. Convergence of vertical spreads to binary call. This table calculates the value of a vertical spread (VS) with parameter n, where n is the number of call spreads and its reciprocal is the spread between strikes. C is a European call. As n increases the vertical spread converges to a binary call (BC) with strike \$105, with analytical value \$0.292384.

n	$VS(n)$	$= n\left[C(105) - C\left(105 + \dfrac{1}{n}\right)\right]$
1	0.276446	$= C(105) - C(106)$
2	0.284331	$= 2[C(105) - C(105.50)]$
3	0.286997	$= 3[C(105) - C(105.33)]$

Clearly, the binary call replicating strategy in Eq. (11) is impractical. To remedy this, we use a technique called Richardson extrapolation which has been previously employed for option pricing (see, e.g., Geske and Johnson (1984)). Given a set of approximations indexed by a parameter (e.g., step size), Richardson extrapolation is a technique for guessing the value when the parameter is infinitesimal. We illustrate the approach for binary calls with the following example for FX options assuming constant interest rates and volatility. Suppose $F = S = \$100$, $K = \$105$, $r = r_f = 4\%$, $\sigma = 20\%$, and $T = 0.25$ years. Then the exact Black (1976) model value of the binary call is \$0.292384. Define $VS(n)$ as the value of n vertical call spreads involving strikes 105 and $105 + n^{-1}$, $n = 1, 2, 3$. Again using Black's model, Table 1 indicates the speed of convergence of $VS(n)$ to the correct value.

While the vertical spread values are slowly converging, five-decimal-place accuracy can be obtained by using the following three-point Richardson extrapolation:[k]

$$VS^{1-2-3} \equiv 0.5 \times VS(1) - 4 \times VS(2) + 4.5 \times VS(3).$$

---

[j]See Chriss and Ong (1995) for a discussion of this result.
[k]See Marchuk and Shaidurov (1983), p. 24, for a derivation of Richardson weights.

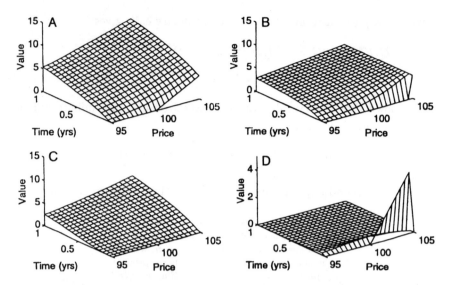

Fig. 3.   Static hedge for an up-and-out call ($K = 100$, $H = 105$). Panel A shows a European call struck at 100. Panel B shows five up-and-in bonds with strike 105, which pay \$1 each at expiration if the barrier 105 has been hit by then. Panel C shows 0.9524 call struck at 110.25. The total hedge is shown in Panel D, which has been scaled up.

Thus the value of a binary call is well approximated[1] by the following simple portfolio of standard calls:

$$BC(105) \approx VS^{1-2-3} = 6C(105) - 0.5C(106) + 8C(105.50) - 13.5C(105.33).$$

Figure 3 shows the value of the components of the static hedge for an up-and-out call with strike $K = \$100$, barrier $H = \$105$, and initial maturity of one year. The standard call struck at \$100 shown in Panel A has both intrinsic and time value along the barrier (105). Panel B is the $H - K = 5$ up-and-in bonds, which match the intrinsic value of the call along the barrier. Panel C is of $KH^{-1} = 0.9524$ call struck at $H^2K^{-1} = \$110.25$, which match the time value of the call along the barrier. When Panels B and C are subtracted from Panel A, the result is shown in Panel D, indicating zero value along the barrier and the call payoff at maturity.

3. Multiple Barrier Options

In this section, we discuss complex barrier options involving multiple barriers.[m] Although more complex specifications are possible, we simply assume that the volatility of the underlying is henceforth a deterministic function of time, as in Black's (1976) model; i.e., $\sigma(F, t) = \sigma(t)$ for all $F > 0$ and $t \in [0, T]$.

---

[1]The approximation deteriorates near expiration when prices are near the strike.
[m]Partial barrier options may be statically hedged using a portfolio of standard and compound options. A discussion of this can be obtained from the authors.

3.1. *Double knockout calls*

Consider a call option that has two barriers,[n] so that the call knocks out if either barrier is hit. We assume that the initial forward price and strike are both between the two barriers. There is a parity relation between this double knockout call (O^2C) and a double knock in call (I^2C), which knocks in if either barrier is hit:

$$O^2C(K, L, H) = C(K) - I^2C(K, L, H), \qquad (12)$$

where K is the strike, L is the lower barrier, and H is the higher barrier. We will again focus on replicating the payoffs of a double knockout call using static portfolios of standard options.

On its surface, a double knockout call $O^2C(K, L, H)$ appears to be a combination of a $DOC(K, L)$ and a $UOC(K, H)$. The payoff of the $O^2C(K, L, H)$ is zero if either barrier is hit and the standard call payoff at expiry if neither barrier is hit. A portfolio of a call knocking out at the lower barrier and a call knocking out at the higher barrier would give these payoffs, so long as the knockout of one option also knocked out the other. The additional specification is necessary as otherwise the surviving option contributes value at the other's barrier.

To construct the replicating portfolio for the $O^2C(K, L, H)$, we begin as before by purchasing a standard call $C(K)$ to provide the desired payoff at expiry. We will then attempt to zero out the value at each barrier separately. If we knew in advance that the forward price reaches the lower barrier L before it reaches the higher barrier H, then our previous analysis of a down-and-out call implies that the value of the call $C(K)$ can be nullified along the barrier L by initially selling KL^{-1} puts struck at L^2K^{-1}. Thus, the replicating portfolio under this assumption would be

$$O^2C(K, L, H) \approx C(K) - KL^{-1}P(L^2K^{-1}). \qquad (13)$$

Alternatively, if we knew in advance that the forward price reaches the higher barrier H first, then from Eq. (10) the replicating portfolio would instead be

$$O^2C(K, L, H) \approx C(K) - KH^{-1}C(H^2K^{-1})$$
$$- (H - K)[2BC(H) + H^{-1}C(H)]. \qquad (14)$$

Because we don't know in advance which barrier will be hit first, we try combining the two portfolios:

$$O^2C(K, L, H) \approx C(K) - DIC(K, L) - UIC(K, H)$$
$$= C(K) - KL^{-1}P(L^2K^{-1}) - KH^{-1}C(H^2K^{-1})$$
$$- (H - K)[2BC(H) + H^{-1}C(H)]. \qquad (15)$$

---

[n] Double barrier calls and puts have been priced analytically in Kunitomo and Ikeda (1992).

The problem with this portfolio is that each written-in call contributes (negative) value at the other's barrier. For example, if the forward price reaches H first, then the $DIC(K, L) = KL^{-1}P(L^2K^{-1})$ contributes (negative) value along H. Thus, we need to add securities to the portfolio in an effort to zero out value along each barrier. Along the barrier H, the negative influence of the KL^{-1} puts struck at L^2K^{-1} can be offset by buying LH^{-1} calls struck at H^2KL^{-2}. To cancel the negative influence of the $UIC(K, H)$ along the barrier L, we will need to extend PCS to binary calls.

Recall that a binary call (put) is a cash-or-nothing option that pays \$1 if the stock price is above (below) a strike price K, and zero otherwise. Similarly, a gap call (put) is an asset-or-nothing option that pays the stock price S if it is above (below) a strike price K, and zero otherwise. The following parity relations are easily shown:

$$GC(K) = K \cdot BC(K) + C(K), \quad GP(K) = K \cdot BP(K) - P(K).$$

Since binary options may be synthesized out of standard options, these parity relations imply that the same is true for gap options. The Appendix proves the following symmetry result, relating values of binary options to gap options.

Binary Put–Call Symmetry: *Given frictionless markets, no arbitrage, zero drift, and deterministic volatility, the following relationships hold:*

$$K^{1/2}BC(K) = GP(H)H^{-1/2}, \quad H^{1/2}BP(H) = GC(K)K^{-1/2}, \quad (16)$$

where the geometric mean of the binary call strike K and the binary put strike H is the forward price F:

$$(KH)^{1/2} = F.$$

Armed with this result, we can cancel the negative influence of the $UIC(K, H)$ in Eq. (15) along the barrier L. Thus, our first layer approximation for the double knockout call value is

$$\begin{aligned}
O^2C(K, L, H) = C(K) &- L^{-1}(KP(L^2K^{-1}) - HP(L^2KH^{-2})) \\
&- H^{-1}(KC(H^2K^{-1}) - LC(H^2KL^{-2})) \\
&- (H - K)[2BC(H) + H^{-1}C(H) - 2L^{-1}GP(L^2H) \\
&- L^{-1}P(L^2H^{-1})].
\end{aligned} \quad (17)$$

Although Eq. (17) is a better approximation than Eq. (15), the added options still contribute value at the other's barrier. Thus, we need to continue to subtract or add options, noting that each additional layer of hedge at one barrier creates an error at the other barrier. As a result, the replicating portfolio for a double knockout call can be written as an infinite sum:

$$O^2C(K, L, H) = C(K) - \sum_{n=0}^{\infty} [L^{-1}(HL^{-1})^n(KP(L^2K^{-1}(LH^{-1})^{2n})$$

$$- HP(K(LH^{-1})^{2(n+1)}))$$

$$+ H^{-1}(LH^{-1})^n(KC(H^2K^{-1}(HL^{-1})^{2n})$$

$$- LC(K(HL^{-1})^{2(n+1)})) + 2(H - K)(HL^{-1})^n$$

$$\times [BC(H(HL^{-1})^{2n}) - L^{-1}GP(L(LH^{-1})^{2n+1})]$$

$$+ (H - K)[H^{-1}(LH^{-1})^nC(H(HL^{-1})^{2n}) - L^{-1}(HL^{-1})^n$$

$$\times P(L(LH^{-1})^{2n+1})]]. \tag{18}$$

Note that the options in the infinite sum are all initially out-of-the-money. Furthermore, as n increases, the number of options held and the options' moneyness both decrease exponentially. As a result, for large n, the options' contribution to the sum becomes minimal. Thus we can get a good approximation to the option value with a small value of n. Table 2 shows a typical example. With $F = K = 100$, barriers at $L = 95$ and $H = 105$, $r = r_f = 4\%$, $\sigma = 20\%$, and $T = 0.25$, five-decimal-place accuracy has occurred by summing the values for $n = 0, 1, 2$. The value for $n = \infty$ is obtained from the analytic solution by Kunitomo and Ikeda (1992).

Figure 4 graphs the value of second-layer hedge, i.e., $n = 1$ in Eq. (18), for a double knockout call option. Notice that the value along both barriers is very close to zero.

In general, bringing in the barriers of a double knockout call reduces both its value and the number of options needed to achieve a given accuracy.

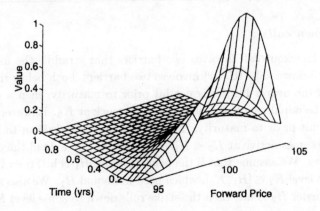

Fig. 4. Second layer static hedge for a double knockout call option with barriers at 95 and 105 and strike at 100.

Table 2. Convergence of replicating portfolio to double knockout call (O^2C) value. The values are generated by using the static hedging portfolio for $O^2C(K, L, H)$ for increasing values of N.

$$O^2C(K, L, H)$$

$$= C(K) - \sum_{n=0}^{\infty} [L^{-1}(HL^{-1})^n(KP(L^2K^{-1}(LH^{-1})^{2n})$$

$$- HP(K(LH^{-1})^{2(n+1)}))$$

$$+ H^{-1}(LH^{-1})^n(KC(H^2K^{-1}(HL^{-1})^{2n})$$

$$- LC(K(HL^{-1})^{2(n+1)}))$$

$$+ 2(H - K)(HL^{-1})^n[BC(H(HL^{-1})^{2n})$$

$$- L^{-1}GP(L(LH^{-1})^{2n+1})]$$

$$+ (H - K)[H^{-1}(LH^{-1})^n C(H(HL^{-1})^{2n})$$

$$- L^{-1}(HL^{-1})^n P(L(LH^{-1})^{2n+1})]] .$$

C and P are European calls and puts, respectively; K is the strike price; L and H are lower and upper barriers, respectively; BC is a binary call; GP is a gap put; r_f is the foreign interest rate and r is the domestic interest rate; σ is the volatility of the underlying asset; and T is the time to maturity of the option. The parameters for the option are initial forward price $F = 100$, $K = 100$, barriers at $L = 95$ and $H = 105$, $r = r_f = 4\%$, $\sigma = 20\%$, and $T = 0.25$. The value for $N = \infty$ is given by the analytic solution of Kunitomo and Ikeda (1992).

N	Value of Replicating Portfolio
0	0.074763
1	0.007781
2	0.007746
3	0.007746
⋮	⋮
∞	0.007744

3.2. Roll-down calls

A double knockout call involves two barriers that straddle the initial spot. In contrast, a roll-down call (RDC)[o] involves two barriers, both below the initial spot and strike. If the nearer barrier is not hit prior to maturity, then a roll-down call has the same terminal payoff as a standard call struck at K_0. However, if the nearer barrier H_1 is hit prior to maturity, then the strike is rolled down to it, and a new out-barrier becomes active at $H_2 < H_1$. For later use, we extend the definition of a RDC as follows. We assume that if the nearer barrier H_1 is hit, then the strike rolls down to some level $K_1 \in [H_1, K_0]$, which need not equal H_1. We also assume that if the farther barrier H_2 is hit, then the strike rolls down to some level $K_2 \in [H_2, K_1]$

[o]For a discussion of roll-down calls and roll-up puts, see Gastineau (1994).

and a new out-barrier becomes active further down at $H_3 < H_2$. This process repeats an arbitrary number of times.

Let $H_1, \ldots, H_n$ be a decreasing sequence of positive barrier levels set below the initial forward price, $F > H_1$. Similarly, let $K_0, \ldots, K_n$ be a decreasing sequence of strikes, with $K_i \geq H_i$, $i = 1, \ldots, n$. Then at initiation, the extended roll-down call can be decomposed into down-and-out calls as

$$ERDC(K_i, H_i) = DOC(K_0, H_1) + \sum_{i=1}^{n} [DOC(K_i, H_{i+1}) - DOC(K_i, H_i)]. \quad (19)$$

This representation is model-independent. To obtain a standard roll-down call, we set $n = 1$ and $K_1 = H_1$. For any n, the hedge works as follows. If the underlying never hits the barrier H_1, then the $DOC(K_0, H_1)$ provides the desired payoff $(F_T - K_0)^+$, and the knockout calls in the sum cancel each other. If the barrier H_1 is hit, then $DOC(K_0, H_1)$ vanishes, as does the written $DOC(K_1, H_1)$. Thus the position when $F = H_1$ may be rewritten as

$$ERDC(K_i, H_i) = DOC(K_1, H_2) + \sum_{i=2}^{n} [DOC(K_i, H_{i+1}) - DOC(K_i, H_i)]. \quad (20)$$

This is analogous to our initial position. In between any two barriers H_i and H_{i+1}, the $DOC(K_i, H_{i+1})$ provides the desired payoff if the next barrier is never hit, but the DOCs in the sum roll down the strike to K_{i+1} if this barrier is hit.

When PCS holds at each barrier, the extended roll-down call value at initiation, for $F > H_1$, is given by

$$ERDC(K_i, H_i) = C(K_0) - K_0 H_1^{-1} P(H_1^2 K_0^{-1})$$
$$+ \sum_{i=1}^{n} [K_i H_i^{-1} P(H_i^2 K_i^{-1}) - K_i H_{i+1}^{-1} P(H_{i+1}^2 K_i^{-1})]. \quad (21)$$

The replicating strategy is as follows. At any time, we are always holding a call struck at or above the highest untouched barrier and puts struck at or below this barrier. Thus, if the forward price never reaches this barrier, the call provides the desired payoff at expiry, and the puts expire worthless. Each time the forward price touches a barrier H_i for the first time, we sell the call struck at K_{i-1} and buy back $K_{i-1} H_i^{-1}$ puts struck at $H_i^2 K_{i-1}^{-1}$; sell $K_i H_i^{-1}$ puts struck at $H_i^2 K_i^{-1}$ and buy the call struck at K_i. PCS guarantees that both transitions are self-financing.

As previously mentioned, the standard roll-down call is the special case of Eq. (21) with $n = 1$ and $K_1 = H_1$. Figure 5 illustrates the replication procedure for a standard roll-down call with initial strike $K_0 = \$100$, rolled-down strike $K_1 = H_1 = \$95$, and final out-barrier $H_2 = \$90$. Panel A shows the value of the replicating portfolio before the first barrier is hit. If the forward hits the first barrier H_1, then the portfolio is costlessly revised to $C(H_1) - H_1 H_2^{-1} P(H_2^2 H_1^{-1})$. Panel B

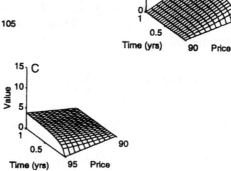

Fig. 5. The two part static hedge for a roll-down call ($K_0 = 100$, $K_1 = H_1 = 95$, $H_2 = 90$). Panel A shows the replicating portfolio when the price is above the first barrier, 95. Panel B shows the replicating portfolio after the barrier 95 has been hit. The two portfolios are identical along the barrier of 95, and the second is worth zero along the lower barrier 90. Panel C is a different perspective of Panel B, showing more clearly value of the portfolio at 95.

shows the value of this new portfolio for prices below \$95. The revised portfolio has zero value along the knockout barrier $H_2 = 90$ as required. Panel C is just Panel B with a different orientation, showing that the value of the two portfolios match along the first barrier $H_1 = 95$.

3.3. *Ratchet calls*

A ratchet call is an extended roll-down call, with strikes set at the barriers, which never knocks out completely. This is accomplished by having the only purpose of the lowest barrier be to ratchet down the strike. This suggests that we can create a static hedge for a ratchet call once we account for this difference.

To synthesize a ratchet call with initial strike K_0, we set the strikes K_i in the $ERDC(K_i, H_i)$ equal to the barriers H_i, $i = 1, \ldots, n-1$. To deal with the fact that an extended roll-down call knocks out completely if the forward reaches H_n, while the ratchet call rolls down the strike to H_n, we replace the last spread of down-and-out calls $[DOC(H_n, H_{n+1}) - DOC(H_n, H_n)]$ in Eq. (19) with a down-and-in call $DIC(H_n, H_n)$. Thus, a model-independent valuation of a ratchet call, using barrier calls, is

$$RC(K_0, H_i) = DOC(K_0, H_1) + \sum_{i=1}^{n-1}[DOC(H_i, H_{i-1}) - DOC(H_i, H_i)]$$

$$+ DIC(H_n, H_n), \quad F > H_1. \tag{22}$$

Substituting in the model-free results, $DOC(K, H) = C(K) - DIC(K, H)$ and $DIC(H, H) = P(H)$ simplifies the result to

$$RC(K_0, H_i) = DOC(K_0, H_1) + \sum_{i=1}^{n-1}[P(H_i) - DIC(H_i, H_{i+1})]$$

$$+ P(H_n), \quad F > H_1. \tag{23}$$

When PCS holds at each barrier, a ratchet call can be represented in terms of standard options as

$$RC(K_0, H_i) = C(K_0) - K_0 H_1^{-1} P(H_1^2 K_0^{-1}) + \sum_{i=1}^{n-1}[P(H_i) - H_i H_{i+1}^{-1} P(H_{i+1}^2 H_i)]$$

$$+ P(H_n), \quad F > H_1. \tag{24}$$

Hedging with this replicating portfolio is analogous to the extended roll-down call hedge: the position held is changed at every barrier, and the transitions are self-financing. Comparing Eq. (24) with its counterpart for an extended roll-down call allows us to capture the value of removing an out-barrier at H_{n+1}. Setting $K_i = H_i$ in Eq. (21) and comparing with Eq. (24) implies that the value of removing this barrier is given by $H_n H_{n+1}^{-1}$ puts struck at $H_{n+1}^2 H_n^{-1}$.

3.4. Lookback calls

A floating strike lookback call (LC) is similar to a ratchet call, except that there is a continuum of rolldown barriers extending from the initial forward price to the origin, so that the strike price is the minimum price over the option's life. A ratchet call with $K_0 = F$, $H_n = 0$ undervalues a lookback because the active strike is always at or above that of a lookback. Thus, a model-free lower bound for a lookback call is

$$LC \geq RC(F, H_i)$$

$$= DOC(F, H_i) + \sum_{i=1}^{n-1}[P(H_i) - DIC(H_i, H_{i+1})] + P(H_n). \tag{25}$$

When PCS holds at each barrier, this lower bound can be expressed in terms of standard options:

$$LC \geq C(F) - FH_1^{-1}P(H_1^2 F^{-1})$$

$$+ \sum_{i=1}^{n-1}[P(H_i) - H_i H_{i+1}^{-1} P(H_{i+1}^2 H_i^{-1}) + P(H_n). \tag{26}$$

The portfolio of standard options undervalues the lookback because the call held is always struck at or above the lookback. By adding more strikes, we obtain a tighter bound. Since the underlying's prices are actually discrete, one possibility is to set the barriers at each possible level.

To obtain an upper bound on the value of a lookback call, we may use an *extended* ratchet call, which ratchets the strike down to the next barrier each time a new barrier is crossed. When the last positive barrier is touched, the strike is ratcheted down to zero. Thus, a model-free upper bound in terms of down-and-in bonds is

$$LC \leq C(H_1) - P(H_1)$$

$$+ \sum_{i=1}^{n-1} [(H_i - H_{i+1})DIB(H_i)] + H_n DIB(H_n). \qquad (27)$$

Intuitively, when each barrier H_i is reached for the first time, the down-and-in bonds ratchet down the delivery price of the synthetic forward $C(H_1) - P(H_1)$ by $H_i - H_{i+1}$ dollars.

When PCS holds at each barrier, it can be used to represent the down-and-in bonds in terms of standard options. In particular, using an argument analogous to that in the Appendix for an up-and-in bond, a down-and-in bond can be replicated using the following static portfolio of binary and standard puts: $DIB(H) = 2BP(H) - H^{-1}P(H)$. Richardson extrapolation may again be used to efficiently represent the binary puts in terms of standard puts.[P]

We can modify the above bounds for both a forward-start and a backward-start lookback call. Let 0 be the valuation date and let T_1 be the start date of the lookback period. In the backward-start case $(T_1 < 0)$, the underlying has some minimum-to-date, m, which is in between two barriers $H_{\hat{i}}$ and $H_{\hat{i}+1}$ for some $\hat{i}$. The lower bound is thus a ratchet call with initial strike $H_{\hat{i}}$ and barriers H_i, where $i = \hat{i}+1, \ldots, n-1$. Similarly, the upper bound is an extended ratchet call with initial strike $H_{\hat{i}+1}$ and barriers $H_i, i = \hat{i}+1, \ldots, n-1$. Because ratchet calls and extended ratchet calls can be replicated with standard options, we have bounded the lookback call in terms of static portfolios of standard options.

In the forward-start case $(T_1 > 0)$, we use the fact that the formula for a backward-start lookback call is linearly homogeneous in the current spot/forward price and the minimum to date. At T_1, the minimum is S_{T_1}, so the lookback call value at T_1 may be written as $c(\cdot)S_{T_1}$, for some function $c(\cdot)$ independent of S_{T_1}. Thus, for a forward-start LC, we should initially hold $ce^{-\delta T_1}$ units of the underlying. Moving forward through time, the dividends received are reinvested back into the security, bringing the number of units held up to c by time T_1. At T_1, the c shares can be sold for proceeds just sufficient to initiate the approximating strategy described above.

[P]Richardson extrapolation may also be used to enhance convergence of the lower and upper bounds of a lookback call by extrapolating down the distance between barriers. A discussion of this can be obtained from the authors.

4. Nonzero Carrying Costs

The previous results were derived assuming that the drift of the underlying was zero (under the martingale measure). This assumption is natural for options on futures, but strained somewhat for options on the spot. In this section, we relax the assumption of zero drift. Although we are no longer able to obtain exact static hedges for options on the spot, we can develop tight bounds on option values using static hedges. Bowie and Carr (1994) give the bounds for single barrier options, so we concentrate on multiple barrier calls. For concreteness, we deal with options on spot foreign exchange (FX), assuming constant interest rates for simplicity. Then, interest rate parity links forward prices $(F(t))$ and spot prices $(S(t))$ of FX by

$$F(t) = S(t)e^{(r-r_f)(T-t)}, \quad t \in [0,T], \tag{28}$$

where r is the domestic rate and r_f is the foreign rate. Thus, when the spot hits a flat barrier H, the forward hits a time-dependent barrier $H(t) \equiv He^{(r-r_f)(T-t)}$.

4.1. *Double knockout calls*

When the drift of the underlying is not zero, a double knockout call on the spot with flat barriers L and H is equivalent to a double knockout call on the forward price with time-dependent barriers $L(t) = Le^{(r-r_f)(T-t)}$ and $H(t) = He^{(r-r_f)(T-t)}$. with $t \in [0,T]$. We can give flat upper and lower bounds on these time-dependent barriers. If $r > r_f$,[q]

$$L \leq L(t) \leq \overline{L} \equiv Le^{(r-r_f)T} \quad H \leq H(t) \leq \overline{H} \equiv He^{(r-r_f)T}.$$

Double knockout options increase in value as the out-barriers are moved farther apart. Thus for the double knockout call on the forward,[r] we can write

$$O^2 C_f(K, \overline{L}, H) \leq O^2 C_f(K, L(t), H(t)) \leq O^2 C_f(K, L, \overline{H}). \tag{29}$$

Furthermore, by definition, the double knockout on the forward with time-dependent barriers is the same as the double knockout on the spot with flat barriers:

$$O^2 C_f(K, L(t), H(t)) = O^2 C_s(K, L, H). \tag{30}$$

Combining Eqs. (29) and (30) allows us to bound the value for a double knockout call on the spot between the values of two double knockout calls on the forward:

$$O^2 C_f(K, \overline{L}, H) \leq O^2 C_s(K, L, H) \leq O^2 C_f(K, L, \overline{H}). \tag{31}$$

As we know how to replicate each of the two bounds with a static portfolio, we have upper and lower bounds on the double knockout call on the spot. Figure 6 indicates how the bounds vary with the interest rate differential.

---

[q]The details for hedging multiple barrier calls when $r < r_f$ are left as an exercise for the reader.
[r]Since interest rates are constant, results for options on forwards also hold for options on futures.

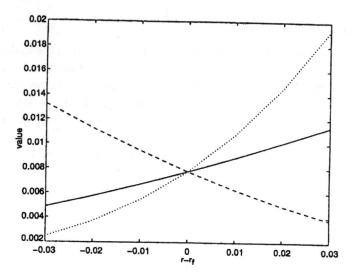

Fig. 6. Synthesizing a double knockout call with cost of carry. Value of the upper (dashed line) and lower (dotted line) bound static hedges for a double knockout call ($K = 100$, lower barrier 95, and upper barrier 105) compared with the analytical value (solid line). The foreign interest rate (r_f) is fixed at 4% and the domestic interest rate (r) varies from 1% to 7%. For $r < r_f$ the lower bound is the upper bound and vice versa.

4.2. *Roll-down calls*

Recall that under zero drift and with PCS holding at every barrier, an extended roll-down call (ERDC) was synthesized out of standard call and put options in Eq. (21).

When the drift of the underlying is not zero, an ERDC on spot with flat strikes K_i, $i = 0, \ldots, n$ and barriers H_i, $i = 1, \ldots, n$ is equivalent to an ERDC on the forward price with time-dependent strikes $K_{it} \equiv K_i e^{(r-r_f)(T-t)}$ and barriers $H_{it} \equiv H_i e^{(r-r_f)(T-t)}$:

$$ERDC_s(K_i, H_i) = ERDC_f(K_{it}, H_{it}) \,.$$

We can give flat upper and lower bounds on these time-dependent quantities. If $r > r_f$,

$$K_i \leq K_{it} \leq \overline{K}_i \equiv K_i e^{(r-r_f)T} \,, \quad t \in [0, T], i = 0, \ldots, n \,,$$

$$H_i \leq H_{it} \leq \overline{H}_i \equiv K_i e^{(r-r_f)T} \,, \quad t \in [0, T], i = 1, \ldots, n \,.$$

The $ERDC_f$ value is decreasing in the level of the strikes and increasing in the level of the barriers. Thus, we can place bounds on $ERDC_f(K_{it}, H_{it})$:

$$ERDC_f(\overline{K}_i, H_i) \leq ERDC_f(K_{it}, H_{it}) \leq ERDC_f(K_i, \overline{H}_i) \,. \tag{32}$$

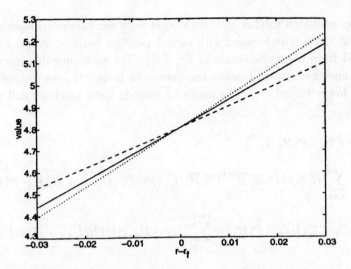

Fig. 7. Synthesizing a roll-down call with cost of carry. Value of the upper (dashed line) and lower (dotted line) bound static hedges for a roll-down call ($K_0 = 100, K_1 = H_1 = 95, H_2 = 90$) compared with the analytical value (dotted line). The foreign interest rate is fixed at 4% and the domestic interest rate varies between 1% and 7%. When the domestic rate is below the foreign rate, the lower bound becomes the upper bound and vice versa.

From Sec. 3.2 we can create static hedge portfolios for the upper and lower bounds, and the tightness of these bounds for a standard roll-down call ($n = 1$, $K_1 = H_1$) is shown in Fig. 7.

4.3. *Ratchet calls*

A ratchet call on the spot is a special case of an extended roll-down call on the spot created by setting the strikes K_i equal to the barriers H_i and removing the last knockout barrier. As a result, the bounds for a ratchet call on the spot are determined similarly to an extended roll-down call. The lower bound is a ratchet call that ratchets every time the flat barrier H_i is hit to a strike of $\overline{H}_i$. The higher bound is an extended ratchet call on the forward that ratchets every time the flat barrier $\overline{H}_i$ is hit to a strike of H_i:

$$RC_f(\overline{H}_i, H_i) \leq RC_s(H_i \leq RC_f(H_i, \overline{H}_i). \tag{33}$$

4.4. *Lookback calls*

Consider a ratchet call on spot with the initial strike K_0 set at the initial spot price and the final rung H_n set at the origin. As in the case of zero carrying costs, this ratchet call undervalues a lookback due to the discreteness of the rungs. As a result, the lower bound for a ratchet call on spot is also a lower bound for a lookback call on spot. An upper bound for a lookback call on spot can be obtained from an extended ratchet call on spot, which ratchets the strike to the next lower barrier.

However, an extended ratchet call on the spot with flat barriers is equivalent to an extended ratchet call on forward with time-dependent barriers. A lower bound can be obtained from a generalization of Eq. (27). For each time-dependent barrier, H_{it}, each time the forward reaches the flat upper bound $\overline{H}_i$, we ratchet the strike to the flat lower bound H_i. The resulting bounds for a lookback call on spot at issuance are

$$C(F) - FH_1^{-1}P(H_1^2\overline{H}_0^{-1})$$

$$+ \sum_{i=1}^{n-1}[\overline{H}_iH_i^{-1}P(H_i^2\overline{H}_i^{-1}) - \overline{H}_iH_{i+1}^{-1}P(H_{i+1}^2\overline{H}_i^{-1})] + \overline{H}_nH_n^{-1}P(H_n^2\overline{H}_n^{-1})$$

$$\leq LC_s \leq C(H_1) - P(H_1) + \sum_{i=1}^{n-1}[(H_1 - H_{i+1})DIB(\overline{H}_i)] + H_nDIB(\overline{H}_n).$$

Using this approach, bounds for forward and backward start lookback calls can also be obtained.

5. Conclusion

The concept of hedging exotic options with a static portfolio of standard instruments simplifies the risk management of exotic options in several ways. First, when compared with dynamically rebalancing with the underlying, the static portfolio is easier to construct initially and to maintain over time. Instead of continuously monitoring the underlying and trading with every significant price change, the hedger can place contingent buy and sell orders with start/stop prices at the barriers. Second, when compared with offsetting the risk using another path-dependent option, the investor uses instruments with which he is familiar, the risks are better understood, and the markets are more liquid.

A static hedge can exactly replicate the payoffs of the path-dependent option when carrying costs are zero; and a pair of static hedges can bracket the payoffs when nonzero carrying costs are introduced. These techniques apply to many path-dependent options, which are related in that their payoffs depend on whether one or more barriers are crossed.

The fundamental result underpinning the creation of our replicating portfolios is put–call symmetry. By using this simple formula, we can engineer simple portfolios to mimic the values of standard options along barriers. The result is an extension of put–call parity to different strike prices which provides further insight into the relation between put and call options.

The main extension to this line of research would involve relaxation of the zero drift and symmetry conditions. Just as bounds were developed when drift is nonzero, perhaps tight bounds can be developed when volatility structures display asymmetry with sufficient stationarity. In the interests of brevity, this extension is left for future research.

Appendix

Put–call symmetry

Let $F(t)$ be the forward price at $t \in [0, T]$ of the underlying for delivery in T years. Let $\sigma(F(t), t)$ be the local volatility rate of the forward price as a function of the forward price $F(t)$ and time t. Under the martingale measure,[s] the forward price process is

$$\frac{dF(t)}{F(t)} = \sigma(F(t), t)\, dW(t)\,. \tag{A.1}$$

Let $B(0)$ be the price at time 0 of a bond paying one dollar in T years and let $C(0, K, T)$ and $P(0, K, T)$ be the initial value of a European call and put struck at K and maturing in T years. Let $G_c(K, T) \equiv B(0)^{-1}C(0, K, T)$ and $G_p(K, T) \equiv B(0)^{-1}P(0, K, T)$ be the respective forward values quoted at 0 of these options for delivery in T years. We now show that both forward values satisfy the following forward partial differential equation (pde):

$$\frac{\sigma^2(K, T)K2}{2} \frac{\partial 2G}{\partial K^2}(K, T) = \frac{\partial G}{\partial T}(K, T)\,, \quad K > 0,\ T > 0\,. \tag{A.2}$$

In contrast to Black (1976) backward pde, this pde indicates how (forward) option values change with the strike and maturity, holding the initial time and underlying forward price fixed. The above result and its proof below are essentially due to Dupire (1994).

To prove the forward pde for a call, we begin with the standard result that the forward price of a call is given by its expected payoff under the equivalent martingale measure:

$$G_c(K, T) = \int_K^\infty (F(T) - K)_p(F(T), T; F(0), 0)\, dF(T)\,, \tag{A.3}$$

where $p(F(T), T; F(0), 0)$ is the transition density of the forward price, indicating the probability density of the forward price being at $F(T)$ at time T, given that it is at $F(0)$ at time 0. The Kolmogorov forward equation governing this density is

$$\frac{1}{2} \frac{\partial^2}{\partial K^2}[\sigma^2(K, T)K^2 p(K, T; F(0), 0)] = \frac{\partial}{\partial T} p(K, T; F(0), 0)\,, \quad K > 0,\ T > 0\,. \tag{A.4}$$

Differentiating (A.3) twice with respect to K gives

$$\frac{\partial^2 G_c(K, T)}{\partial K^2} = p(K, T; F(0), 0)\,. \tag{A.5}$$

Substituting into the Kolmogorov equation gives

$$\frac{1}{2} \frac{\partial^2}{\partial K^2}\left[\sigma^2(K, T)K^2\frac{\partial^2 G_c(K, T)}{\partial K^2}\right] = \frac{\partial}{\partial T} \frac{\partial^2 G_c(K, T)}{\partial K^2}\,, \quad K > 0,\ T > 0\,. \tag{A.6}$$

---

[s]We use a pure discount bond maturing at T as the numeraire.

Integrating twice with respect to K gives the desired result. The same proof applies to European puts. It is easily verified that Black (1976) formulas for calls and puts satisfy the above equation with $\sigma^2(K,T) = \sigma^2$.

The forward call value $G_c(K,T)$ is the unique solution of Eq. (A.2) subject to the following boundary conditions:

(a) $G_c(K,0) = \max[F(0) - K, 0]$, $K > 0$;
(b) $\lim_{K\uparrow\infty} G_c(K,T) = 0$, $T > 0$;
(c) $\lim_{K\downarrow 0} G_c(K,T) = F(0)$, $T > 0$.

Similarly, the forward put value $G_p(K,T)$ is the unique solution of Eq. (A.2) subject to the following boundary conditions:

(a) $G_p(K,0) = \max[K - F(0), 0]$, $K > 0$;
(b) $\lim_{K\uparrow\infty} G_p(K,T) \sim K$, $T > 0$;
(c) $\lim_{K\downarrow 0} G_p(K,T) = 0$, $T > 0$.

Let $u_c(x,T) \equiv G_c(K,T)(KF(0))^{-1/2}$ and $u_p(x,T) \equiv G_p(K,T)(KF(0))^{-1/2}$ be *normalized* call and put forward values, respectively, written as functions of $x \equiv \ln(K/F(0))$ and maturity T. Then, the normalized values both solve the following pde:

$$\frac{v^2(x,T)}{2}\frac{\partial^2 u}{\partial x^2}(x,T) - \frac{v^2(x,T)}{8}u(x,T) = \frac{\partial u}{\partial T}(x,T), \quad x \in (-\infty,\infty), \ T > 0,$$

$$(\text{A.7})$$

where $v(x,T) \equiv \sigma(F(0)e^x, T)$ is the volatility expressed as a function of x and T.

The normalized forward call value $u_c(x,T)$ is the unique solution of Eq. (A.7) subject to the following boundary conditions:

(a) $u_c(x,0) = \max[e^{-x/2} - e^{x/2}, 0]$, $x \in \mathfrak{R}$;
(b) $\lim_{x\uparrow\infty} u_c(x,T) = 0$, $T > 0$;
(c) $\lim_{x\downarrow-\infty} u_c(x,T) = e^{-x/2}$, $T > 0$.

Similarly, the normalized forward put value $u_p(x,t)$ is the unique solution of Eq. (A.7) subject to the following boundary conditions:

(a) $u_p(x,0) = \max[e^{x/2} - e^{-x/2}, 0]$, $x \in \mathfrak{R}$;
(b) $\lim_{x\uparrow\infty} u_p(x,T) \sim e^{x/2}$, $T > 0$;
(c) $\lim_{x\downarrow-\infty} u_p(x,T) = 0$, $T > 0$.

Our symmetry condition is that $v^2(x,T) = v^2(-x,T)$ for all $x \in \mathfrak{R}$ and for all $T > 0$. Given this condition, it is easy to see that $u_c(x,T)$ and $u_p(-x,T)$ satisfy the same boundary value problems and are therefore equal:

$$u_c(x,T) = u_p(-x,T), \quad \text{for all } x \in \mathfrak{R} \text{ and for all } T > 0.$$

Reverting to forward prices gives

$$G_c(K_c, T)(K_c F(0))^{-1/2} = G_p(K_p, T)(K_p F(0))^{-1/2},$$

where $(K_c K_p)^{1/2} = F(0)$. Multiplying both sides by $F(0)^{1/2} B(0)$ gives the desired result:

$$C(0, K_c, T)K_c^{-1/2} = P(0, K_p, T)K_p^{-1/2}. \qquad \square$$

Binary put–call symmetry

Assuming that volatility is a function of time alone, the payoff and values for binary calls and puts, and gap calls and puts with time T until maturity, and strike H can be written as

Payoff at time T	Value at time 0
$BC = 1(F(T) > H)$	$BC = B(0)N(d_2)$
$GC = F(T)1(F(T) > H)$	$GC = B(0)F(0)N(d_1)$
$BP = 1(F(T) < H)$	$BP = B(0)N(-d_2)$
$GP = F(T)1(F(T) < H)$	$GP = B(0)F(0)N(-d_1),$

where

$$N(d) \equiv \int_{-\infty}^{d} \frac{e^{-z^2/2}}{\sqrt{2\pi}} \, dz$$

is the standard normal distribution function,

$$d_1 \equiv \frac{\ln\left(\dfrac{F(0)}{H}\right) + \dfrac{\bar{\sigma}^2 T}{2}}{\bar{\sigma}\sqrt{T}}, \qquad d_2 = d_1 - \bar{\sigma}\sqrt{T},$$

and

$$\bar{\sigma}^2 \equiv \frac{1}{T}\int_0^T \sigma^2(t) \, dt.$$

It may be verified by direct substitution that

$$K^{1/2}BC(K) = GP(H)H^{-1/2}, \qquad H^{1/2}BP(H) = GC(K)K^{-1/2}, \qquad \text{(A.8)}$$

where the geometric mean of the binary (gap) call strike K and the gap (binary) put strike H is the forward price F: $(KH)^{1/2} = F$.

Up-and-in bond

We can rewrite an up-and-in bond as a combination of an up-and-in binary call and an up-and-in binary put:

$$UIB(H) = UIBC(H) + UIBP(H). \qquad \text{(A.9)}$$

However, a $UIBC(H)$ is the same as a standard $BC(H)$, as it has to knock in to have positive value. We can expand the $UIBP(H)$ into its components:

$$UIB(H) = BC(H) + \lim_{n\uparrow\infty} n[UIP(H,H) - UIP(H - n^{-1}, H)]. \qquad (A.10)$$

We can apply PCS:

$$UIB(H) = BC(H) + \lim_{n\uparrow\infty} n[C(H) - (H - n^{-1})H^{-1}C(H^2(H - n^{-1})^{-1})], \quad (A.11)$$

or equivalently

$$UIB(H) \approx BC(H) + H^{-1}\lim_{n\uparrow\infty} C(H^2(H - n^{-1})^{-1})$$

$$+ \lim_{n\uparrow\infty} n[C(H) - C(H + n^{-1})]. \qquad (A.12)$$

The approximation error is $O(n^{-2})$. The final term can now be rewritten as a binary call and so

$$UIB(H) = 2BC(H) + H^{-1}C(H). \qquad (A.13)$$

$\square$

References

David Bates, *The crash premium: Option pricing under asymmetric processes, with applications to options on Deutschemark futures*, working paper, University of Pennsylvania (1988).

David Bates, *The crash of 87: Was it expected? The evidence from options markets*, J. Finance **46** (1991) 1009–1044.

Fischer Black, *The pricing of commodity contracts*, J. Financial Economics **3** (1976) 167–179.

Fischer Black and Myron Scholes, *The pricing of options and corporate liabilities*, The J. Political Economy **81** (1973) 637–659.

Jonathon Bowie and Peter Carr, *Static simplicity*, Risk **7** (1994) 45–49.

Neil Chriss, *Black–Scholes and beyond: Modern option pricing* (Irwin Professional Publishing, Burr Ridge, IL, 1996).

Neil Chriss and Michael Ong, *Digitals diffused*, Risk **8** (1995) 56–59.

Emanuel Derman, Deniz Ergener, and Iraj Kani, *Forever hedged*, Risk **7** (1994) 139–145.

Bruno Dupire, *Pricing with a smile*, Risk **7** (1994) 18–20.

Gary Gastineau, *Roll-up puts, roll-down calls, and contingent premium options*, The J. Derivatives **1** (1994) 40–43.

Robert Geske and Herbert Johnson, *The American put option valued analytically*, J. Finance **39** (1984) 1511–1524.

Naoto Kunitomo and Masayuki Ikeda, *Pricing options with curved boundaries*, Math. Finance (1992) 275–298.

Gurii Marchuk and Vladimir Shaidurov, *Difference methods and their extrapolations* (Springer-Verlag, NY, 1983).

Robert Merton, *Theory of rational option pricing*, Bell J. Economics and Management Science **4** (1973) 141–183.

Reprinted from Int. J. Theor. Appl. Finance
2(1) (1998) 17–42

CLOSED FORM FORMULAS FOR EXOTIC OPTIONS AND THEIR LIFETIME DISTRIBUTION

RAPHAËL DOUADY

*C.M.L.A., Ecole Normale Supérieure, 61 av. du Pdt. Wilson,
94235 Cachan Cedex, France
E-mail: rdouady@cmla.ens-cachan.fr*

Received 7 July 1998
Revised 8 September 1998

We first recall the well-known expression of the price of barrier options, and compute double barrier options by the mean of the iterated mirror principle. The formula for double barriers provides an intraday volatility estimator from the information of high-low-close prices. Then we give explicit formulas for the probability distribution function and the expectation of the exit time of single and double barrier options. These formulas allow to price time independent and time dependent rebates. They are also helpful to hedge barrier and double barrier options, when taking into account variations of the term structure of interest rates and of volatility. We also compute the price of rebates of double knock-out options that depend on which barrier is hit first, and of the BOOST, an option which pays the time spent in a corridor. All these formulas are either in closed form or double infinite series which converge like $e^{-\alpha n^2}$.

Keywords: Double barrier options, closed forms, first hitting time.

JEL classification: G130.

1. Introduction

Exit times of Ito processes from various domains appear to be a recurrent issue in option pricing and hedging. Some options explicitly involve their lifetime in their pay-off (time dependent rebates, etc.). Moreover, hedging — hence pricing — a non-European option, the lifetime of which is thus uncertain, requires the knowledge of its lifetime distribution, in order to properly take into account the term structure of volatility and/or of interest rates.

In this paper, we first compute the value of a (single) barrier option, assuming constant interest rate and volatility, and the probability distribution of its lifetime. The method is a standard technique based on the reflection principle. Then we deduce the expected lifetime and its Laplace transform, which provides all its momenta. This gives a formula to price time independent and time dependent rebates, or rebates that are paid at a postponed date. Time independent rebates are also called "American digital options". Single barrier options are widely studied in the finance literature. See [28] where prices of barrier options are computed, and [22, Chap. 6] where the probability of passing a barrier is given as an exercise. Various books now provide the formula of single barrier options. See for instance [10, 35, 25]. Formulas for exit times distribution and expectation go back to the first half of the century, with [23, 12]. One of the best references concerning the exit time of Brownian and Ornstein–Uhlenbeck processes from variously shaped domains is [7] (see also [26]). An interesting history of the first passage time of a Brownian motion through a barrier can be found in [33, Chap. 1].

In the second part, we compute the price of double barrier options and the probability distribution of their lifetime as a "quasi-closed form" formula: we use an iterated reflection principle, which again is not new (it goes back to the 50's), but provides extremely fast converging series, the nth term being of the order of $\exp(-\alpha n^2)$. In practice, no more than 2 or 3 terms have a significant value. The series we get is different from that obtained by Geman and Yor [16] and it converges much faster. A methodology similar to ours is used in a recent paper by He *et al.* [20] to price double lookback options. From the double barrier option formula, we deduce an intraday volatility estimator using only the daily "high-low-close" information.

In the third part, we provide the value of rebates of double knock-out options. In order to provide the value of rebates that depend on which barrier is hit first, we also compute the distribution of the first hitting time of a barrier, conditionally to the fact that the other one has not been crossed before.

In the last part, we again give a quasi-closed form formula for the price of the *BOOST* option, using Laplace transforms. We recall that the *BOOST* can be considered as the rebate of a double knock-out option which is proportional to the number of days it stayed in (see Sec. for a precise definition). This option is described by Chesney *et al.* in their book [10].

2. Notations

Throughout this paper, $W_t = W(t, \omega)$, $(t, \omega) \in \mathbb{R}_+ \times \Omega$ will denote a standard Brownian motion under a given probability measure $\mathcal{P}$ on Ω. That is

$$W_0 = 0, \quad \text{Var } W_t = t.$$

The Gaussian distribution density and cumulative distribution, respectively g and N, are defined by

$$g(x) = \frac{1}{\sqrt{2\pi}} \exp\left(-\frac{x^2}{2}\right) \quad N(x) = \int_{-\infty}^{x} g(s)\, ds \tag{2.1}$$

The underlying of all the options considered here will be an asset S_t the price of which follows a "geometric Brownian motion" driven by the Brownian motion W_t.

Letting μ denote the (constant) risk-neutral drift[a] and σ the (constant) volatility; we thus assume that S_t satisfies the diffusion equation:

$$\frac{dS_t}{S_t} = \mu\, dt + \sigma\, dW_t\,, \qquad (2.2)$$

so that one has

$$S_t = S_0 \exp\left(\left(\mu - \frac{\sigma^2}{2}\right) t + \sigma\, W_t\right).$$

The reader is aware not to confuse the probability of the *intersection* of two events A and B:

$$\mathcal{P}(A \text{ and } B) = \mathcal{P}(A \cap B)$$

with the *conditional* probability of A upon the condition that B is realised:

$$\mathcal{P}(A\,|\,B) = \frac{\mathcal{P}(A \cap B)}{\mathcal{P}(B)}\,.$$

The acronym P.D.F. stands for "Probability Density Function", while P.D.E. means "Partial Differential Equation". The P.D.F. $\psi_A(x)$ of the random variable X *constrained by the event A* is defined by

$$\mathcal{P}(X \in [x, x + dx] \text{ and } A) = \psi_A(x)\, dx\,.$$

In particular, one has

$$\int_{-\infty}^{\infty} \psi_A(x)\, dx = \mathcal{P}(A)\,.$$

Generally speaking, we reserve the letter ψ for P.D.F. with respect to the x variable, while φ will be used for those with respect to the t variable. The letter τ will always denote a stochastic hitting time.

In all closed form formulas provided here for option prices, the option nominal is set to 1.

3. Single Barriers

The results of this section are not new,[b] but they will be useful for the next one. We adopt Lamberton–Lapeyre's methodology [22] based on the reflection principle (see also Revuz–Yor [29, p. 101]).

---

[a]That is, we assume that $\mathcal{P}$ is the risk-neutral probability and that $\mu = r_d - r_f$, where r_d is the domestic (currencies) or refinancing (equities) rate and r_f is the foreign (currencies) or dividend (equities) rate.

[b]They were already known in the 70's by finance people (Merton [24]), and in the 50's (and perhaps even before) by probabilists (Levy [23]).

3.1. *The mirror principle for drift-free processes*

This "trick" allows one to compute the probability of crossing the barrier. Let $h > 0$ be the barrier level and T be the time horizon. We define the process W_t' by

$$W_t' = W_t \qquad \text{if} \quad W_s < h \quad \forall s < t,$$

$$W_t' = 2h - W_t \quad \text{if} \quad \exists s < t, \quad W_s = h.$$

The process W_t' is again a standard Brownian motion (it is a martingale with independent increments and volatility 1). It has the following property:

$$\forall T > 0 \quad \left(W_T = x < h \text{ and } \max_{[0,T]} W_t \geq h \right) \iff W_T' = 2h - x. \qquad (3.1)$$

Therefore, $\max_{[0,T]} W_t \geq h$, if and only if $W_T \geq h$ or $W_T' \geq h$, and these two events are incompatible. As a consequence, we may add their probabilities:

$$P\left(\max_{[0,T]} W_t \geq h \right) = 2 P(W_T \geq h) = 2 N\left(-\frac{h}{\sqrt{T}} \right). \qquad (3.2)$$

Let τ_h be the first time at which W_t hits the barrier h:

$$\tau_h = \min\{t > 0, \ W_t = h\} \quad (\text{possibly} > T \text{ or even} = \infty).$$

From (3.2), we deduce that

$$P(\tau_h \leq T) = 2 N\left(-\frac{h}{\sqrt{T}} \right). \qquad (3.3)$$

This equation provides, by a simple derivation with respect to t, the P.D.F. of τ_h:

$$\varphi_h(t) = \frac{\partial}{\partial T} P(\tau_h \leq T) = \frac{h}{\sqrt{2\pi t^3}} \exp\left(-\frac{h^2}{2t} \right).$$

3.2. *Processes with drift*

The case of processes with a constant drift will be solved by the mean of Girsanov theorem. Let S_t be the price at time t of the underlying asset. We set

$$X_t = \frac{1}{\sigma} \log \frac{S_t}{S_0} = \lambda t + W_t, \quad \lambda = \frac{\mu}{\sigma} - \frac{\sigma}{2}.$$

Thanks to Girsanov theorem,[c] X_t is a standard Brownian motion under the probability $\mathcal{P}^\lambda$ which has the following density with respect to $\mathcal{P}$:

$$\frac{d\mathcal{P}^\lambda}{d\mathcal{P}}|_t = \exp\left(-\lambda X_t + \frac{\lambda^2 t}{2}\right)$$

We now know, owing to (3.1), that, if[d] $x < h$:

$$\mathcal{P}^\lambda\left(X_T \in [x, x+dx] \text{ and} \max_{[0,T]} X_t < h\right)$$
$$= \frac{dx}{\sqrt{2\pi T}}\left(\exp\left(-\frac{x^2}{2T}\right) - \exp\left(-\frac{(2h-x)^2}{2T}\right)\right),$$

hence

$$\mathcal{P}\left(X_T \in [x, x+dx] \text{ and} \max_{[0,T]} X_t < h\right) = \psi_h^\lambda(x, T)\, dx, \qquad (3.4)$$

with

$$\psi_h^\lambda(x, T) = \frac{1}{\sqrt{2\pi T}}\left(\exp\left(-\frac{(x-\lambda T)^2}{2T}\right) - \exp\left(2\lambda h - \frac{(2h+\lambda T - x)^2}{2T}\right)\right) \qquad (3.5)$$

In particular,

$$\mathcal{P}\left(\max_{[0,T]} x_t < h\right) = \int_{-\infty}^h \psi_h^\lambda(x, T)\, dx$$
$$= N\left(\frac{h}{\sqrt{T}} - \lambda\sqrt{T}\right) - e^{2\lambda h} N\left(-\frac{h}{\sqrt{T}} - \lambda\sqrt{T}\right). \qquad (3.6)$$

Let τ_h^λ be the first time X_t hits the barrier h (still assumed positive). We know with certainty that $X_{\tau_h^\lambda} = h$. Therefore, the P.D.F. (under probability $\mathcal{P}$) of τ_h^λ is given by

$$\varphi_h^\lambda(t) = \frac{d\mathcal{P}}{d\mathcal{P}^\lambda}|_{X_t=h}\, \varphi_h(t) = \frac{h}{\sqrt{2\pi t^3}}\, \exp\left(\lambda h - \frac{\lambda^2 t}{2} - \frac{h^2}{2t}\right). \qquad (3.7)$$

[c]We use here a very simple form of this theorem, and one may also apply (again in a simple form) Cameron–Martin formula:

$$\mathcal{P}^\lambda\left(X_t \in [x, x+dx]\right) = \frac{1}{\sqrt{2\pi t}} \exp\left(-\frac{x^2}{2t}\right) dx$$
$$\mathcal{P}\left(X_t \in [x, x+dx]\right) = \frac{1}{\sqrt{2\pi t}} \exp\left(-\frac{(x-\lambda t)^2}{2t}\right) dx,$$

thus

$$\frac{d\mathcal{P}^\lambda}{d\mathcal{P}} = \exp\left(\frac{(x-\lambda t)^2 - x^2}{2t}\right) = \exp\left(-\lambda x + \frac{\lambda^2 t}{2}\right).$$

[d]In the sequel, dx is meant to be an infinitesimal length of interval. Equalities should be understood as valid at the limit $dx \to 0$, when divided by dx:

$$\mathcal{P}\left(X \in [x, x+dx]\right) = \psi(x)\, dx \iff \lim_{dx \to 0} \frac{1}{dx}\mathcal{P}\left(X \in [x, x+dx]\right) = \psi(x).$$

One can easily check that

$$\varphi_h^\lambda(T) = -\frac{d}{dT}\, \mathcal{P}\left(\max_{[0,T]} x_t < h\right).\tag{3.8}$$

3.3. *Price of (single) barrier options*

We here consider only knock-in and knock-out options that can knock in or out all along their life, like currency options. We exclude those for which the barrier is only active at expiration ("European" barrier), which are actually combinations of binary options and vanilla ones, like limited caps, etc., and those barriers that apply in a "time window". For options which only check whether the closing price is beyond the barrier, we refer the reader to [5, 6].

Let H denote an upper barrier and L a lower one. The following relation always holds:

$$\text{Knock-In}\quad+\quad\text{Knock-Out}\quad=\quad\text{Vanilla}$$

provided they all have the same strike, maturity and (except for the vanilla one) barrier. We shall therefore compute only knock-out options. The price of an "Up-and-Out Call" will be $\mathrm{UOC}(S, K, H)$, while that of an "Down-and-Out Put" will be $\mathrm{DOP}(S, K, L)$, etc. There is no special relation between Puts and Calls.[e] The price of an Up-and-Out option with trigger H, the pay-off of which at the expiration date T is $f(S_T)$, is provided by the knowledge of the function ψ_h^λ:

$$\mathrm{UO}_f(S_0, H) = e^{-rT}\int_{-\infty}^{h}\psi_h^\lambda(x, T)\, f(S_0\exp(\sigma x))dx,$$

where $h = \frac{1}{\sigma}\log\frac{H}{S_0}$. Replacing the process X_t by $-X_t$ yields the price of Down-and-Out options.

In the following formulae, $r = r^d$ denotes the domestic (refinancing) interest rate, and μ is the risk-neutral drift of S_t that is $\mu = r^d - r^f$, where r^f is the dividend rate earned by the underlying (or foreign interest rate for a currency).

For easier notations, we set $S_0 = S$ and

$$h = \frac{1}{\sigma}\log\frac{H}{S},\quad k = \frac{1}{\sigma}\log\frac{K}{S},\quad \ell = \frac{1}{\sigma}\log\frac{L}{S},$$

$$\lambda = \frac{\mu}{\sigma} - \frac{\sigma}{2},\quad \lambda' = \frac{\mu}{\sigma} + \frac{\sigma}{2}.$$

For Up-and-Out options:

$$\alpha = e^{2\lambda h},\quad \alpha' = e^{2\lambda' h} = \frac{\alpha H^2}{S^2},$$

---

[e]Some approximations are available for "symmetric like" structures, for instance $P(S, K, H) \simeq$ cst.$C(S, K', L)$ with $K' = S^2/K$ and $L = S^2/H$. The fail from exact equality comes from interest rates and from a possible volatility smile (which is not taken into account in this study). See [9, 15].

$$\begin{cases} d_1 = \lambda'\sqrt{T} - k/\sqrt{T} \\ d_2 = \lambda\sqrt{T} - k/\sqrt{T} \\ d_3 = \lambda'\sqrt{T} - h/\sqrt{T} \\ d_4 = \lambda\sqrt{T} - h/\sqrt{T} \end{cases} \begin{cases} d_5 = -\lambda\sqrt{T} - h/\sqrt{T} \\ d_6 = -\lambda'\sqrt{T} - h/\sqrt{T} \\ d_7 = -\lambda\sqrt{T} - (2h-k)/\sqrt{T} \\ d_8 = -\lambda'\sqrt{T} - (2h-k)/\sqrt{T} \end{cases}$$

For Down-and-Out options:

$$\alpha = e^{2\lambda\ell}, \quad \alpha' = e^{2\lambda'\ell} = \frac{\alpha L^2}{S^2},$$

$$\begin{cases} d_1 = \lambda'\sqrt{T} - k/\sqrt{T} \\ d_2 = \lambda\sqrt{T} - k/\sqrt{T} \\ d_3 = \lambda'\sqrt{T} - \ell/\sqrt{T} \\ d_4 = \lambda\sqrt{T} - \ell/\sqrt{T} \end{cases} \begin{cases} d_5 = -\lambda\sqrt{T} - \ell/\sqrt{T} \\ d_6 = -\lambda'\sqrt{T} - \ell/\sqrt{T} \\ d_7 = -\lambda\sqrt{T} - (2\ell-k)/\sqrt{T} \\ d_8 = -\lambda'\sqrt{T} - (2\ell-k)/\sqrt{T} \end{cases}$$

For simplicity, we denote $N_i = N(d_i)$. Up-and-Out Calls and Down-and-Out Puts (with $L < K < H$ otherwise these are void options) are given by:

$$\mathrm{UOC}(S,K,H) = e^{(\mu-r)T}S(N_1 - N_3 - \alpha'(N_6 - N_8))$$
$$- e^{-rT}K(N_2 - N_4 - \alpha(N_5 - N_7))$$

$$\mathrm{DOP}(S,K,L) = e^{-rT}K(N_4 - N_2 - \alpha(N_7 - N_5))$$
$$- e^{(\mu-r)T}S(N_3 - N_1 - \alpha'(N_8 - N_6))$$

Down-and-Out Calls and Up-and-Out Puts depend on the location of the trigger with respect to the strike:

$$\mathrm{DOC}(S,K,L \le K) = e^{(\mu-r)T}S(N_1 - \alpha'(1 - N_8)) - e^{-rT}K(N_2 - \alpha(1 - N_7))$$

$$\mathrm{DOC}(S,K,L \ge K) = e^{(\mu-r)T}S(N_3 - \alpha'(1 - N_6)) - e^{-rT}K(N_4 - \alpha(1 - N_5))$$

$$\mathrm{UOP}(S,K,H \ge K) = e^{-rT}K(1 - N_2 - \alpha N_7) - e^{(\mu-r)T}S(1 - N_1 - \alpha'N_8)$$

$$\mathrm{UOP}(S,K,H \le K) = e^{-rT}K(1 - N_4 - \alpha N_5) - e^{(\mu-r)T}S(1 - N_3 - \alpha'N_6)$$

Remark 1. These formulae should only be considered as a basis to price barrier options. Not only do they take into account flat and non-stochastic interest rate and volatility, even in this case, barrier options involve violent discontinuities and uneven behavior with respect to volatility changes, that make them a real challenge to hedge. The impact of transaction costs will depend on the size of the book. We also warn the reader of the unbounded delta Δ of "irregular" options[f] which

[f]$\mathrm{DOC}(S,K,L \le K)$ and $\mathrm{UOP}(S,K,H \ge K)$ are "regular" knock-out options, but $\mathrm{DOC}(S,K,L \ge K)$ and $\mathrm{UOP}(S,K,H \le K)$ are irregular: the trigger is in the money and they contain an "American bet" which causes hedging difficulties close to maturity. The same holds for the "reverse knock-out" $\mathrm{UOC}(S,K,H)$ and $\mathrm{DOP}(S,K,L)$.

makes a pure "Black–Scholes" dynamic hedging strategy inapplicable, because of "liquidity holes", and causes irreducible risk.[g] Hedging difficulties are thoroughly analysed in [34] and an attempt to optimise the sharing of the hedge between static and dynamic is discussed in [3].

3.4. *Rebates and American digitals*

Often, knock-out options include a rebate delivered to the buyer as a compensation when the underlying hits the barrier. If the rebate is paid just after knocking out, its price per unit face value equals the expectation of the discount factor, under the condition that the underlying crosses the barrier. If, independently of the date the option is knocked out, the rebate is paid at expiry, then its price is simply equal to the probability of hitting the barrier, discounted by e^{-rT}, where r is the rate applicable on this period. In this latter case, the Up-and-Out Rebate is given by Eq. (3.6):

$$\text{UOR}_{\text{end}}(S, K, H) = e^{-rT} \left(1 - \mathcal{P}\left(\max_{[0,T]} X_t < h \right) \right)$$

$$= e^{-rT} N\left(\lambda\sqrt{T} - \frac{h}{\sqrt{T}} \right) + e^{2\lambda h - rT} N\left(-\frac{h}{\sqrt{T}} - \lambda\sqrt{T} \right) \quad (3.9)$$

and the Down-and-Out Rebate, by changing the sign of X_t:

$$\text{DOR}_{\text{end}}(S, K, L) = e^{-rT} N\left(\frac{\ell}{\sqrt{T}} - \lambda\sqrt{T} \right) + e^{2\lambda\ell - rT} N\left(\frac{\ell}{\sqrt{T}} + \lambda\sqrt{T} \right). \quad (3.10)$$

When the rebate is paid just after knocking out, and still assuming constant interest rate, one has:

$$\text{UOR}_{\text{KO}}(S, K, H) = \int_0^T e^{-rt} \varphi_h^\lambda(t)\, dt$$

$$= e^{(\lambda - \sqrt{\lambda^2 + 2r})h} N\left(\sqrt{(\lambda^2 + 2r)\, T} - \frac{h}{\sqrt{T}} \right)$$

$$+ e^{(\lambda + \sqrt{\lambda^2 + 2r})h} N\left(-\frac{h}{\sqrt{T}} - \sqrt{(\lambda^2 + 2r)\, T} \right)$$

and

$$\text{DOR}_{\text{KO}}(S, K, L) = \int_0^T e^{-rt} \varphi_{-\ell}^{-\lambda}(t)\, dt$$

$$= e^{(\lambda + \sqrt{\lambda^2 + 2r})\ell} N\left(\sqrt{(\lambda^2 + 2r)\, T} + \frac{\ell}{\sqrt{T}} \right)$$

$$+ e^{(\lambda - \sqrt{\lambda^2 + 2r})\ell} N\left(\frac{\ell}{\sqrt{T}} - \sqrt{(\lambda^2 + 2r)\, T} \right). \quad (3.11)$$

[g]I remember a trader giving an order to his broker: "Sell 100,000 shares!" and the broker: "But to whom, sir?".

Remark 2. When taking $T = +\infty$, the rebate price provides the Laplace transform of the exit time distribution (considering the interest rate r as the variable):

$$\int_0^\infty e^{-rt}\varphi_h^\lambda(t)\,dt = \exp\left(\left(\lambda - \sqrt{\lambda^2 + 2r}\right)h\right).$$

One knows that the coefficients of the Taylor expansion of the Laplace transform at 0 provide the momenta of the distribution.

Remark 3. The price of a constant rebate is also the value of the "American digital" option which pays a fixed amount as soon as the spot crosses a given trigger.

3.5. *Delta of (single) barrier options*

From the formulas giving the prices of barrier options, one can easily deduce, through a simple differentiation with respect to the spot, the value of the Δ. Again, we compute them only for knock-out options (g is the standard Gaussian P.D.F.):

$$\Delta_{\text{UOC}} = e^{(\mu-r)T}\left(N_1 - N_3 + 2\alpha'\frac{r}{\sigma^2}\left(N_6 - N_8\right)\right)$$

$$- 2\,e^{-rT}\left(\alpha\,\frac{\lambda K}{\sigma S}\left(N_5 - N_7\right) + \frac{H - K}{S\sigma\sqrt{T}}\,g(d_4)\right).$$

For symmetry reasons, the Δ of the Down-and-Out Put has the same expression, for both options are the two sides of the same analytic solution of Black–Scholes P.D.E., vanishing along the barrier:

$$\Delta_{\text{DOP}} = e^{(\mu-r)T}\left(N_1 - N_3 + 2\alpha'\frac{r}{\sigma^2}\left(N_6 - N_8\right)\right)$$

$$- 2\,e^{-rT}\left(\alpha\,\frac{\lambda K}{\sigma S}\left(N_5 - N_7\right) - \frac{K - L}{S\sigma\sqrt{T}}\,g(d_4)\right).$$

As previously for prices, Down-and-Out Calls and Up-and-Out Puts deltas depend on whether the strike is above or below the barrier:

$$\Delta_{\text{DOC},K\geq L} = e^{(\mu-r)T}\left(N_1 + 2\frac{r}{\sigma^2}\,\alpha'\left(1 - N_8\right)\right) - 2\,e^{-rT}\frac{\lambda K}{\sigma S}\,\alpha(1 - N_7)$$

$$\Delta_{\text{DOC},K\leq L} = e^{(\mu-r)T}\left(N_3 + 2\frac{r}{\sigma^2}\,\alpha'\left(1 - N_6\right)\right) - 2\,e^{-rT}\frac{\lambda K}{\sigma S}\,\alpha\left(1 - N_5\right)$$

$$+ 2\,e^{-rT}\frac{L - K}{S\sigma\sqrt{T}}\,g(d_4)$$

$$\Delta_{\text{UOP},K\leq H} = 2\,e^{-rT}\frac{\lambda K}{\sigma S}\,\alpha N_7 - e^{(\mu-r)T}\left(1 - N_1 + 2\frac{r}{\sigma^2}\,\alpha'N_8\right)$$

$$\Delta_{\text{UOP},K\geq H} = 2\,e^{-rT}\frac{\lambda K}{\sigma S}\,\alpha N_5 - e^{(\mu-r)T}\left(1 - N_3 + 2\frac{r}{\sigma^2}\,\alpha'N_6\right)$$

$$- 2\,e^{-rT}\frac{K - H}{S\sigma\sqrt{T}}\,g(d_4)$$

Remark 4. Note that the term $g(d_4)$ appears when the option pay-off presents a discontinuity at the barrier. When the option is a regular knock out — like a Down-and-Out Call with a strike above the barrier — this term cancels out. The same cancellation occurs in the Δ of standard Calls and Puts, but not in that of binary options.

3.6. *Expectation of the exit time*

The distribution of the exit time $\varphi_h^\lambda(t)$ is information worthy of note. Indeed, in a stochastic interest rate framework, it provides an approximation of the hedge splitting over the different forwards of the underlying which is widely adopted by traders (especially in the commodity market). In this case, the sensitivity of a barrier option to one particular forward price (at a date before the option maturity) is equal to the probability density of knocking out at this date multiplied by the Δ of the option on the barrier at this precise date. Therefore, one can deduce how to display the hedge over the various maturities. In a first attempt, one can adjust the "duration" of this distribution (its definition is a straightforward extent of Macauley's one for bonds) to the expectation of the exit time, or better, to the conditional expectation of the exit time assuming the option is knocked out, the hedge being dispatched between a combination of forward contracts that globally has this duration and an amount of forward contracts at maturity, according to the probability of remaining until expiration, given by Eq. (3.6). We here provide the expectation of the exit time, unconditional or conditioned by a maximum maturity T:

$$\mathbf{E}(\tau_{h,T}^\lambda) = \frac{h}{\lambda} - e^{2\lambda h} \left(T + \frac{h}{\lambda}\right) N\left(-\frac{h}{\sqrt{T}} - \lambda\sqrt{T}\right) + \left(T - \frac{h}{\lambda}\right) N\left(\frac{h}{\sqrt{T}} - \lambda\sqrt{T}\right)$$

$$\mathbf{E}\left(\tau_h^\lambda \,|\, \tau_h^\lambda < T\right) = \frac{h\left(N\left(\lambda\sqrt{T} - \frac{h}{\sqrt{T}}\right) - e^{2\lambda h} N\left(-\frac{h}{\sqrt{T}} - \lambda\sqrt{T}\right)\right)}{\lambda\left(N\left(\lambda\sqrt{T} - \frac{h}{\sqrt{T}}\right) + e^{2\lambda h} N\left(-\frac{h}{\sqrt{T}} - \lambda\sqrt{T}\right)\right)}$$

Remark 5. Note that the unconditional expectation of the exit time has the following simple expression:

$$\mathbf{E}(\tau_h^\lambda) = \frac{h}{\lambda}$$

It tends to $+\infty$ as λ tends to 0 and is infinite if $\lambda < 0$.

Let $\Delta_H(t)$ be the forward value of the option Δ at date t assuming the spot hits the barrier at this time ($S_t = H$). The distribution of hedges will, up to the first order, depend on the expectation:

$$\mathbf{E}\left(\tau_h^\lambda \, \Delta_H(\tau_h^\lambda) \,|\, \tau_h^\lambda < T\right).$$

Let $\pi^* = \mathcal{P}(\max_{[0,T]} X_t < h)$ be the probability of the option not being knocked-out. One has

$$\mathbf{E}\left(\tau_h^\lambda \, \Delta_H(\tau_h^\lambda) \,|\, \tau_h^\lambda < T\right) = \frac{1}{1-\pi^*} \int_0^T t \, \Delta_H(t) \, \varphi_h^\lambda(t) \, dt \,.$$

This integral cannot, in general, be computed in closed form, but standard numerical techniques like Gauss–Legendre (see [27]) allow very fast and accurate evaluations of a one-dimensional integral.

This formula should be used as follows: assume that hedging the barrier option requires one to buy an amount Δ of underlying (spot) in a constant interest rate framework. Then, when interest rates vary, one should split Δ between an amount $\pi^* \Delta / B(0, T)$ of forward contracts on the underlying at date T (where $B(0, T) = e^{-rT}$ is the discount factor on the interval $[0, T]$) and quantities $\delta(t_i)$ of forward contracts at dates $t_1, \ldots, t_n < T$ such that

$$\sum \delta(t_i) B(0, t_i) = (1 - \pi^*) \Delta \,,$$

and

$$\frac{\sum \delta(t_i) B(0, t_i) \, \Delta_H(t_i) \, t_i}{(1 - \pi^*) \Delta} = \mathbf{E}\left(\tau_h^\lambda \, \Delta_H(\tau_h^\lambda) \,|\, \tau_h^\lambda < T\right) \,.$$

Such a hedge will at least protect against a parallel shift of the yield curve (in case of a stock) and both yield curves (in case of a currency).

4. Double Barriers

4.1. *The iterated mirror principle*

Let us first compute the probability that a standard (drift-free) Brownian motion W_t remains in a corridor $[\ell, h]$ (with $\ell < 0 < h$) during the period $[0, T]$. Denote by $\delta = h - \ell$ the corridor width and by $\tau_{h,\ell}$ the first time W_t hits one of the two barriers (possibly $\tau_{h,\ell} > T$). We also define

$$W_\tau^T \begin{cases} W^{\tau_{h,\ell}} & (= h \text{ or } \ell) & \text{if} \quad \tau_{h,\ell} < T \,, \\ W_T & \text{if} \quad \tau_{h,\ell} \geq T \,. \end{cases}$$

According to the mirror principle, one has:[h]

$$\mathcal{P}\left(W_T = x \text{ and } \tau_{h,\ell} \geq T\right) = \mathcal{P}\left(W_T = x\right) - \mathcal{P}(W_T = x \text{ and } W_\tau^T = h)$$

$$- \mathcal{P}\left(W_T = x \text{ and } W_\tau^T = \ell\right)$$

$$= \mathcal{P}(W_T = x) - \mathcal{P}\left(W_T = 2h - x \text{ and } W_\tau^T = h\right)$$

$$- \mathcal{P}\left(W_T = 2\ell - x \text{ and } W_\tau^T = \ell\right)$$

[h] Here, we write for simplicity $W_T = x$ for $W_T \in [x, x + dx]$.

Then

$$\mathcal{P}\left(W_T = 2h - x \text{ and } W_\tau^T = h\right) = \mathcal{P}(W_T = 2h - x)$$

$$- \mathcal{P}\left(W_T = 2h - x \text{ and } W_\tau^T = \ell\right)$$

$$= \mathcal{P}(W_T = 2h - x)$$

$$- \mathcal{P}\left(W_T = x - 2\delta \text{ and } W_\tau^T = \ell\right)$$

We end into the following expression for the distribution density of W_T under the condition that W_t remains inside the corridor for all $t \in [0, T]$:

$$\psi_{h,\ell,T}(x) = \lim_{dx \to 0} \frac{1}{dx} \mathcal{P}(W_T \in [x, x + dx] \text{ and } \tau_{h,\ell} \geq T)$$

$$= \mathbf{1}_{[\ell,h]} \sum_{n=-\infty}^{+\infty} (g_T(x + 2n\delta) - g_T(2h - x + 2n\delta)), \qquad (4.1)$$

where $g_T(x)$ denotes the centered Gaussian distribution density of standard deviation $\sqrt{T}$:

$$g_T(x) = \frac{1}{\sqrt{2\pi T}} \exp\left(-\frac{x^2}{2T}\right).$$

One may easily check that $\psi_{h,\ell,T}(\ell) = \psi_{h,\ell,T}(h) = 0$.

4.2. *Double knock-out calls*

The notations are the same as those for single barriers:

$$h = \frac{1}{\sigma} \log \frac{H}{S}, \quad k = \frac{1}{\sigma} \log \frac{K}{S}, \quad \ell = \frac{1}{\sigma} \log \frac{L}{S},$$

$$\lambda = \frac{\mu}{\sigma} - \frac{\sigma}{2}, \quad \lambda' = \frac{\mu}{\sigma} + \frac{\sigma}{2}, \quad \delta = h - \ell.$$

We first assume that the strike K is between the barriers L and H. Then the price of the Double Knock-Out Call is

$$\text{DKOC}(S, K, L, H) = e^{-rT} \int_k^h (S e^{\sigma x} - K)\psi_{h,\ell,T}^\lambda(x) \, dx,$$

where the distribution density $\psi_{h,\ell,T}^\lambda(x)$ is given by Girsanov theorem:

$$\psi_{h,\ell,T}^\lambda(x) = e^{-\frac{1}{2}\lambda^2 T + \lambda x} \psi_{h,\ell,T}(x). \qquad (4.2)$$

Let us set, for any $n \in \mathbb{Z}$:

$$I_n(u, k, h, \delta) = \int_k^h e^{-\frac{1}{2}u^2 T + ux} g_T(x + 2n\delta) \, dx$$

$$= e^{-2nu\delta} \int_k^h g_T(x + 2n\delta - u) \, dx$$

$$= e^{-2nu\delta} \left(N\left(\frac{h + 2n\delta}{\sqrt{T}} - u\sqrt{T}\right) - N\left(\frac{k + 2n\delta}{\sqrt{T}} - u\sqrt{T}\right) \right)$$

and

$$J_n(u, k, h, \delta) = \int_k^h e^{-\frac{1}{2}u^2 T + ux} g_T(2h - x + 2n\delta) \, dx$$

$$= e^{2u(n\delta + h)} \left(N\left(\frac{2h - k + 2n\delta}{\sqrt{T}} + u\sqrt{T} \right) - N\left(\frac{h + 2n\delta}{\sqrt{T}} + u\sqrt{T} \right) \right).$$

These series tend towards 0 like a Gaussian: for reasonable values of K, L and H, they numerically vanish for $|n| \geq 4$, and one has

$$\mathrm{DKOC}(S, K, L, H) = e^{(\mu - r)T} S \sum_{n=-\infty}^{\infty} (I_n(\lambda', k, h, \delta) - J_n(\lambda', k, h, \delta))$$

$$- e^{-rT} K \sum_{n=-\infty}^{\infty} (I_n(\lambda, k, h, \delta) - J_n(\lambda, k, h, \delta)).$$

If $K < L$, then the lower bound of the integrals must be set to ℓ and the formula becomes

$$\mathrm{DKOC}(S, K, L, H) = e^{(\mu - r)T} S \sum_{n=-\infty}^{\infty} (I_n(\lambda', \ell, h, \delta) - J_n(\lambda', \ell, h, \delta)),$$

$$- e^{-rT} K \sum_{n=-\infty}^{\infty} (I_n(\lambda, \ell, h, \delta) - J_n(\lambda, \ell, h, \delta)),$$

with

$$I_n(u, \ell, h, \delta) = \int_\ell^h e^{-\frac{1}{2}u^2 T + ux} g_T(x + 2n\delta) \, dx$$

$$= e^{-2nu\delta} \left(N\left(\frac{h + 2n\delta}{\sqrt{T}} - u\sqrt{T} \right) - N\left(\frac{h + (2n-1)\delta}{\sqrt{T}} - u\sqrt{T} \right) \right),$$

and

$$J_n(u, \ell, h, \delta) = \int_\ell^h e^{-\frac{1}{2}u^2 T + ux} g_T(2h - x + 2n\delta) \, dx$$

$$= e^{2u(n\delta + h)} \left(N\left(\frac{h + (2n+1)\delta}{\sqrt{T}} + u\sqrt{T} \right) - N\left(\frac{h + 2n\delta}{\sqrt{T}} + u\sqrt{T} \right) \right).$$

4.3. *Double knock-out puts*

As for the Calls, the price of a Double Knock-Out Puts depends on the upper barrier location with respect to the strike. One has

$$\mathrm{DKOP}(S, K, L, H) = e^{-rT} K \sum_{n=-\infty}^{\infty} (I_n(\lambda, \ell, k, \delta) - J_n(\lambda, \ell, k, \delta))$$

$$- e^{(\mu - r)T} S \sum_{n=-\infty}^{\infty} (I_n(\lambda', \ell, k, \delta) - J_n(\lambda', \ell, k, \delta)),$$

where if $L < K < H$,

$$I_n(u, \ell, k, \delta) = \int_\ell^k e^{-\frac{1}{2}u^2 T + ux} \, g_T(x + 2n\delta) \, dx$$

$$= e^{-2nu\delta} \left(N\left(\frac{k + 2n\delta}{\sqrt{T}} - u\sqrt{T} \right) - N\left(\frac{h + (2n - 1)\delta}{\sqrt{T}} - u\sqrt{T} \right) \right)$$

$$J_n(u, \ell, k, \delta) = \int_\ell^k e^{-\frac{1}{2}u^2 T + ux} \, g_T(2h - x + 2n\delta) \, dx$$

$$= e^{2u(n\delta + h)} \left(N\left(\frac{h + (2n + 1)\delta}{\sqrt{T}} + u\sqrt{T} \right) \right.$$

$$\left. - N\left(\frac{2h - k + 2n\delta}{\sqrt{T}} + u\sqrt{T} \right) \right)$$

and if $K \geq H$,

$$\mathrm{DKOP}(S, K, L, H) = e^{-rT} K \sum_{n=-\infty}^\infty (I_n(\lambda, \ell, h, \delta) - J_n(\lambda, \ell, h, \delta))$$

$$- e^{(\mu - r)T} S \sum_{n=-\infty}^\infty (I_n(\lambda', \ell, h, \delta) - J_n(\lambda', \ell, h, \delta))$$

Remark 6. The price of barrier, and even more double barrier options is very sensitive to the term structure of interest rates and to the shape of the volatility surface (the "smile"). If one needs to take them into account, then, generally speaking, no closed form formula is available but, because of the Dirichlet boundary conditions, the value of the option is efficiently computed through a θ-scheme with meshes precisely on the barriers (see [27]).

4.4. *Intraday volatility estimator*

Data bases are usually settled daily, but they provide a very useful information: the highest quotation of the day, the lowest, the opening and the closing. One can apply the present computations to deduce the joint distribution of the triple $\left(\max_{t \leq T} S_t, \min_{t \leq T} S_t, S_T \right)$ and therefore an estimator of the intraday volatility. Such an estimation may, for instance, allow one to optimise the frequency of dynamic hedging.

As previously, we assume that

$$\frac{dS}{S} = \mu \, dt + \sigma \, dW \,,$$

so that

$$S_t = S_0 \exp \sigma X_t \,,$$

$$X_t = \lambda t + W_t \,, \quad \lambda = \frac{\mu}{\sigma} - \frac{\sigma}{2} \,.$$

Let $T > 0$. One has

$$P\left(X_T \in [x, x + dx] \text{ and } \max_{t \leq T} X_t \leq h, \min_{t \leq T} X_t \geq \ell\right) = \psi_{h,\ell,T}^\lambda(x)\, dx,$$

thus the joint P.D.F. of $\left(\max_{t \leq T} X_t, \min_{t \leq T} X_t, X_T\right)$ is

$$\frac{\partial^2}{\partial h \partial \ell} \psi_{h,\ell,T}^\lambda(x)$$

$$= e^{-\frac{\lambda^2 T}{2} + \lambda x} \frac{\partial^2}{\partial h \partial \ell} \psi_{h,\ell,T}(x)$$

$$= 1_{\ell < x < h}\, e^{-\frac{\lambda^2 T}{2} + \lambda x} \sum_{n=-\infty}^{+\infty} \frac{\partial^2}{\partial h \partial \ell} \left(g_T(x + 2n\delta) - g_T(2h - x + 2n\delta)\right)$$

$$= 1_{\ell < x < h}\, e^{-\frac{\lambda^2 T}{2} + \lambda x} \sum_{n=-\infty}^{+\infty} 4n((n+1)g_T''(2h - x + 2n\delta) - ng_T''(x + 2n\delta)), \quad (4.3)$$

where

$$g_T''(x) = \frac{d^2}{dx^2}\, g_T(x) = \frac{x^2 - T}{\sqrt{2\pi T^5}}\, \exp\left(-\frac{x^2}{2T}\right).$$

Given a time series:

$$(S_{\text{open}}(t_i),\, S_{\text{high}}(t_i),\, S_{\text{low}}(t_i),\, S_{\text{close}}(t_i)) \quad i = 1, \ldots, N$$

one computes

$$h_i = \log \frac{S_{\text{high}}(t_i)}{S_{\text{open}}(t_i)}, \quad \ell_i = \log \frac{S_{\text{low}}(t_i)}{S_{\text{open}}(t_i)}, \quad x_i = \log \frac{S_{\text{close}}(t_i)}{S_{\text{open}}(t_i)},$$

and finds the maximum likelihood values of λ and σ such that the P.D.F. of $(h_i/\sigma, \ell_i/\sigma, x_i/\sigma)$ be given by formula (4.3).

Remark 7.   One may also assume that λ is given as the risk-neutral drift and perform the maximum likelihood estimation only on σ, but in a market with a strong trend, this might lead to hedging errors.

5. Corridors and Rebates of Double Barrier Options

5.1. *Corridors*

A *Corridor* is an option that pays a fixed amount at a given date, provided the underlying remained within a range $[L, H]$ all the time until expiration. According to the previous analysis, its price is given by

$$\text{Corridor}(S, L, H, T) = e^{-rT} \sum_{n=-\infty}^{\infty} (I_n(\lambda, \ell, h, \delta) - J_n(\lambda, \ell, h, \delta)). \quad (5.1)$$

Remark 8. The expression giving the price of a Corridor corresponds to the term in K in the Double Knock-Out Calls and Puts, when K is outside the range $[L, H]$. An intuitive way to understand this is to see a Corridor as an extreme Double Knock-Out Put with an infinitely high strike and an infinitely small amount accordingly, or to differentiate the option value with respect to K.

5.2. *Symmetric rebates*

If a rebate is included in a double knock-out option, then in the case it is paid at the expiration date T, whenever the option is knocked out, it is formally equivalent to the complement of a Corridor option to the zero-coupon of maturity T. The price (per unit) of a Rebate Double Knock-Out option, paid at the end, is thus given by

$$\mathrm{RDKO}_{\mathrm{end}}(S, L, H, T) = e^{-rT}\left(1 - \sum_{n=-\infty}^{\infty}(I_n(\lambda, \ell, h, \delta) - J_n(\lambda, \ell, h, \delta))\right).$$

In the case it is paid just when the option is killed, then its price is given in the next section, Remark 9 and also in Sec. .

5.3. *Asymmetric rebates*

We here compute separately the value of the upper rebate, which is paid only if the upper barrier is hit first, and of the lower one, paid only if the lower barrier is hit first. As previously, we let τ_h^λ (resp. τ_ℓ^λ) be the first time the process $X_t = \lambda t + W_t$ reaches the level h (resp. ℓ) and $\tau_{h,\ell}^\lambda = \min(\tau_h^\lambda, \tau_\ell^\lambda)$.

Let $\varphi_{h,\ell,+}^\lambda(t)$ (resp. $\varphi_{h,\ell,-}^\lambda(t)$) be the P.D.F. of τ_h^λ (resp. τ_ℓ^λ) conditioned by $\tau_h^\lambda < \tau_\ell^\lambda$ (resp. $\tau_\ell^\lambda < \tau_h^\lambda$), that is

$$\varphi_{h,\ell,+}^\lambda(t) = \frac{1}{dt}\,\mathcal{P}\left(\tau_h^\lambda \in [t, t+dt] \text{ and } \tau_h^\lambda < \tau_\ell^\lambda\right),$$

$$\varphi_{h,\ell,-}^\lambda(t) = \frac{1}{dt}\,\mathcal{P}\left(\tau_\ell^\lambda \in [t, t+dt] \text{ and } \tau_\ell^\lambda < \tau_h^\lambda\right).$$

The sum of these two functions is the P.D.F. of the exit time $\tau_{h,\ell}^\lambda$:

$$\varphi_{h,\ell,+}^\lambda(t) + \varphi_{h,\ell,-}^\lambda(t) = \frac{1}{dt}\,\mathcal{P}\left(\tau_{h,\ell}^\lambda \in [t, t+dt]\right)$$

$$= -\frac{d}{dt}\int_\ell^h \psi_{h,\ell,t}^\lambda(x)\,dx,$$

where $\psi_{h,\ell,t}^\lambda$ is defined by (4.2). We know that $\psi_{h,\ell,t}^\lambda$ satisfies the Kolmogorov equation:

$$\frac{d}{dt}\psi_{h,\ell,t}^\lambda = \frac{1}{2}\frac{d^2}{dx^2}\psi_{h,\ell,t}^\lambda - \lambda\frac{d}{dx}\psi_{h,\ell,t}^\lambda. \tag{5.2}$$

Therefore

$$\varphi_{h,\ell,+}^{\lambda}(t) + \varphi_{h,\ell,-}^{\lambda}(t) = -\frac{1}{2}\left(\frac{d\psi_{h,\ell,t}^{\lambda}}{dx}(h) - \frac{d\psi_{h,\ell,t}^{\lambda}}{dx}(\ell)\right) + \lambda\left(\psi_{h,\ell,t}^{\lambda}(h) - \psi_{h,\ell,t}^{\lambda}(\ell)\right)$$

$$= \frac{1}{2}\left(\frac{d\psi_{h,\ell,t}^{\lambda}}{dx}(\ell) - \frac{d\psi_{h,\ell,t}^{\lambda}}{dx}(h)\right)$$

because $\psi_{h,\ell,t}^{\lambda}(h) = \psi_{h,\ell,t}^{\lambda}(\ell) = 0$. It is not difficult to see, by perturbing separately h and ℓ, that one can identify each term:

$$\varphi_{h,\ell,+}^{\lambda}(t) = -\frac{1}{2}\frac{d\psi_{h,\ell,t}^{\lambda}}{dx}(h), \quad \varphi_{h,\ell,-}^{\lambda}(t) = \frac{1}{2}\frac{d\psi_{h,\ell,t}^{\lambda}}{dx}(\ell).$$

One has

$$\frac{d\psi_{h,\ell,t}^{\lambda}}{dx}(\ell) = e^{-\frac{\lambda^2 t}{2}+\lambda x}\left(\lambda\psi_{h,\ell,t}(\ell) + \frac{d}{dx}\psi_{h,\ell,t}(\ell)\right)$$

$$= e^{-\frac{\lambda^2 t}{2}+\lambda x}\frac{d}{dx}\psi_{h,\ell,t}(\ell)$$

$$= -\frac{2}{t}e^{-\frac{\lambda^2 t}{2}+\lambda\ell}\sum_{n=-\infty}^{+\infty}(\ell+2n\delta)g_t(\ell+2n\delta)$$

and

$$\frac{d\psi_{h,\ell,t}^{\lambda}}{dx}(h) = -\frac{2}{t}e^{-\frac{\lambda^2 t}{2}+\lambda h}\sum_{n=-\infty}^{+\infty}(h+2n\delta)g_t(h+2n\delta).$$

The price of the Upper Rebate Double Knock-Out option, paid at the end, can now be computed by replacing, for each term, h by $h+2n\delta$ in (3.8) and (3.6):

$$\text{URDKO}_{\text{end}}(S, L, H, T) = e^{-rT}\int_0^T \varphi_{h,\ell,+}^{\lambda}(t)\,dt$$

$$= \sum_{n=-\infty}^{+\infty}\left(e^{-rT-2\lambda n\delta}\,N\left(-\frac{h+2n\delta}{\sqrt{T}}+\lambda\sqrt{T}\right)\right.$$

$$\left. + e^{-rT+2\lambda(h+n\delta)}\,N\left(-\frac{h+2n\delta}{\sqrt{T}}-\lambda\sqrt{T}\right)\right)$$

If paid at Knock-Out time, one has, setting $\rho = \sqrt{\lambda^2 + 2r}$:

$$\text{URDKO}_{\text{KO}}(S, L, H, T) = \int_0^T e^{-rt}\varphi_{h,\ell,+}^{\lambda}(t)\,dt$$

$$= \sum_{n=-\infty}^{+\infty}\left(e^{-\rho(h+2n\delta)+\lambda h}\,N\left(-\frac{h+2n\delta}{\sqrt{T}}+\rho\sqrt{T}\right)\right.$$

$$\left. + e^{\rho(h+2n\delta)+\lambda h}\,N\left(-\frac{h+2n\delta}{\sqrt{T}}-\rho\sqrt{T}\right)\right) \qquad (5.3)$$

Symmetrically, the Lower Rebate Double Knock-Out, paid at the end, is given by:

$$\text{LRDKO}_{\text{end}}(S, L, H, T) = e^{-rT} \int_0^T \varphi_{h,\ell,-}^{\lambda}(t)\, dt$$

$$= \sum_{n=-\infty}^{+\infty} \left(e^{-rT-2\lambda n\delta}\, N\left(\frac{\ell + 2n\delta}{\sqrt{T}} - \lambda\sqrt{T}\right) \right.$$

$$\left. + e^{-rT+2\lambda(\ell+n\delta)}\, N\left(\frac{\ell + 2n\delta}{\sqrt{T}} + \lambda\sqrt{T}\right) \right)$$

and, if paid at Knock-Out time, one has, with the same value of ρ:

$$\text{LRDKO}_{\text{KO}}(S, L, H, T) = \int_0^T e^{-rt}\, \varphi_{h,\ell,-}^{\lambda}(t)\, dt$$

$$= \sum_{n=-\infty}^{+\infty} \left(e^{-\rho(\ell+2n\delta)+\lambda\ell}\, N\left(\frac{\ell + 2n\delta}{\sqrt{T}} - \rho\sqrt{T}\right) \right.$$

$$\left. + e^{\rho(\ell+2n\delta)+\lambda\ell}\, N\left(\frac{\ell + 2n\delta}{\sqrt{T}} + \rho\sqrt{T}\right) \right) \tag{5.4}$$

Remark 9. Adding up (5.3) and (5.4) provides the price of a symmetric rebate which is paid as soon as the option is knocked out.

6. The BOOST

The BOOST (Banking on Over All STability) is an option that was introduced into the market by Société Générale in 1994. It is characterised by a corridor $[L, H]$. As soon as one of the barriers L or H is reached, the option ends and the buyer receives an amount proportional to the *time* (or to the number of days if discretised) the underlying spent inside the corridor since its issuance. Therefore, after having "lived" for some time, the BOOST has a price made of two parts. The first one is not model dependent and corresponds to the time already spent in the corridor, while the second one is the risk-neutral expectation of the remaining lifetime. For security reasons, a limit to the lifetime has been imposed. If the underlying stays in the corridor until the limit, then the BOOST pays, at expiry, the amount corresponding to this maximum time (as if exit were forced at this date, if it did not happen before). One can also see the BOOST as the time increasing rebate of a double knock-out option.

There are at least three ways to get a closed form formula for the BOOST. One of them is to compute the Green function of the Black–Scholes P.D.E., but the series one gets this way converge slowly. The second is to integrate the value of a Corridor Option with respect to its maturity. This method leads to an integral which is not formally computable by standard means (although we here show that a solution exists). In this section, we give another method to get the BOOST price, through Laplace transforms.[i] This formula is a double infinite series, but which converges extremely fast (like $e^{-\alpha n^2}$: the nth term measures the probability that a Brownian path alternately hits the upper and the lower barriers n times, similarly to double knock-out options). Practically, three or four terms are enough to get a 10^{-6} accuracy. The same computation provides the value of the rebate of a double barrier option which is paid as soon as the option knocks out.

6.1. *Exit time of a brownian motion from a corridor*

Let us consider the following random process:

$$X_t = \lambda t + W_t,$$

where W_t is a standard Brownian motion and λ a scalar.

Let h and ℓ denote the barriers and assume $\ell < 0 < h$. As in previous sections, we set $\delta = h - \ell$ and

$$\tau_h^\lambda = \inf\{t, X_t = h\},$$

$$\tau_\ell^\lambda = \inf\{t, X_t = \ell\},$$

$$\tau_{h,\ell}^\lambda = \inf(\tau_h^\lambda, \tau_\ell^\lambda).$$

We first evaluate

$$T_\lambda(h, \ell) = \mathbf{E}(\tau_{h,\ell}^\lambda)$$

Let

$$p = \mathcal{P}(\tau_h^\lambda < \tau_\ell^\lambda).$$

Thanks to the fundamental theorem of martingales, one has

$$\mathbf{E}(X_{\tau_{h,\ell}^\lambda}) = \lambda T_\lambda(h, \ell) = hp + \ell(1 - p).$$

On the other hand, for any σ, the process

$$Y_t^\sigma = e^{\sigma W_t - \frac{1}{2}\sigma^2 t}$$

is a martingale. Taking $\sigma = -2\lambda$, and considering the stopping time $\tau_{h,\ell}^\lambda$, we get

$$Y_t^{-2\lambda} = e^{-2\lambda X_t}$$

and

$$\mathbf{E}(Y_{\tau_{h,\ell}^\lambda}^{-2\lambda}) = 1 = pe^{-2\lambda h} + (1 - p)e^{-2\lambda \ell}$$

---

[i]It is not obvious that the series coming from the Green function and that given by the following method converge to the same value. This has been shown by S. Eddé in a personal communication from MUREX. It is left as an exercise to the reader to show that integrating the corridor formula leads to the same value.

thus

$$p = \frac{e^{-2\lambda\ell} - 1}{e^{-2\lambda\ell} - e^{-2\lambda h}} .$$

Consequently

$$T_\lambda(h, \ell) = \frac{(e^{-2\lambda\ell} - 1)h + (1 - e^{-2\lambda h})\ell}{\lambda(e^{-2\lambda\ell} - e^{-2\lambda h})} .$$

Remark 10. If $\lambda \to 0$, one finds the simple formula:

$$T_0(h, \ell) = -h\ell .$$

6.2. *Distribution of $\tau_{h,\ell}^\lambda$ and BOOST without time limit*

We now compute the Laplace transform of the distribution of $T_{h,\ell}^\lambda$. For every $r > 0$, one has

$$\mathbf{E}(e^{-r\tau_{h,\ell}^\lambda}) = p\, A_r + (1 - p)\, B_r ,$$

where

$$A_r = \mathbf{E}(e^{-r\tau_h^\lambda} \,|\, \tau_h^\lambda < \tau_\ell^\lambda) \quad B_r = \mathbf{E}(e^{-r\tau_\ell^\lambda} \,|\, \tau_\ell^\lambda < \tau_h^\lambda) .$$

But if the number $\sigma \in \mathbf{R}$ satisfies

$$\sigma\lambda + \frac{1}{2}\sigma^2 = r , \tag{6.1}$$

then

$$Y_t^\sigma = e^{\sigma X_t - rt} ,$$

and

$$\mathbf{E}(Y_{\tau_{h,\ell}^\lambda}^\sigma) = 1 = p\, e^{\sigma h} A_r + (1 - p)\, e^{\sigma\ell} B_r .$$

Let us first consider the case where

$$r \neq -\frac{1}{2}\lambda^2 ,$$

and denote by σ_1 and σ_2 the two solutions to the Eq. (6.1)

$$\sigma_1 = -\lambda + \sqrt{\lambda^2 + 2r} ,$$

$$\sigma_2 = -\lambda - \sqrt{\lambda^2 + 2r} .$$

The conditional expectations A_r and B_r are given by the linear system:

$$\begin{cases} p\, e^{\sigma_1 h} A_r + (1 - p)\, e^{\sigma_1 \ell} B_r = 1 \\ p\, e^{\sigma_2 h} A_r + (1 - p)\, e^{\sigma_2 \ell} B_r = 1 \end{cases}$$

leading to

$$
\begin{cases}
A_r = \dfrac{e^{\lambda h}\ \sinh(\sqrt{\lambda^2 + 2r}\,|\ell|)}{p\sinh(\sqrt{\lambda^2 + 2r}\,(h - \ell))} \\[3mm]
B_r = \dfrac{e^{\lambda \ell}\ \sinh(\sqrt{\lambda^2 + 2r}\,h)}{(1 - p)\sinh(\sqrt{\lambda^2 + 2r}\,(h - \ell))}.
\end{cases}
$$

These expressions provide the Laplace transform of $T^{\lambda}_{h,\ell}$:

$$
\mathbf{E}(e^{-r\tau^{\lambda}_{h,\ell}}) = \frac{e^{\lambda \ell}\sinh(\sqrt{\lambda^2 + 2r}\,h) - e^{\lambda h}\sinh(\sqrt{\lambda^2 + 2r}\,\ell)}{\sinh(\sqrt{\lambda^2 + 2r}\,\delta)}. \qquad (6.2)
$$

By differentiating the Laplace transform with respect to r, we obtain the value of the BOOST without time limit:

$$
\mathbf{E}(\tau^{\lambda}_{h,\ell}\,e^{-r\tau^{\lambda}_{h,\ell}}) = -\frac{\partial}{\partial r}\mathbf{E}(e^{-r\tau^{\lambda}_{h,\ell}}).
$$

Letting $\rho = \sqrt{\lambda^2 + 2r}$ we get the formula

$$
\mathbf{E}(\tau^{\lambda}_{h,\ell}e^{-r\tau^{\lambda}_{h,\ell}})
$$

$$
= \frac{\delta\left(e^{\lambda \ell}\sinh\rho h - e^{\lambda h}\sinh\rho\ell\right)\cosh\rho\delta - (h\,e^{\lambda \ell}\cosh\rho h - \ell\,e^{\lambda h}\cosh\rho\ell)\sinh\rho\delta}{\rho\sinh^2\rho\delta}
$$

$$
(6.3)
$$

and, when $\rho \to 0$, that is, if $r = -\frac{1}{2}\lambda^2$, then

$$
\mathbf{E}(\tau^{\lambda}_{h,\ell}e^{-r\tau^{\lambda}_{h,\ell}}) = \frac{\delta}{3}(he^{\lambda \ell} - \ell e^{\lambda h}) - \frac{h^3 e^{\lambda \ell} - \ell^3 e^{\lambda h}}{3\delta}.
$$

Assume now that S_t is a process that satisfies the following diffusion equation:

$$
\frac{dS}{S} = \mu\,dt + \sigma\,dW,
$$

where μ and σ are *constants* and W_t a standard Brownian motion. We set

$$
X_t = \frac{1}{\sigma}\log\frac{S_t}{S_0},
$$

and we have

$$
dX = \lambda\,dt + dW, \qquad \lambda = \frac{\mu}{\sigma} - \frac{\sigma}{2}.
$$

The process S_t respectively crosses the barriers L and H, $L < S_0 < H$, if X_t respectively crosses h and ℓ given by

$$
\ell = \frac{1}{\sigma}\log\frac{L}{S_0}, \qquad h = \frac{1}{\sigma}\log\frac{H}{S_0}.
$$

By inputting these values of λ, h and ℓ in formula (6.3), we get the value of the unlimited BOOST written on S_t with barriers L and H.

Remark 11. Assume that the process Z_t satisfies the following diffusion equation:

$$dZ = \mu \, dt + \sigma \, dW \,,$$

with $Z_0 = z$. The value of the unlimited BOOST written on Z_t with barriers h and ℓ, $\ell < x < h$, is obtained by replacing in the previous formula h by $\dfrac{h-z}{\sigma}$, ℓ by $\dfrac{\ell-z}{\sigma}$ and λ by $\dfrac{\mu}{\sigma}$.

Remark 12. The formula (6.2) provides the price of the (symmetric) rebate of a double knock-out option with infinite maturity, that would be paid when knocking out. See Sec. for the rebate of double barrier options with finite maturity.

6.3. *BOOST with a time limit*

Let $M > 0$ and

$$\tau_M = \inf\left(\tau_{h,\ell}^\lambda, M\right)$$

be the stopping time of a BOOST with time limit M. Our aim is to compute the value of such a BOOST, that is

$$\mathbf{E}(e^{-r\tau_M}\,\tau_M) = \int_0^M t\,e^{-rt} P_{h,\ell}^\lambda(t)\,dt + M\,e^{-rM}\int_M^\infty P_{h,\ell}^\lambda(t)\,dt\,,$$

where $\varphi_{h,\ell}^\lambda$ is the distribution density of $\tau_{h,\ell}^\lambda$.

We shall apply the following Laplace transform inversion formula to compute the function $P_{h,\ell}^\lambda$:

$$\tilde{f}(x) = \int_0^\infty e^{-xt} f(t)\,dt = \frac{\sinh\left(\alpha\sqrt{x}\right)}{\sinh\left(\beta\sqrt{x}\right)}$$

when

$$f(t) = \frac{2\pi}{\beta^2}\sum_{n=1}^\infty (-1)^n\, n\, e^{-\frac{n^2\pi^2 t}{\beta^2}}\sin\frac{n\pi\alpha}{\beta}\,.$$

This formula is a limit case of Heaviside expansion theorem on Laplace transforms (see [1, p. 1021, 29.2.20]): let $a_1,\ldots,a_n \in \mathbb{C}$ and P be a polynomial of degree less than n, then the inverse Laplace transform of the rational function $P(x)/Q(x)$, where $Q(x) = \prod_{k=1}^n (x - a_k)$, is

$$\sum_{k=1}^n \frac{P(a_k)}{Q'(a_k)}\,\exp(a_k t)$$

applied to the product expansion (see [1, p. 85, 4.5.68]):

$$\sinh z = z\prod_{k=1}^\infty \left(1 + \frac{z^2}{k^2\pi^2}\right)$$

From the identity

$$(\widetilde{e^{-ct}f})(s) = \tilde{f}(s+c),$$

and from Eq. (6.2), we get

$$\varphi_{h,\ell}^{\lambda}(t) = \frac{\pi}{\delta^2} e^{-\frac{\lambda^2}{2}t} \sum_{n=1}^{\infty} (-1)^{n-1} n\, e^{-\frac{n^2\pi^2 t}{2\delta^2}} \left(e^{\lambda\ell} \sin\frac{n\pi h}{\delta} - e^{\lambda h}\sin\frac{n\pi\ell}{\delta}\right).$$

In order to compute $\mathbf{E}(e^{-r\tau_M}\,\tau_M)$, we observe that the previous series for $\varphi_{h,\ell}^{\lambda}(t)$ converges very fast (like e^{-an^2}) provided t is not too small. As we already know the value of the unlimited BOOST, it is sufficient to compute its difference with the limited one. Let I_M be defined by

$$\mathbf{E}(\tau_M e^{-r\tau_M}) = \mathbf{E}(\tau_{h,\ell}^{\lambda} e^{-r\tau_{h,\ell}^{\lambda}}) - I_M.$$

One has

$$I_M = \int_M^{\infty} (t\,e^{-rt} - M\,e^{-rM})\varphi_{h,\ell}^{\lambda}(t)\,dt,$$

and if we set $u_n = \dfrac{n^2\pi^2}{2\delta^2} + \dfrac{\lambda^2}{2}$, then

$$\int_M^{\infty} t\,e^{-rt}\varphi_{h,\ell}^{\lambda}(t)\,dt$$

$$= \frac{\pi}{\delta^2}\sum_{n=1}^{\infty}(-1)^n\, n\,\frac{1+(u_n+r)M}{(u_n+r)^2}\,e^{-(u_n+r)M}\left(e^{\lambda\ell}\sin\frac{n\pi h}{\delta} - e^{\lambda h}\sin\frac{n\pi\ell}{\delta}\right)$$

$$M\,e^{-rM}\int_M^{\infty}\varphi_{h,\ell}^{\lambda}(t)\,dt$$

$$= \frac{\pi}{\delta^2}\sum_{n=1}^{\infty}(-1)^n\, n\,\frac{M}{u_n}\,e^{-(u_n+r)M}\left(e^{\lambda\ell}\sin\frac{n\pi h}{\delta} - e^{\lambda h}\sin\frac{n\pi\ell}{\delta}\right)$$

We conclude that

$$I_M = \frac{\pi}{\delta^2}\sum_{n=1}^{\infty}(-1)^n\,\frac{rM(1+\frac{r}{u_n})-1}{(u_n+r)^2}\,e^{-(u_n+r)M}\left(e^{\lambda\ell}\sin\frac{n\pi h}{\delta} - e^{\lambda h}\sin\frac{n\pi\ell}{\delta}\right).$$

Remark 13. This expression for I_M is not singular when $\frac{\lambda^2}{2}+r=0$.

Remark 14. By modifying the drift λ, one can compute the value of a BOOST with barriers that exponentially depend on time: $H(t)=H_0 e^{\theta t}$, $L(t)=L_0 e^{\theta t}$.

Remark 15. When $M\to\infty$, the first term of the series provides the following equivalent for I_M:

$$I_M \sim \frac{\pi^2}{\delta^2}\,\frac{1-rM(1+\frac{r}{u_1})}{(u_1+r)^2}\,e^{-(u_1+r)M}\left(e^{\lambda\ell}\sin\frac{\pi h}{\delta} - e^{\lambda h}\sin\frac{\pi\ell}{\delta}\right)$$

with

$$u_1 = \frac{\pi^2}{2\delta^2} + \frac{\lambda^2}{2}$$

this equivalent is already a very good approximation when $M > \delta^2$.

Remark 16. If $M > \dfrac{u_1}{r(u_1 + r)}$, and in particular, if $M > \dfrac{1}{r}$, then $I_M < 0$ which means that the "unlimited" BOOST is cheaper than the limited one. This phenomenon comes from the fact that the BOOST reaches such a value that postponing the payment date by one day costs more than the day value.

6.4. *Symmetric rebate of double barrier options*

The price of rebates of double barrier options which are paid as soon as the option is knocked out is given by the expectation of the discount factor $e^{-r\tau_M}$:

$$\text{RDKO}_{\text{KO}}(S, L, H) = \mathbf{E}(e^{-r\tau_M}) = \int_0^M e^{-rt}\varphi_{h,\ell}^{\lambda}(t)\, dt\,,$$

where M denotes the option maturity. As for the BOOST, we shall compute its value by its complement to the "infinite" case given by formula (6.2). One has

$$\mathbf{E}(e^{-r\tau_M}) = \mathbf{E}(e^{-r\tau_{h,\ell}^{\lambda}}) - J_M\,,$$

with

$$J_M = \int_M^\infty e^{-rt}\,\varphi_{h,\ell}^{\lambda}(t)\, dt$$

$$= \frac{\pi}{\delta^2}\sum_{n=1}^{\infty}(-1)^n\, n\,\frac{e^{-(u_n+r)M}}{u_n+r}\left(e^{\lambda\ell}\sin\frac{n\pi h}{\delta} - e^{\lambda h}\sin\frac{n\pi\ell}{\delta}\right),$$

where, as previously,

$$u_n = \frac{n^2\pi^2}{2\delta^2} + \frac{\lambda^2}{2}\,.$$

It is left to the reader to check that this formula is consistent with the results of Sec. (see footnote 9).

Acknowledgments

The author wishes to thank Marc Yor and Albert Shiryaev for helpful discussions and bibliographical indications. He is also grateful to the teams of Société Générale and of Murex who implemented and checked all the formulas in this article.

References

[1] M. Abramowitz, I. S. Stegun, *Handbook of Mathematical Functions*, Dover, New York (1965) (1st Ed.) (1972) (9th Ed.). References are given with respect to the 9th edition.

[2] M. Avellaneda, *Capped-Delta valuation: managing stop-loss risk in the hedging of barrier options*, working paper, Courant Institute, New York Univ. (1996).

[3] M. Avellaneda, M. Levy and A. Paras, *Pricing and hedging derivative securities in markets with uncertain volatilities*, Appl. Math. Finance **2** (1995) 73–88.

[4] C. Bardos, R. Douady and A. V. Fursikov, *Static hedging of barrier options with a smile: An inverse problem*, Preprint C.M.L.A., Ecole Normale Sup. (1998).

[5] M. Broadie, P. Glasserman and S. Kou, *A continuity correction for discrete barrier options*, preprint, Columbia Univ. (1996).

[6] M. Broadie, P. Glasserman and S. Kou, *Connecting discrete and continuous path-dependent options*, preprint, Columbia Univ. (1996).

[7] A. Borodin and P. Salminen, *Handbook of Brownian Motion: Facts and Formulae*, Birkhauser, Berlin (1996).

[8] P. Carr and K. Ellis, *Non-standard valuation of barrier options*, preprint Cornell Univ. (1995).

[9] P. Carr, K. Ellis and V. Gupta, *Static hedging of exotic options*, preprint Cornell Univ. (1996).

[10] M. Chesney, B. Marois and R. Wojakowski, *Les options de change*, Economica, Paris, (1995).

[11] J. C. Cox and M. Rubinstein, *Option markets*, Prentice Hall, Englewood Cliffs NJ (1985).

[12] J. L. Doob, *Stochastic Processes*, J. Wiley and Sons, New York (1953).

[13] R. Douady, *Calcul du BOOST avec tendance et limite, taux et volatilité constants*, preprint, Société Générale (1994).

[14] R. Douady, *Optionsà limite et à double limite*, preprint, Société Générale (1995).

[15] N. El Karoui, *Evaluation et couverture des options exotiques*, working paper, Univ. Paris VI (1997).

[16] H. Geman and M. Yor, *Pricing and hedging double barrier options: a probabilistic approach*, working paper, ESSEC Finance Dept. and Univ. Paris VI (1996).

[17] I. V. Girsanov, *On transforming a certain class of stochastic processes by absolutely continuous substitution of measures*, Theory of Probability and Appl. **5** (1960) 285–301.

[18] I. Gikhman and A. Skorokhod, *Introduction to the Theory of Random Processes*, Mir, Moscow, (1977) (Russian), (1980) (French). English translation by Saunders, Philadelphia (1979).

[19] J. M. Harrison, *Brownian Motions and Stochastic Flows in Networks*, J. Wiley and Sons, New York (1985) (re-ed. Krieger (1990)).

[20] H. He, B. Keirstead and J. Rebholz *Double lookbacks*, Math. Finance (July 1998).

[21] N. Kunimoto and M. Ikeda, *Pricing options with curved boundaries*, Math. Finance **2**(4) (Oct. 1992) 275–298.

[22] D. Lamberton and B. Lapeyre, *Calcul stochastique appliqué à la finance*, Ellipse, Paris (1992).

[23] P. Lévy, *Processus stochastiques et mouvement brownien*, Gauthier-Villars, Paris (1953), 2nd Ed. (1965).

[24] R. C. Merton, *Theory of rational option pricing*, Bell J. Econ. Management Sci. **4** (1973) 141–183.

[25] I. Nelken, *Handbook of Exotic Options: Instruments, Analysis and Applications*, Irwin, Chicago IL (1995).

[26] N. U. Prabhu, *Stochastic Processes*, McMillan, New York (1965).

[27] W. Press, A. Teukolsky, W. Vetterling and B. Flannery, *Numerical Recipes in C*, Cambridge Univ. Press (1992) 2nd Ed. (1996).

[28] E. Reiner and M. Rubinstein, *Breaking down the barriers*, Risk Mag. **4** (Sept. 1991) 28–35.

[29] D. Revuz and M. Yor, *Continuous Martingales and Brownian Motions*, Springer-Verlag Berlin (1990).

[30] D. Rich, *The mathematical foundation of barrier option pricing theory*, AFOR **7** (1994) 267–312.

[31] L. C. G. Rogers and S. E. Satchell, *Estimating variance from high, low and closing prices*, Annals of Appl. Probability, **1** n°4, p. 504–512.

[32] M. Rubinstein, *Exotic options*, Finance Working Paper n°220, U. C. Berkeley, (1991).

[33] V. Seshadri, *The Inverse Gaussian Distribution*, Oxford Univ. Press, New York (1993).

[34] N. Taleb, *Dynamic Hedging* J. Wiley and Sons, New York (1997).

[35] P. Zhang, *Exotic options: A guide to the second generation options*, World Scientific Press, Singapore (1997).

Reprinted from J. Financial and Quantitative Analysis
33(3) (1998) 409–422

ASIAN OPTIONS, THE SUM OF LOGNORMALS, AND THE RECIPROCAL GAMMA DISTRIBUTION*

MOSHE ARYE MILEVSKY

*Schulich School of Business, York University,
4700 Keele Street, Ontario, M3J 1P3, Canada*

STEVEN E. POSNER

*Marsh & McLennan Securities Corp.,
114 West 47th Street, New York, NY 10036, USA*

Arithmetic Asian options are difficult to price and hedge as they do not have closed-form analytic solutions. The main theoretical reason for this difficulty is that the payoff depends on the *finite sum* of correlated lognormal variables, which is not lognormal and for which there is no recognizable probability density function. We use elementary techniques to derive the probability density function of the *infinite sum* of correlated lognormal random variables and show that it is reciprocal gamma distributed, under suitable parameter restrictions. A random variable is reciprocal gamma distributed if its inverse is gamma distributed. We use this result to approximate the *finite sum* of correlated lognormal variables and then value arithmetic Asian options using the reciprocal gamma distribution as the state-price density function. We thus obtain a closed-form analytic expression for the value of an arithmetic Asian option, where the cumulative density function of the gamma distribution, $G(d)$ in our formula, plays the exact same role as $N(d)$ in the Black–Scholes/Merton formula. In addition to being theoretically justified and exact in the limit, we compare our method against other algorithms in the literature and show our method is quicker, at least as accurate, and, in our opinion, more intuitive and pedagogically appealing than any previously published result, especially when applied to high yielding currency options.

1. Introduction and Motivation

There are two basic types of Asian options, the arithmetic Asian option and the geometric Asian option. The geometric Asian option has a payoff function that depends on the *product* of the prices of the underlying security at fixed points in its life. The payoff from an arithmetic Asian option is a function of the *sum* of the prices of the underlying security at fixed points in its life.

*This project was supported by a financial grant from the Social Sciences and Humanities Research Council of Canada and by the York University Research Authority. The authors acknowledge the helpful comments and suggestions of the seminar participants at the Fields Institute at the University of Toronto, the 8th Annual Derivative Securities Conference at Boston University, New York University, and York University, as well as Marco Avellaneda, Peter Carr, George Chacko, Sanjiv Ranjan Das, Aron Gottesman, Dilip Madan, Tom Salisbury, Marti Subrahmanyam, Hans Tuenter, Jonathan Karpoff (the editor), and Robert Jarrow (associate editor and referee). The usual disclaimer applies.

Geometric Asian options are relatively easy to price and hedge with known closed-form solutions in the Black–Scholes (1973) and Merton (1973) framework, (see for example, Turnbull and Wakeman (1991) and Kemna and Vorst (1990)), but this type of averaging is relatively uncommon and not used in practice. On the other hand, arithmetic Asian options are the most common type of Asian option but are quite difficult to price.

The main theoretical reason for this difficulty is that in the standard Black–Scholes environment, security prices are lognormally distributed. Consequently, the geometric Asian option is characterized by the correlated *product* of lognormal random variables, which is also, conveniently, lognormally distributed. As a result, the state-price density function is lognormal and hence the analytic no-arbitrage value of the option can be easily obtained using risk-neutral expectations. In contrast, the arithmetic Asian option depends on the finite *sum* of correlated lognormal random variables, which is clearly not lognormally distributed and for which there is no recognizable closed-form probability density function.

As a result, a variety of techniques have been developed in the literature to analyze arithmetic Asian options. Generally, they can be classified as follows:

(i) Monte Carlo simulations with variance reduction techniques (see Haykov (1993), Boyle (1977), Corwin *et al.* (1996), Kemna and Vorst (1990)).

(ii) Binomial trees and lattices with efficiency enhancements (see Hull and White (1993) or Neave and Turnbull (1994)).

(iii) The PDE approach (see Dewynne and Wilmott (1995), Rogers and Shi (1995), and Alziary, Decamps and Koehl (1997)).

(iv) General numerical methods (see Carverhill and Clewlow (1990), Curran (1994), and Nielsen and Sandman (1996)).

(v) Pseudo-analytic characterizations (see Geman and Yor (1993), Yor (1993), Kramkov and Mordecky (1994), as well as Ju (1997) and Chacko and Das (1997) for some recent extensions).

(vi) Analytic approximations that produce closed-form expressions (see Turnbull and Wakeman (1991), Levy (1992), Vorst (1992), Vorst (1996), and Bouaziz, Briys, and Crouhy (1994)).

Our work is in the spirit of the latter two classes. In this paper we use elementary techniques to obtain a probability density function for the *infinite sum* of correlated lognormal random variables and show that it is reciprocal gamma distributed (defined in Sec. 2). With this result, we are able to approximate the *finite sum* of correlated lognormal variables and then value arithmetic Asian options using the reciprocal gamma distribution as the state-price density function. We obtain a closed form analytic expression for the arithmetic Asian option where the cumulative density function of the Gamma distribution, $G(d)$ in our formula, plays the exact same role as $N(d)$ in the Black–Scholes formula. In addition to being theoretically justified and precise in the limit, when we compared our model against other algorithms in the literature, our method is quicker, at least as accurate, and in our opinion, more intuitive than any previously published result, especially when applied to high yielding currency options.

The remainder of the paper is organized as follows. Section 2 introduces notation and terminology, Sec. 3 (Theorem 1) derives the conditions under which the sum of correlated lognormal random variables converges to the reciprocal gamma distribution. Section 4 uses the reciprocal gamma as the state-price density to obtain a closed form analytic expression (Theorem 2) for the value of an arithmetic mean Asian call option. As well, Sec. 4 presents hedging parameters (greeks) for the arithmetic mean Asian call option. Section 5 provides a numerical example comparing our approximation to previous literature on arithmetic mean Asian options and Sec. 6 concludes the paper with some directions for further research. All proofs are in the Appendices.

2. Notation and Terminology

We assume the standard risk-neutral Geometric Brownian Motion (GBM) model,

$$dS_t = (r - q)S_t \, dt + \sigma S_t \, dB_t \iff$$

$$S_t = S_0 \exp\left\{ \left(r - q - \frac{1}{2}\sigma^2\right) t + \sigma B_t \right\}, \tag{1}$$

where r is the domestic interest rate; σ is the volatility; B_t is the Brownian motion driving the stochastic differential equation; and time t is measured in units of years. For equities, q is the dividend yield and for foreign exchange; S_t denotes the price of one unit of the foreign currency at time t; and q is the foreign risk-free rate. The process defined by Eq. (1) implies that the (one plus) total return from the stock, over any time period, is lognormally distributed.

2.1. Sums, means, and moments

The arithmetic Asian option is defined and structured over a finite interval of time denoted by $[0, T]$. The price of the underlying security is observed on N different occasions during the life of the option. The observation times are denoted by $0 \leq \tau(1) < \tau(2) < \cdots < \tau(N) \leq T$; which is general enough to include unequal time increments, forward-start options and continuous observations. We are currently at time $t \leq T$, which can also be an observation time, and define $j := \min\{i\}, t \leq \tau(i)$, which denotes the index counter for the next observation time. We then define

$$I_{[t,\tau(j),T]} := \frac{1}{S_t} \sum_{i=j}^{N} S_{\tau(i)}, \tag{2}$$

which represent the arithmetic sum of $N - j + 1$ *discrete* prices scaled by the current price. The middle subscript of $I_{[t,\tau(j),T]}$ indicates that, although the current time is t, the next averaging time is $\tau(j) \geq t$. Analogously, we define the arithmetic sum (integral) of *continuous* prices scaled by the current price, by

$$I_{[t,T]} := \frac{1}{S_t} \int_t^T S_u du = \frac{1}{S_t} \int_t^T \exp\left\{\left(r - q - \frac{1}{2}\sigma^2\right)t + \sigma B_u\right\} du. \tag{3}$$

We naturally dispense with the formality of $\tau(j)$ in the subscript of $I_{[t,T]}$, when integrating continuous prices, since, by definition, $t = \tau(j)$.

In a similar fashion, define the arithmetic mean of $N - j + 1$ *discrete* prices to be

$$A_{[t,\tau(j),T]} := \frac{1}{N-j+1} \sum_{i=j}^N S_{\tau(i)} = \frac{S_t}{N-j+1} I_{[t,\tau(j),T]}. \tag{4}$$

In continuous time,

$$A_{[t,T]} := \frac{1}{T-t} \int_t^T S_u du = \frac{S_t}{T-t} I_{[t,T]}. \tag{5}$$

Once again we drop $\tau(j)$ in the subscript to simplify notation when dealing with continuous averaging. Technically, at any time $t, 0 \leq t \leq T$, we can decompose the stochastic variable $A_{[0,\tau(1),T]}$ into the observed and unobserved parts,

$$A_{[0,\tau(1),T]} = \frac{t}{T} A_{[0,\tau(j-1),t]} + \frac{T-t}{T} A_{[t,\tau(j),T]}, \tag{6}$$

where $A_{[0,\tau(j-1),t]} = (\frac{1}{j-1}) \sum_{i=1}^{j-1} S_{\tau(i)}$ denotes the average observed to date and where the last averaging time was $\tau(j-1)$. An expression for the first two risk-neutral moments of $A_{[t,\tau(j),T]}$ can be found in Turnbull and Wakeman (1991), Nielsen and Sandman (1996) and the appendix of Levy (1992). In continuous time, using the Wiener measure $\mathbf{P}$ and Fubini's Theorem,[a]

$$E^*[A_{[t,T]}] = \int_\Omega \left[\frac{S_t}{T-t} \int_t^T \exp\left\{\left(r - q - \frac{1}{2}\sigma^2\right)u + \sigma B_u\right\} du\right] d\mathbf{P}$$

$$= \begin{cases} S_t \left(\dfrac{\exp\{(r-q)(T-t)\} - 1}{(r-q)(T-t)}\right), & \text{if } r \neq q, \\[2ex] S_t, & \text{if } r = q, \end{cases} \tag{7}$$

and

$$E^*[A_{[t,T]}^2] = \int_\Omega \left[\frac{S_t}{T-t} \int_t^T \exp\left\{\left(r - q - \frac{1}{2}\sigma^2\right)u + \sigma B_u\right\} du\right]^2 d\mathbf{P}$$

$$= \begin{cases} \dfrac{2S_t^2}{(T-t)^2} \left(\dfrac{\exp\left\{(2(r-q)+\sigma^2)(T-t)\right\}}{(r-q+\sigma^2)(2r-2q+\sigma^2)}\right. \\[2ex] \quad + \dfrac{1}{(r-q)} \left(\dfrac{1}{2(r-q)+\sigma^2} - \left.\dfrac{\exp\{(r-q)(T-t)\}}{r-q+\sigma^2}\right)\right), & \text{if } r \neq q, \\[2ex] \dfrac{2S_t^2}{(T-t)^2} \left(\dfrac{e^{\sigma^2(T-t)} - 1 - \sigma^2(T-t)}{\sigma^4}\right), & \text{if } r = q. \end{cases} \tag{8}$$

[a]An equivalent expression appears in Hull (1997, Chapter 18).

2.2. *Gamma and reciprocal gamma distributions*

A Gamma distributed random variable X has the probability density function (p.d.f.),

$$X \sim \mathbf{g}(x \mid \alpha, \beta) := \frac{\beta^{-\alpha} \, x^{\alpha-1} \, \exp\left\{-\frac{x}{\beta}\right\}}{\Gamma(\alpha)}, \quad \forall \, x > 0, \tag{9}$$

where $\alpha > 0$ and $\beta > 0$, and $\Gamma(\alpha)$ is the gamma function and satisfies $(\alpha - 1)\Gamma(\alpha - 1) = \Gamma(\alpha)$. The cumulative density function (c.d.f.) of X is denoted $\mathbf{G}(x \mid \alpha, \beta)$ and is available in most statistical software packages and commercial spreadsheets. Three easy-to-derive identities are used when proving some of our results:

$$\mathbf{G}(x|\alpha, \beta) = \mathbf{G}\left(\frac{x}{\beta}|\alpha, 1\right), \tag{10}$$

$$\mathbf{g}(x|\alpha, \beta) = \frac{1}{\beta}\mathbf{g}\left(\frac{x}{\beta}|\alpha, 1\right), \tag{11}$$

$$\mathbf{g}(x|\alpha, \beta) = \frac{x}{\beta(\alpha - 1)}\mathbf{g}(x|\alpha - 1, \beta), \quad \forall \, \alpha > 1. \tag{12}$$

The random variable $Y = \frac{1}{X}$ is reciprocal gamma distributed with c.d.f. and p.d.f.:

$$\mathbf{G_R}(y \mid \alpha, \beta) = \Pr(Y \leq y) = \Pr(1/Y > 1/y) \tag{13}$$

$$= 1 - \mathbf{G}(1/y \mid \alpha, \beta),$$

$$\mathbf{g_R}(y \mid \alpha, \beta) := \frac{\mathbf{g}(1/y \mid \alpha, \beta)}{y^2}, \quad \forall \, y > 0. \tag{14}$$

Using this duality, we will express all option pricing formulas using $\mathbf{G}(\cdot \mid \alpha, \beta)$ as opposed to $\mathbf{G_R}(\cdot \mid \alpha, \beta)$. The moments of the Reciprocal Gamma distribution are

$$M_i := E[Y^i] = E\left[\frac{1}{X^i}\right] = \frac{1}{\beta^i(\alpha - 1)(\alpha - 2)\cdots(\alpha - i)},$$

$$i = 1, 2, 3, \dots, \tag{15}$$

and are positive provided that $a > i$. We can invert the parameters α and β in terms of the first two (by definition) positive moments, so that

$$\alpha = \frac{2M_2 - M_1^2}{M_2 - M_1^2}, \qquad \beta = \frac{M_2 - M_1^2}{M_2 \, M_1}. \tag{16}$$

The purpose of this inversion is that we can compute the first two risk-neutral moments of the payoff function and then match the moments to the reciprocal gamma distribution, and then solve for the parameters that specify the p.d.f. In other words, if we want to approximate a random variable by the reciprocal gamma distribution, we can use Eq. (16) to locate the appropriate density parameters, provided that the first two moments are available.

3. Limiting Distribution

We now present the theoretical justification for using the reciprocal gamma distribution as the state-price density when pricing Asian options.

Theorem 1. *If* $r - \frac{1}{2}\sigma^2 - q < 0$ *in Eq. (1), then*

$$I_\infty := \lim_{T \to \infty} I_{[0,T]} \sim \mathbf{g_R}(\cdot \mid \alpha, \beta)\,,$$

which implies that

$$\Pr[I_\infty \geq w] := \lim_{T \to \infty} \Pr[I_{[0,T]} \geq w] = \lim_{T \to \infty} \Pr\left[\left(I_{[0,T]}\right)^{-1} \leq \frac{1}{w}\right]$$

$$= \mathbf{G}\left(\frac{1}{w} \mid \alpha, \beta\right),$$

with parameters $\alpha = 1 + 2(q - r)/\sigma^2 > 0$ *and* $\beta = \sigma^2/2$.

Proof of Theorem 1. See Appendix A. □

Despite the elegance and simplicity of the above theorem, the quantity $I_{[0,T]}$, which is the sum of correlated lognormals, converges to the reciprocal gamma distribution when both $N \to \infty$ and $T \to \infty$. Unfortunately, both of these conditions are unlikely to occur in practice. First, Asian (and all traded) options have a finite lifetime. Second, Asian options always have discrete observation periods and thus $N \ll \infty$.

Can we approximate the true distribution of $I_{[0,\tau(1),T]}$, and $A_{[0,\tau(1),T]}$ by the reciprocal gamma for finite values of N and T? Indeed, we argue that as a result of Theorem 1, it is a superior approximation in the finite case. First, it is possible to show that the c.d.f of the random variable $A_{[0,\tau(1),T]}$ is closer to the reciprocal gamma distribution than the lognormal distribution in the Kolmogorov–Smirnoff sense.[b] Second, we have compared arithmetic Asian option prices from different analytic state-price density approximations and Monte Carlo simulations to conclude (Table 1) that the reciprocal gamma approximation provides the closest fit for a wide range of (negative drift) parameter values.

---

[b]In particular, if we let $\mathbf{F}(x)$ denote the c.d.f. of the "true" random variable $A_{[0,T]}(n)$, and let $\mathbf{L_N}(x)$ denote the c.d.f. of the lognormal distribution and all three random variables have the same M_1 and M_2, then
$$\max_x |\mathbf{F}(x) - \mathbf{G_R}(x)| \leq \max_x |\mathbf{F}(x) - \mathbf{L_N}(x)|\,,$$
which we interpret to mean that the reciprocal gamma distribution is a better fit than the lognormal distribution. A detailed derivation is available from the authors upon request.

4. Option Pricing Using Risk-Neutral Valuation

The arithmetic Asian call option is a function of eight explicit parameters and is denoted by

$$ASC_t := ASC_t(S_t, A_{[0,\tau(j-1),t]}, K, t, T, \sigma, r, q),\tag{17}$$

where S_t is the current stock price, $A_{[0,\tau(j-1),t]}$ is the average observed to date, K is the strike price, $t \geq 0$ is the time (in years) since the start of the option, $T - t$ is the time remaining to maturity. The payoff at maturity from an arithmetic Asian call option is $\max[A_{[0,\tau(1),T]} - K, 0]$. By the fundamental theorem of option pricing, the no-arbitrage value of the Asian call option is the discounted expectation of the payoff using the risk-neutral measure:

$$ASC_t = e^{-r(T-t)} E^*[\max(A_{[0,\tau(1),T]} - K, 0)].\tag{18}$$

The following theorem finds the price of the Asian option by evaluating this expectation under the risk-neutral measure, induced by the reciprocal gamma distribution.

Theorem 2. *The no-arbitrage value of an arithmetic Asian option is*

$$ASC_t = e^{-r(T-t)} \frac{T-t}{T} E^*[A_{[t,\tau(j),T]}]$$

$$\cdot \mathbf{G}\left(\frac{T-t}{TK - t A_{[0,\tau(j-1),t]}} \bigg| \alpha - 1, \beta\right)$$

$$- e^{-r(T-t)}\left(K - \frac{t}{T} A_{[0,\tau(j-1),t]}\right)$$

$$\cdot \mathbf{G}\left(\frac{T-t}{TK - t A_{[0,\tau(j-1),t]}} \bigg| \alpha, \beta\right),\tag{19}$$

when $A_{[0,\tau(j-1),t]} < (T/t)K$, *and*

$$ASC_t = e^{-r(T-t)} \frac{T-t}{T} E^*[A_{[t,\tau(j),T]}] - e^{-r(T-t)}\left(K - \frac{t}{T} A_{[0,\tau(j-1),t]}\right)\tag{20}$$

when $A_{[0,\tau(j-1),t]} \geq (T/t)K$.

In particular, for continuous averaging, $E^*[A_{[t,T]}] = S_t((e^{(r-q)(T-t)} - 1)/((r - q)(T - t)))$, *and so*

$$ASC_t = \frac{e^{-q(T-t)} - e^{-r(T-t)}}{(r-q)T} S_t \cdot \mathbf{G}\left(\frac{T-t}{TK - t A_{[0,t]}} \bigg| \alpha - 1, \beta\right)$$

$$- e^{-r(T-t)}\left(K - \frac{t}{T} A_{[0,t]}\right) \cdot \mathbf{G}\left(\frac{T-t}{TK - t A_{[0,t]}} \bigg| \alpha, \beta\right)\tag{21}$$

and, therefore, at $t = 0$,

$$ASC_0 = \frac{e^{-qT} - e^{-rT}}{(r-q)T} S_0 \mathbf{G}\left(\frac{1}{K}\bigg|\alpha - 1, \beta\right) - e^{-rT} K \mathbf{G}\left(\frac{1}{K}\bigg|\alpha, \beta\right). \quad (22)$$

Proof of Theorem 2. See Appendix B. □

We emphasize that α, β are themselves functions of S_t, r, q, σ^2, and $T - t$, via Eq. (16). Equation (22) is remarkably similar to the original Black–Scholes (1973) equation, where $\mathbf{G}(1/K|\alpha - 1, \beta)$ plays the role of $\mathbf{N}(d_1)$ and $\mathbf{G}(1/K|\alpha, \beta)$ plays the role of $\mathbf{N}(d_2)$ which is the probability of maturing in the money, in the risk-neutral world. By Theorem 1, Eq. (21) converges to the true option pricing formula when $T \to \infty$ and $r - q < \frac{1}{2}\sigma^2$. Thus, for example, when pricing (long-dated) Asian options on foreign exchange with a high interest rate relative to the domestic interest rate, our methodology should perform at its best. The intuition for the case when $A_{[0,\tau(j-1)],t]} \geq (T/t)K$ is that if the average observed to date $A_{[0,\tau(j-1)],t]}$ is high enough, the option will expire in the money with probability one.

4.1. *Hedging parameters and the greeks*

For continuous averaging, taking first and second derivatives of Eq. (21) involves substantial algebra and the use of identities (10)–(12), but the final expression is simplified to

$$\Delta := \begin{cases} \dfrac{e^{-q(T-t)} - e^{-r(T-t)}}{(r-q)T} \mathbf{G}\left(\dfrac{T-t}{TK - tA_{[0,t]}}\bigg|\alpha - 1, \beta\right), & A_{[0,t]} < \dfrac{T}{t}K \\[4mm] \dfrac{e^{-q(T-t)} - e^{-r(T-t)}}{(r-q)T}, & A_{[0,t]} \geq \dfrac{T}{t}K, \end{cases} \quad (23)$$

and

$$\Gamma := \begin{cases} \dfrac{e^{-r(T-t)}(T-t)}{TS_t^2} \mathbf{g}\left(\dfrac{T-t}{TK - tA_{[0,t]}}\bigg|\alpha, \beta\right), & A_{[0,t]} < \dfrac{T}{t}K \\[4mm] 0, & A_{[0,t]} \geq \dfrac{T}{t}K. \end{cases} \quad (24)$$

Qualitatively, when the option is guaranteed to finish in-the-money, the formula for delta gives an explicit expression for how to "unwind the hedge" independently of the spot S_t, and is gamma neutral. If the option's moneyness is uncertain at time t, then the delta and the gamma depend crucially on the spot and the current average $A_{[0,t]}$.

5. Numerical Comparisons

Table 1 compares values for FX arithmetic Asian call options from our reciprocal gamma formula to the results from (i) a Monte Carlo simulation, with standard

Table 1. Theoretical model values for arithmetic Asian call options on foreign exchange, alternative strike, volatility, averaging frequency, maturity, and drift parameter values; relative error comparison with Monte Carlo simulation.

Panel A. Risk Free = 6%; FX Price = 100; Foreign Interest Rate = 18%; Forward Price = 88.7; Maturity = 1 Year, Weekly Averaging, N = 52 Observations, Start Next Week

Vol.	Strike	M.C.	S.Error	Vorst	Levy	ModGeom	RG
10%	80	13.306	0.001	13.3060	13.3055	13.3060	13.3046
10%	90	4.508	0.003	4.5325	4.5116	4.5329	4.4992
10%	100	0.409	0.002	0.4174	0.4001	0.4176	0.4112
20%	80	13.623	0.004	13.6578	13.6433	13.6619	13.6015
20%	90	6.218	0.008	6.2737	6.2350	6.2836	6.1931
20%	100	2.032	0.007	2.0502	2.0113	2.0595	2.0339
30%	80	14.571	0.010	14.6486	14.6343	14.6748	14.5070
30%	90	8.103	0.013	8.1682	8.1431	8.2072	8.0499
30%	100	3.961	0.012	3.9695	3.9417	4.0074	3.9490
40%	80	15.883	0.016	15.9769	16.0060	16.0543	15.7608
40%	90	10.029	0.018	10.0763	10.1090	10.1732	9.9302
40%	100	5.970	0.017	5.9430	5.9718	6.0348	5.9260
Average Deviation from Monte Carlo:				0.636%	0.566%	1.004%	0.434%

Panel B. Risk Free = 6%; FX Price = 100; Foreign Interest Rate = 18%; Forward Price = 69.8; Maturity = 3 Years, Monthy Averaging, N = 36 Observations, Start Next Month

Vol.	Strike	M.C.	S.Error	Vorst	Levy	ModGeom	RG
10%	70	11.405	0.002	11.4337	11.4128	11.4354	11.3980
10%	85	2.186	0.005	2.2966	2.1862	2.3041	2.1863
10%	100	0.108	0.001	0.1214	0.0972	0.1230	0.1128
20%	70	12.503	0.009	12.6749	12.5777	12.7118	12.4466
20%	85	4.901	0.012	5.0884	4.9245	5.1422	4.8790
20%	100	1.510	0.008	1.5828	1.4609	1.6194	1.5334
30%	70	14.324	0.019	14.5741	14.5124	14.7231	14.1810
30%	85	7.608	0.019	7.8025	7.7033	7.9787	7.5205
30%	100	3.791	0.016	3.8578	3.7556	4.0061	3.7849
40%	70	16.394	0.029	16.6092	16.7582	16.9819	16.1113
40%	85	10.289	0.029	10.3865	10.5253	10.7965	10.0683
40%	100	6.368	0.026	6.3243	6.4295	6.6962	6.2560
Average Deviation from Monte Carlo:				3.093%	1.924%	5.065%	1.268%

errors in brackets, (ii) the Vorst (1992) method, (iii) the Levy (1992) method, and (iv) the modified geometric method, described in Levy and Turnbull (1992).[c] We include the average relative error for each method as a means of comparison.

[c]For comparison purposes, we selected pricing algorithms that are at the same level of computational complexity as the methodology we propose. Further comparisons with alternative parameter values are available from the authors upon request.

Table 1. (*Continued*)

Panel C. Risk Free = 6%; FX Price = 100; Foreign Interest Rate = 18%; Forward Price = 54.88; Maturity = 5 Years, Monthly Averaging, N = 60 Observations, Start Next Month

Vol.	Strike	M.C.	S.Error	Vorst	Levy	ModGeom	RG
10%	60	11.051	0.002	11.0932	11.0632	11.1026	11.0457
10%	75	2.615	0.005	2.7868	2.6230	2.8267	2.6125
10%	90	0.222	0.002	0.2682	0.2064	0.2827	0.2315
20%	60	12.077	0.010	12.3128	12.1875	12.4147	12.0215
20%	75	5.276	0.012	5.5592	5.3340	5.7059	5.2440
20%	90	1.962	0.009	2.1213	1.9231	2.2351	1.9805
30%	60	13.791	0.020	14.1053	14.0759	14.4308	13.6374
30%	75	7.902	0.021	8.1771	8.0917	8.8509	7.7834
30%	90	4.395	0.018	4.5426	4.4279	4.8850	4.3549
40%	60	15.730	0.033	15.9349	16.2940	16.6696	15.4097
40%	75	10.477	0.032	10.5720	10.9187	11.3796	10.1843
40%	90	7.010	0.029	6.9698	7.2533	7.7264	6.8087
Average Deviation from Monte Carlo:				4.581%	2.335%	9.131%	1.463%

Panel D. Risk Free = 6%; FX Price = 100; Foreign Interest Rate = 4%; Forward Price = 102.0; Maturity = 1 Year, Weekly Averaging, N = 52 Observations, Start Next Week

Vol.	Strike	M.C.	S.Error	Vorst	Levy	ModGeom	RG
10%	95	6.083	0.003	6.0826	6.0874	6.0838	6.0734
10%	100	2.735	0.004	2.7258	2.7341	2.7280	2.7282
10%	105	0.884	0.003	0.8678	0.8748	0.8696	0.8847
20%	95	7.722	0.008	7.7155	7.7436	7.7284	7.6901
20%	100	4.933	0.009	4.9058	4.9385	4.9206	4.9132
20%	105	2.935	0.008	2.8886	2.9201	2.9027	2.9303
30%	95	9.677	0.014	9.6444	9.7267	9.6898	9.6099
30%	100	7.137	0.014	7.0725	7.1608	7.1206	7.0897
30%	105	5.130	0.013	5.0354	5.1228	5.0825	5.1049
40%	95	11.716	0.020	11.6291	11.8118	11.7378	11.5941
40%	100	9.336	0.020	9.2053	9.3963	9.3178	9.2394
40%	105	7.355	0.019	7.1854	7.3758	7.2964	7.2884
Average Deviation from Monte Carlo:				0.990%	0.396%	0.482%	0.521%

Levy = JIMF (1992), Vorst = IRFA (1992), M.C. = Monte Carlo Simulation, N = 200,000, ModGeom = Modified Geometric Method, RG = Reciprocal Gamma Distribution

The results are consistent with Theorem 1. The table illustrates that the RG methodology is robust for a wide variety of volatility, strike, averaging and maturity values. Specifically, in the negative drift case, for the three maturities analyzed, the average RG price was the closest to the Monte Carlo and consistently within the standard error. Furthermore, in the case where the drift was positive, the

RG method performed quite reasonably, compared to the other techniques, and dominated the Vorst (1992) method.

6. Conclusion

Knowledged practitioners and academics are well aware that the finite sum of correlated lognormal random variables is not lognormal. This issue arises in portfolio theory, asset pricing, and in the field of derivative securities. In particular, there are many options whose payoffs depend on the finite sum of correlated lognormal variables. The most common such security is the arithmetic Asian option where the payoff is based on the path taken by the underlying security during its life. The arithmetic average is a scaled sum of correlated lognormal random variables for which there is no recognizable density function. Consequently, there is no closed form expression for valuing arithmetic Asian options. In this paper we use elementary techniques to obtain a density function for the *infinite sum* of correlated lognormal random variables and show that it is reciprocal gamma distributed when the underlying diffusion has a negative drift. We then approximate the finite sum of correlated lognormal variables and value arithmetic Asian options using the reciprocal gamma distribution, together with moment matching, as the state-price density function. Our contribution is that we obtain a closed form analytic expression for the no-arbitrage value of the arithmetic Asian option, where the cumulative density function of the gamma distribution, $G(d)$ in our formula, plays the exact same role as $N(d)$ in the Black–Scholes formula. Further research by the authors, Posner and Milevsky (1998), expands this methodology to exotic options on more general underlying processes by locating suitable finite-time approximations to the state-price density.

Appendix A: Proof of Theorem 1

The quantity I_∞ and its relationship to the gamma distribution has been studied in the actuarial literature by Dufresne (1990), Yor (1993), and DeSchepper, Teunen and Goovaerts (1994), as well as in the economics literature by Merton (1975) and Majumdar and Radner (1991).

In this appendix, we provide an elementary proof of this result by showing that $P[I_\infty^{-1} \leq 1/w]$ has a functional form that is identical to $\mathbf{G}(1/w|1+2(q-r)/\sigma^2, \sigma^2/2)$ and, thus, by definition, I_∞ is reciprocal gamma distributed and I_∞^{-1} is gamma distributed. Define the diffusion process:

$$dW_t := ((q - r + \sigma^2) W_t - 1)dt - \sigma W_t dB_t, \tag{25}$$

where $r > 0, q > 0, \sigma > 0$, and $W_0 = w > 0$ is the value of the process at time $t = 0$. The solution to the stochastic differential equation is

$$W_t = e^{(q-r+\frac{1}{2}\sigma^2)t - \sigma B_t} \left[w - \int_0^t e^{(r-q-\frac{1}{2}\sigma^2)s + \sigma B_s} ds \right]. \tag{26}$$

See the textbook by Karatzas and Shreve ((1992), Chapter 5.6) for proof and details. Note that we constructed the arbitrary diffusion in Eq. (25), so that the integral in Eq. (26) will exactly equal the quantity we are interested in, namely $I_{[0,t]}$. Furthermore, since $e^{(q-r+\frac{1}{2}\sigma^2)t-\sigma B_t} \geq 0$, we have for any t^*

$$W_{t^*} \leq 0 \quad \Longleftrightarrow \quad w \leq \int_0^{t^*} e^{(r-q-\frac{1}{2}\sigma^2)s-\sigma B_s} ds \,. \tag{27}$$

Also note that, once the process W_t crosses zero, it never crosses zero again. This results from the non-decreasing monotonicity of the integral $\int_0^{t^*} e^{(r-q-\frac{1}{2}\sigma^2)s-\sigma B_s} ds$ with respect to t^*. Consequently, the probability that W_t crosses zero, prior to some time t^*, is equivalent to the probability that $W_{t^*} \leq 0$ i.e.,

$$\Pr[\inf\{W_s; 0 \leq s \leq t^*\} \leq 0] = \Pr[W_{t^*} \leq 0]\,. \tag{28}$$

Hence we have for $t^* = \infty$

$$\Pr[\inf\{W_s; 0 \leq s \leq \infty\} \leq 0] = \Pr[W_\infty \leq 0]$$

$$= \Pr\left[w \leq \int_0^\infty e^{(r-q-\frac{1}{2}\sigma^2)s+\sigma B_s} ds\right]$$

$$= \Pr\left[\left(\int_0^\infty e^{(r-q-\frac{1}{2}\sigma^2)s+\sigma B_s} ds\right)^{-1} \leq \frac{1}{w}\right]$$

$$= \Pr\left[I_\infty^{-1} \leq \frac{1}{w}\right]\,. \tag{29}$$

To complete the proof, the next lemma finds the $\Pr[\inf\{W_s; 0 \leq s \leq \infty\} \leq 0]$ for arbitrary values of w, thus obtaining an expression for $\Pr[I_\infty^{-1} \leq 1/w]$. See Milevsky (1998) for a generalization of this technique and it application to other density functions.

Lemma 3. $\Pr[\inf\{W_s; 0 \leq s \leq \infty\} \leq 0] = \mathbf{G}(1/w|1 + 2(q-r)/\sigma^2, \sigma^2/2)$.

Proof of Lemma. We use a *scale function* to compute the probability that the diffusion W_t, will exit the interval $(0, \infty)$. Defining $Y_t = f(W_t)$, by Ito's lemma,

$$df(W_t) = f'(W_t)dW_t + \frac{1}{2}f''(W_t)d\langle W\rangle_t\,, \tag{30}$$

where $d\langle W\rangle_t = \sigma^2 W_t^2 dt$. Thus,

$$df(W_t) = \left[f'(W_t)\left((q-r+\sigma^2)W_t - 1\right) + \frac{1}{2}f''(W_t)\sigma^2 W_t^2\right]dt$$

$$+ f'(W_t)\sigma W_t dB_t\,. \tag{31}$$

The diffusion $Y_t = f(W_t)$ will be a martingale if and only if

$$f'(W_t)((q-r+\sigma^2)W_t - 1) + \frac{1}{2}f''(W_t)\sigma^2 W_t^2 = 0\,, \tag{32}$$

or, equivalently, the *scale* function $f(w)$ must satisfy the ordinary differential equation,

$$f'(w)((q - r + \sigma^2)\, w - 1) = -\frac{1}{2}f''(w)\sigma^2 w^2 , \tag{33}$$

subject to any arbitrary initial condition. The solution to (33) is

$$f(w) = \int_{2/w\sigma^2}^{\infty} z^{\left(\frac{2(q-r)}{\sigma^2}\right)} \exp(-z)dz := \Gamma\left(1 + \frac{2(q-r)}{\sigma^2}, \frac{2}{w\sigma^2}\right), \tag{34}$$

which is the definition of the incomplete gamma function. We have mapped the original diffusion process W_t, into a martingale Y_t, via the scale function $f(w)$. The probability that a martingale with initial position y, defined on a state space $(0, l)$, will ever hit zero, is $1 - y/l$. Since $f(\cdot)$ is monotonic and $f(0) = 0$, the probability that the process W_t ever hits zero is the same as the probability of $f(W_t)$ hitting zero:

$$\Pr[\inf\{W_s; 0 \le s \le \infty\} \le 0]$$

$$= \Pr[\inf\{f(W_s); 0 \le s \le \infty\} \le 0] \tag{35}$$

$$= 1 - \frac{f(w)}{f(\infty)}$$

$$= \frac{\Gamma\left(1 + \frac{2(q-r)}{\sigma^2}, 0\right) - \Gamma\left(1 + \frac{2(q-r)}{\sigma^2}, \frac{2}{w\sigma^2}\right)}{\Gamma\left(1 + \frac{2(q-r)}{\sigma^2}, 0\right)}, \tag{36}$$

which, as per Eq. (34), can be written as

$$\Pr[\inf\{W_s; 0 \le s \le \infty\} \le 0]$$

$$= \frac{\int_0^{2/w\sigma^2} z^{\left(\frac{2(q-r)}{\sigma^2}\right)} \exp(-z)dz}{\Gamma\left(1 + \frac{2(q-r)}{\sigma^2}\right)}$$

$$= \int_0^{1/w} \frac{(\sigma^2/2)^{-\left(1 + \frac{2(q-r)}{\sigma^2}\right)} u^{\left(\frac{2(q-r)}{\sigma^2}\right)} \exp\left\{-\frac{u}{\sigma^2/2}\right\}}{\Gamma\left(1 + \frac{2(q-r)}{\sigma^2}\right)} du , \tag{37}$$

which is equivalent to the functional form of the gamma c.d.f. with parameters $\alpha = 1 + 2(q-r)/\sigma^2$ and $\beta = \sigma^2/2$. Hence, $\Pr[\inf\{W_s; 0 \le s \le \infty\} \le 0] = \mathbf{G}(1/w|1 + 2(q-r)/\sigma^2, \sigma^2/2)$. Finally, $\Pr[I_\infty^{-1} \le 1/w] = \Pr[\inf\{W_s; 0 \le s \le \infty\} \le 0]$, our main result follows. $\qquad\square$

Appendix B: Proof of Theorem

Using Eq. (6), we can re-write ASC_t, in terms of the unobserved average $A_{[t,\tau(j),T]}$ and the observed average to date $A_{[0,\tau(j-1),t]}$, as

$$ASC_t = e^{-r(T-t)} E^* \left[\max \left(\frac{T-t}{T} A_{[t,\tau(j),T]} - \left(K - \frac{t}{T} A_{[0,\tau(j-1),t]} \right), 0 \right) \right]$$

$$= e^{-r(T-t)} \frac{(T-t)}{T} E^* \left[\max \left(A_{[t,\tau(j),T]} - K', 0 \right) \right], \tag{38}$$

with the adjusted strike price $K' = (T K - t A_{[0,\tau(j-1),t]})/(T-t)$, which is applicable to both discrete and continuous-time averaging. We approximate the state-price density for $A_{[t,\tau(j),T]}$ with $g_R(y|\alpha, \beta)$. Hence, using risk-neutral valuation,

$$ASC_t = e^{-r(T-t)} \frac{T-t}{T} \int_{K'}^{\infty} (y - K') \, g_R(y|\alpha, \beta) dy, \tag{39}$$

where the parameters are from Eq. (16). Now, using the change of variables $x = \frac{1}{y}$ we get

$$ASC_t = e^{-r(T-t)} \frac{T-t}{T} \left[\int_0^{1/K'} \left(\frac{1}{x} - K' \right) g \left(x | \alpha, \beta \right) dx \right]$$

$$= e^{-r(T-t)} \frac{T-t}{T} \left[\int_0^{1/K'} \frac{g(x|\alpha, \beta)}{x} dx \right]$$

$$- e^{-r(T-t)} \frac{T-t}{T} \left[K' \int_0^{1/K'} g \left(x | \alpha, \beta \right) dx \right]. \tag{40}$$

Using the identities (10)–(12), as well as (15), we have

$$ASC_t = e^{-r(T-t)} \frac{T-t}{T} \left[\frac{1}{\beta(\alpha-1)} G \left(\frac{1}{K'} \middle| \alpha - 1, \beta \right) \right.$$

$$\left. - K' G \left(\frac{1}{K'} \middle| \alpha, \beta \right) \right]$$

$$= e^{-r(T-t)} \frac{T-t}{T} \left[E^* \left[A_{[t,\tau(j),T]} \right] G \left(\frac{1}{K'} \middle| \alpha - 1, \beta \right) \right.$$

$$\left. - K' G \left(\frac{1}{K'} \middle| \alpha, \beta \right) \right]. \tag{41}$$

This only makes sense for $1/K' > 0$, which means $A_{[0,\tau(j-1),t]} < TK/t$. As $A_{[0,t]} \nearrow (TK)/t$, both $G(1/K'|\cdot, \beta)$ functions become 1. $\qquad \square$

References

B. Alziary, J. Decamps and P. Koehl, "A P.D.E. Approach to Asian Options: Analytical and Numerical Evidence", *Journal of Banking and Finance* **21** (1997) 613–640.

F. Black and M. S. Scholes, "The Pricing of Options and Corporate Liabilities", *Journal of Political Economy* **81** (1973) 637–653.

L. Bouaziz, E. Briys and M. Crouhy, "The Pricing of Forward–Start Asian Options", *Journal of Banking and Finance* **18** (1994) 823–839.

P. Boyle, "Options: A Monte Carlo Approach", *Journal of Financial Economics* **4** (1977) 323–338.

A. Carverhill and L. Clewlow, "Flexible Convolution", *RISK* **5** (April 1990) 25–29.

G. Chacko and S. Das, "Average Interest", Working Paper, Harvard Business School (1997).

J. Corwin, P. Boyle and K. Tan, "Quasi-Monte Carlo Methods Numerical Finance", *Management Science* **42** (June 1996) 926–938.

M. Curran, "Valuing Asian and Portfolio Options by Conditioning on the Geometric Mean", *Management Science* **40** (12, 1994) 1705–1711.

A. DeSchepper, M. Teunen and M. Goovaerts, "An Analytic Inversion of a Laplae Transform Related to Annuities Certain", *Insurance: Mathematics and Economics* **14** (1, 1994) 33–37.

J. Dewynne and P. Wilmott, "Asian Options as Linear Complementarity Problems", *Advances in Futures and Options Research* **8** (1995) 145–173.

D. Durfresne, "The Distribution of a Perpetuity with Applications to Risk Theory and Pension Funding", *Scandinavian Actuarial Journal* **9** (1990) 39–79.

H. Geman and M. Yor, "Bessel Processes, Asian Options and Perpetuities", *Mathematical Finance* **3** (4, Oct. 1993) 349–375.

J. Haykov, "A Better Control Variate for Pricing Standard Asian Options", *Journal of Financial Engineering* **2** (1993) 207–216.

J. Hull and A. White, "Efficient Procedures for Valuing European and American Path Dependent Options", *Journal of Derivatives* **1** (Fall 1993) 21–31.

J. C. Hull, *Options, Futures and Other Derivatives*, Third ed. Englewood Cliffs, NJ: Prentice Hall (1997).

N. Ju, "Fourier Transformation, Martingale, and the Pricing of Average Rate Derivatives", Ph.D. Thesis, Univ. of California–Berkeley (1997).

I. Karatzas and S. E. Shreve, *Brownian Motion and Stochastic Calculus*, 2nd ed. New York, NY: Springer-Verlag (1992).

A. Kemna and A. Vorst, "A Pricing Method for Options Based on Average Values", *Journal of Banking and Finance* **14** (1990) 113–129.

D. Kramkov and E. Mordecky, "Integral Options", *Theory of Probability and Its Applications* **39** (June 1994) 162–171.

E. Levy, "Pricing European Average Rate Currency Options", *Journal of International Money and Finance* **11** (1992) 474–491.

E. Levy and S. Turnbull, "Average Intelligence", *RISK* **5** (Feb. 1992) 53–59.

M. Majumdar and R. Radner, "Linear Models of Economic Survival under Production Uncertainty", *Economic Theory* **1** (1991) 13–30.

R. Merton, "An Asymptotic Theory of Growth Under Uncertainty", *Review of Economic Studies* **42** (1975) 375–393; "Theory of Rational Option Pricing", *Bell Journal of Economics and Management Science* **4** (1973) 141–183.

M. A. Milevsky, "Martingales, Scale Functions and Stochastic Life Annuities", *Insurance: Mathematics and Economics* (forthcoming 1998).

E. Neave and S. Turnbull, "Quick Solutions for Arithmetic Average Options on a Recombining Random Walk", *4th Actuarial Approach for Financial Risks International Colloquium*, (1993) 718–739.

J. A. Nielsen and K. Sandman, "The Pricing of Asian Options under Stochastic Interest Rates", *Applied Mathematical Finance* **3** (1996) 209–236.

S. E. Posner and M. A. Milevsky, "Valuing Exotic Options by Approximating the SPD with Higher Moments", *Journal of Financial Engineering* **7**(2) (June 1998).

L. Rogers and Z. Shi, "The Value of an Asian Option", *Journal of Applied Probability* **32** (1995) 1077–1088.

S. Turnbull and L. Wakeman, "A Quick Algorithm for Pricing European Average Options", *Journal of Financial and Quantitative Analysis* **26** (1991) 377–389.

T. Vorst, "Prices and Hedge Ratios of Average Exchange Rate Options", *International Review of Financial Analysis* **1** (3, 1992) 179–193; "Averaging Options", in I. Nelken, ed. *The Handbook of Exotic Options*, Homewood, IL: Irwin (1996) 175–199.

M. Yor, "From Planar Brownian Windings to Asian Options", *Insurance: Mathematics and Economics* **13** (1993) 23–34.

Reprinted from Rev. Financial Studies
9(1) (1996) 277–300
© Oxford Univ. Press

PRICING AND HEDGING AMERICAN OPTIONS:
A RECURSIVE INTEGRATION METHOD

JING-ZHI HUANG and MARTI G. SUBRAHMANYAM

Stern School of Business, Finance Department, New York University,
44 West 4th Street, Suite 9-190, New York, NY 10012, USA

G. GEORGE YU

Goldman, Sachs & Co., 85, Brood Street, New York,
NY 10004, USA

In this article, we present a new method for pricing and hedging American options
along with an efficient implementation procedure. The proposed method is efficient
and accurate in computing *both* option values and various option hedge parameters.
We demonstrate the computational accuracy and efficiency of this numerical procedure
in relation to other competing approaches. We also suggest how the method can be
applied to the case of any American option for which a closed-form solution exists for
the corresponding European option.

A variety of financial products such as fixed-income derivatives, mortgage-backed
securities, and corporate securities have early exercise or American-style features
that significantly influence their valuation and hedging. Considerable interest exists,
therefore, in both academic and practitioner circles, in methods of valuation and
hedging of American-style options that are conceptually sound as well as efficient
in their implementation.

It has been recognized early in the modern academic literature on options pricing that, in general, closed-form solutions to the problems of valuation and hedging
of American-style options are very difficult, if not impossible, to achieve. Hence,
many researchers have focused on numerical methods to solve the problems of valuation and hedging of such options. These methods can be divided into two broad
categories.

The first approach is a solution to the integral equation, where the option value
is written as the expected value, under the risk neutral probability measure, of the
option payoffs. In the case of American-style options, the computation of the conditionally expected value on a given date is subject to the condition that the option

We acknowledge helpful comments from Menachem Brenner, Ren-Raw Chen, Bent Christensen,
Steven Feinstein, Steven Figlewski, Bin Gao, Bjarne Jensen, Herb Johnson, and Nianqin. We
thank Chi-fu Huang, the former editor, and especially an anonymous referee for many helpful
comments and suggestions that have greatly improved this article. We also thank Franklin Allen,
the executive editor, for editorial suggestions.

has not been exercised previously. This approach is in the spirit of the risk neutral valuation approach suggested by Cox and Ross (1976) and formalized by Harrison and Kreps (1979) and others. It is often implemented in practice in the context of the binomial approach of Cox, Ross, and Rubinstein (1979). A variation on the integral equation approach is suggested by Geske and Johnson (1984), who provide a numerical approximation based on the analytical valuation formulae for options exercisable on discrete dates. Their approach uses the exact analytical expressions for options exercisable at one, two, three, and perhaps four dates, to derive an approximate price of an American-style option that is continuously exercisable.

The second approach is to directly solve the Black and Scholes (1973) partial differential equation, subject to the boundary conditions imposed by the possibility of early exercise. This approach is implemented by using numerical approximation methods. For instance, the finite-difference methods of Brennan and Schwartz (1977), Courtadon (1982), and Schwartz (1977) and the analytical approximation methods of Barone-Adesi and Whaley (1987) and MacMillan (1986) are applications of this approach.

Although both types of methods have been used extensively in the academic literature and in practice, they have some limitations. First, as pointed out by Omberg (1987a), some methods such as the binomial method and the Geske and Johnson (1984) extrapolation scheme may not yield uniform convergence in option prices. Second, many of these numerical methods use a perturbation scheme to compute the hedge parameters. Such a scheme requires fairly intensive computation to obtain accurate results. Third, in some cases, such as the explicit finite-difference scheme, the method may not always converge. Finally, it has been recognized that analytical approximations may have large pricing errors for long-dated options.

These difficulties, as well as the desire for a more elegant mathematical representation, have led to a search for an analytical framework for American-style options. In a sense, this is the continuous-time extension of the Geske and Johnson (1984) approach. This solution is discussed by Kim (1990), Carr, Jarrow, and Myneni (1992) and Jacka (1991). These authors obtain the analytical American option pricing formulae under the assumption of a lognormal diffusion for the underlying security price. Kim and Yu (1993) consider some alternative underlying price processes in the same framework.

Two facets of the literature on the analytical valuation of American-style options should be noted. First, much of the research has focused on *standard* American options on stocks such as simple put or call options. Increasingly there is a need to explore how these formulae can be extended to the cases of other American-style claims, such as exotic options. Second, none of the above studies on these formulae offers clear guidance on how to implement the analytical valuation formulae, in practice, for the efficient computation of American option values. This is problematic due to the fact that the early exercise boundary of an American option is implicitly defined by a complicated path integral and does not have any closed-form solution. Therefore, the determination of the early exercise boundary, a pivotal step in the implementation of these analytical valuation formulae, becomes a difficult task. For this reason, there is need for American-style option formulae that are analytically rigorous, and yet improve upon the computational simplicity and intuitive appeal of the existing numerical methods in the literature.

In this article, we provide a new unified framework for the valuation and hedging of American-style options that is based on a recursive computation of the early exercise boundary. This unified framework can be applied to value and hedge several interesting options: options on stocks with dividends, options on futures, and options on foreign exchange. In particular, we implement an efficient numerical procedure that combines a unified analytical valuation formula and the Geske and Johnson (1984) approximation method to compute option prices and hedge parameters. This procedure involves estimating the early exercise boundary at only a few points and then approximating the entire boundary using Richardson extrapolation, rather than computing the boundary point-wise. An attractive feature of this procedure is that it is robust for a general American-style option contract.

An additional important feature of our approach is that it permits the *direct* implementation of formulae to compute option values as well as option hedge parameters. Specifically, once the exercise boundary is estimated, option values and hedge parameters can be obtained analytically.[a] This feature distinguishes our approach from other existing approaches in the literature, for example, variations of the binomial and the finite-difference methods. These methods use a perturbation scheme to compute the option hedge parameters, and hence, reflect the approximation errors, once again, in the values of the hedge parameters. In our approach, in contract, option prices and hedge parameters are calculated at the same time, given the early exercise boundary. As a result, the computational errors in option prices are not compounded in the calculation of hedge parameters.

It is useful to compare our approach to related work by Broadie and Detemple (1994), Carr and Faguet (1994), and Omberg (1987b). Omberg (1987b) uses an exponential function to approximate the early exercise boundary. Since he does not use the exact analytical valuation formula which imposes the optimal early exercise condition, the optimality and the convergence properties of his approach are somewhat unclear. Broadie and Detemple (1994) provide an upper bound on the value of American options using a lower bound for the early exercise boundary (i.e., an approximate exercise boundary).[b] However, their formula for computing option prices requires the use of parameters determined from regressions. Further, the implementation of their method for computing hedge parameters is not discussed in their paper. Carr and Faguet (1994) develop the "method of lines" based on approximations to the Black and Scholes (1973) partial differential equation (PDE). This involves discretization along either the time or the state-variable dimension,

---

[a]This is in the same sense that the Black and Scholes formula is an analytical formula, although it involves numerical integration of the cumulative normal density function.
[b]They also develop a lower bound on the option value using a capped option written on the same underlying asset.

but not both. They obtain an (approximate) exercise boundary from this discretized PDE, and then calculate the option values and hedge parameters. Similar to Broadie and Detemple (1994), their method involves using some adjustment factors for short maturity option contracts. Thus, both Broadie and Detemple (1994) and Carr and Faguet (1994) would seem to require a recalibration of the regression coefficients and adjustment factors to be applicable to nonstandard options.

This article is organized as follows. In Sec. 1, we present the unified analytical valuation formulae for option prices and hedge ratios, for the case of American put options with a proportional cost of carrying the underlying security. Section 2 presents a numerical implementation procedure for the formulae. Section 3 applies this procedure to two American option pricing problems: the first, the case of stock options, and the second, the case of quanto options. We demonstrate the speed and the accuracy of our method by comparing its numerical results with those of existing methods in the literature. Section 4 concludes the article.

1. A Unified Valuation Formula for American Put Options

In this section, we present a unified analytical valuation formula for American options on an underlying security that has an instantaneous, proportional cost of carry at a rate b. As will be clear later, depending on the characterization of b and its value, we can have many commonly traded financial contracts as special cases. For expositional purposes, we choose American put options for analysis, although our method can be applied to almost any American option for which an equivalent European-style valuation formula exists in closed form.

Assume that the capital markets are frictionless and arbitrage-free with continuous trading possibilities. Let S_t denote the underlying security price at time t and k be the exercise price of an American put option expiring at time T. Assume also that the price process $\{S_t;\ t \geq 0\}$ follows a lognormal diffusion with constant return volatility σ and expected return μ:[c]

$$dS_t = \mu S_t\, dt + \sigma S_t\, dZ_t\,, \tag{1}$$

where $\{Z_t;\ t \geq 0\}$ is a one-dimensional standard Brownian motion (defined on some probability space).

As in McKean (1965) and Merton (1973), we define the American put value as the solution to a free boundary problem. We assume that the early exercise boundary for this American put is well-defined, unique, and has a continuous sample path. Let $\mathcal{B} \equiv \{B_t;\ B_t \geq 0, t \in [0,\ T]\}$ denote the optimal early exercise boundary of the American put, and $P(S_t,\ t)$ denote the put value at time t. Function $P(S_t,\ t)$ is $\mathcal{C}^{1,1}$ on $[B_t,\ \infty] \times [0,\ T]$ and $\mathcal{C}^{2,1}$ on $(B_t,\ \infty] \times [0,\ T]$. Then, the arbitrage-free

---

[c]We discuss and illustrate our numerical procedures for the lognormal diffusion process, although the method can be applied to other diffusion processes using the formulae derived by Kim and Yu (1993).

put price, $P(S_t, t)$ satisfies the fundamental Black and Scholes (1973) PDE

$$\frac{\sigma^2 S_t^2}{2} \frac{\partial^2 P}{\partial S^2} + b S_t \frac{\partial P}{\partial S} - rP + \frac{\partial P}{\partial t} = 0, \tag{2}$$

subject to the following boundary conditions:

$$P(S_T, T) = \max[0, K - S_T], \tag{3}$$

$$\lim_{S_t \uparrow \infty} P(S_t, t) = 0, \tag{4}$$

$$\lim_{S_t \downarrow B_t} P(S_t, t) = K - B_t, \tag{5}$$

$$\lim_{S_t \downarrow B_t} \frac{\partial P(S_t, t)}{\partial S_t} = -1, \tag{6}$$

where $r > 0$ denotes the continuously compounded interest rate and is assumed constant. Equation (5) is the "value matching" condition and Eq. (6) is the "super-contact" condition, and they are jointly referred to as the "smooth pasting" conditions. These conditions ensure that the premature exercise of the put option on the early exercise boundary, $\mathcal{B}$, will be optimal and self-financing.

Let $\Psi(S_t; S_0)$ be the risk neutral transitional density function of the underlying security price. We can then derive the following expression for the price of the American put:[d]

$$P_0 = p_0 + rK \int_0^T e^{-rt} \int_0^{B_t} \Psi(S_t; S_0) \, dS_t \, dt$$

$$-\alpha \int_0^T e^{-rt} \int_0^{B_t} S_t \Psi(S_t; S_0) \, dS_t \, dt, \tag{7}$$

where $\alpha = r - b$, $P_0 \equiv P(S_0, 0)$ and

$$p_0 \equiv p(S_0, 0) = e^{-rT} \int_0^K (K - S_T) \Psi(S_T; S_0) \, dS_T \tag{8}$$

is the price of a European put. The ex-expiration date early exercise boundary is given by

$$B_T = \min\left(K, \frac{r}{\alpha} K\right),$$

independent of the underlying security price distribution.[e]

---

[d]This is a direct extension of the results in Carr, Jarrow, and Myneni (1992) and Kim and Yu (1993).

[e]The simple derivation of the ex-expiration date early exercise boundary, B_T, is as follows. First, $B_T \leq K$. In addition, since holding an asset with a cost of carry, b, is equivalent to holding a stock with a constant dividend ratio $\alpha \equiv r - b$, the early exercise decision at the ex-expiration date instant $T - dt$ is justified if $(rK - \alpha B_T)dt \geq 0$, implying $B_T \leq rK/\alpha$. Therefore, $B_T = \min(K, rK/\alpha)$.

Intuitively, the analytical valuation formula in Eq. (7) gives the sources of value of an American put. The first term, the European put price, p_0, is the value of the guaranteed payoffs from the put option. The last two terms represent the early exercise premium of the American put. More specifically, the second term of Eq. (7) is the present value of the benefits from prematurely exercising the American put through the interest gained by receiving the exercise proceeds early. The last term of Eq. (7) captures the costs associated with the early exercise decision through the loss of insurance value, that is, the costs of taking a short position on the underlying security following exercising the put option early. Alternatively, these values can be linked to the expected payoffs from a dynamic trading strategy, suggested by Carr, Jarrow, and Myneni (1992), that converts an American put into an otherwise identical European put.

Under the assumed lognormal security price process in Eq. (1), the pricing formula in Eq. (7) becomes

$$P_0 = p_0 + \int_0^T \left[rK\, e^{-rt}\, N(-d_2(S_0,\, B_t,\, t)) - \alpha S_0\, e^{-\alpha t}\, N(-d_1(S_0,\, B_t,\, t)) \right] dt\,, \quad (9)$$

where p_0 is the Black and Scholes (1973) and Merton (1973) European put option price:

$$p_0 = K\, e^{-rT}\, N(-d_2(S_0,\, K,\, T)) - S_0\, e^{\alpha T}\, N(-d_1(S_0,\, K,\, T))\,, \quad (10)$$

where $N(\cdot)$ is standard normal distribution function, and

$$d_1(x,\, y,\, t) \equiv \frac{\ln(x/y) + (b + \sigma^2/2)t}{\sigma\sqrt{t}}\,,$$

$$d_2(x,\, y,\, t) \equiv d_1(x,\, y,\, t) - \sigma\sqrt{t}\,.$$

It follows from Eqs. (5) and (9) that the time t optimal point on the early exercise boundary, B_t, satisfies the following integral equation:

$$K - B_t = K\, e^{-r\tau_1}\, N(-d_2(B_t,\, K,\, \tau_1)) - B_t\, e^{-\alpha\tau_1}\, N(-d_1(B_t,\, K,\, \tau_1))$$

$$+ \int_t^T \left[rK\, e^{-r\tau_2}\, N(-d_2(B_t,\, B_s,\, \tau_2)) \right.$$

$$\left. - \alpha\, B_t\, e^{-\alpha\tau_2}\, N(-d_1(B_t,\, B_s,\, \tau_2)) \right] ds\,, \quad (11)$$

where $\tau_1 = T - t$ and $\tau_2 = s - t$. In the above equation, the first two terms on the right-hand side represent the value of a European put with a forward price equal to B_t. The last two terms represent the gain due to interest earned and the loss due to the insurance value foregone through early exercise, respectively, which are reflected in the increased value of an American option over an otherwise identical European option.

Since the cost of carry, b, can be specified in quite a general fashion, Eq. (9) unifies the American put valuation formulae for a wide range of American option valuation problems. Some straightforward examples are standard stock options for which the cost of carry is the riskless interest rate, r, less the constant dividend ratio, β, that is, $b \equiv r - \beta$; futures options analyzed in Black (1976), in which $b \equiv 0$; commodities options where there is a convenience yield, $b \equiv r - \gamma$, where γ is the convenience yield; and options on foreign currencies analyzed in Garman and Kohlhagen (1983) and Grabbe (1983), in which the cost of carrying the foreign currency is the domestic riskless interest rate, r, less the foreign riskless interest rate, r^*, that is, $b \equiv r - r^*$.

Clearly, once the early exercise boundary $\mathcal{B}$ is determined, the American put option value can be easily computed using Eq. (9). The valuation formula also offers a simple way to calculate the hedge parameters. Differentiating Eq. (9) yields the following explicit comparative statics results:

delta:
$$\frac{\partial P(S, t)}{\partial S_t} = -e^{-\alpha \tau_1} N(-d_1(S_t, K, \tau_1)) - \int_t^T e^{-\alpha \tau_2}$$
$$\times \left[\frac{rK - \alpha B_s}{B_s \alpha \sqrt{\tau_2}} n(d_1(S_t, B_s, \tau_2)) \right.$$
$$\left. + \alpha N(-d_1(S_t, B_s, \tau_2)) \right] ds, \tag{12}$$

gamma:
$$\frac{\partial^2 P(S, t)}{\partial S_t^2} = \frac{e^{-\alpha \tau_1}}{S_t \sigma \sqrt{\tau_1}} n(d_1(S_t, K, \tau_1))$$
$$+ \frac{1}{\sigma^2 S_t} \int_t^T \frac{e^{-\alpha \tau_2}}{\tau_2} n(d_1(S_t, B_s, \tau_2))$$
$$\times \left[\frac{rK}{B_s} d_1(S_t, B_s, \tau_2) - \alpha d_2(S_t, B_s, \tau_2) \right] ds, \tag{13}$$

vega:
$$\frac{\partial P(S, t)}{\partial \sigma} = S_t \sqrt{\tau_1}\, n(d_1(S_t, K, \tau_1)) e^{-\alpha \tau_1}$$
$$+ S_t \int_t^T e^{-\alpha \tau_2} n(d_1(S_t, B_s, \tau_2))$$
$$\times \left[\frac{rK d_1(S_t, B_s, \tau_2) - \alpha B_s d_2(S_t, B_s, \tau_2)}{\sigma B_s} \right] ds, \tag{14}$$

$$\frac{\partial P(S, t)}{\partial K} = \frac{1}{K} \left[P(S, t) - \frac{\partial P(S, t)}{\partial S_t} S_t \right], \tag{15}$$

theta:
$$\frac{\partial P(S, t)}{\partial \tau_1} = \frac{\sigma^2 S_t^2}{2} \frac{\partial^2 P(S, t)}{\partial S_t^2} + b S_t \frac{\partial P(S, t)}{\partial S_t} - r P(S, t), \tag{16}$$

where $n(\cdot)$ is the standard normal density function.[f] Similar to the option pricing formula in Eq. (9), these hedge parameters also have an intuitive explanation. One can see that all three parameters — delta, gamma, and vega — are equal to their European counterparts plus a term that depends on the early exercise premium. Moreover, given the early exercise boundary, Eq. (12) through (16) provide a *direct* approach to calculate the hedge parameters, while other existing approximation methods based on numerical computation of option hedge ratios have to rely on some scheme of perturbing the option valuation formula.

It is instructive to discuss the reasons why the analytical formulae presented above have some desirable properties in terms of both efficiency and accuracy. Once the early exercise boundary is specified, one can analytically obtain the option values and hedge parameters using Eq. (9) and Eqs. (12) through (16). In contrast, in methods using a perturbation scheme, the option values have to be recomputed for other values of the input variables. Hence, the analytical method is likely to be more efficient compared to the alternatives. In terms of accuracy, the direct computation of option values and hedge parameters restricts the error to that from approximating the early exercise boundary. The competing methods that perturb the option values may compound the approximation error, since they require repetitive calculation of option values, that is, the perturbation methods may enhance error propagation.

2. Implementation

Implementation of the valuation formula in Eq. (9) and the hedge ratio formulae of Eqs. (12) through (16) requires the early exercise boundary as an input. Once the optimal exercise boundary is obtained, calculations of the option prices and sensitivity parameters are straightforward. In the following, we first use a recursive scheme to calculate the exercise boundary and then our approach, which is based on the Richardson extrapolation method, to calculate the option prices and hedge parameters.

2.1. *Optimal exercise boundary*

The optimal exercise boundary $\mathcal{B}$ has no known analytical solution. An approximate exercise boundary, however, can be obtained numerically. A straightforward algorithm is to compute the boundary recursively using Eq. (11).[g] Namely, we divide the interval $[0, T]$ into n subintervals t_i for $i = 0, 1, 2, \ldots, n$, with $t_0 = 0$ and $t_n = T$. This discretization yields, from Eq. (11), n implicit integral equations defining the optimal exercise points, $B_{t_0}, B_{t_1}, \ldots, B_{t_{n-1}}$. Given the boundary condition on expiration that $B_{t_n} = \min(K, rK/\alpha)$, $\mathcal{B}_n \equiv \{B_{t_i}, 0 \leq i \leq n-1\}$ can be computed recursively using the n integral equations. For a large enough n, this algorithm produces a good approximation of the optimal exercise boundary. Given

---

[f]In practice, we need to compute the delta, gamma, and vega only, since the other two comparative statics parameters are explicit functions of these values.
[g]This is in the spirit of Kim (1990).

the discrete optimal boundary $\mathcal{B}_n$, the option price and hedge parameters can then be easily calculated from Eq. (9) and Eqs. (12) through (16) using a numerical integration scheme.

This recursive algorithm is conceptually simple and easy to implement, especially for the valuation problem of short-term American options, where only a few points on the boundary need to be calculated. However, this recursive method can be computationally intensive when an option has a long time to expiration or where the exercise price is a function of time.[h] This is because, in these cases, a fairly large number of points on the boundary may be needed in order to obtain a good approximation of the boundary itself. It would, therefore, be useful to rapidly compute the option value and hedge ratio without approximating the whole early exercise boundary. This calls for a more efficient implementation of the recursive method.

2.2. *Accelerated recursive method*

In the following, we suggest an approach that uses the Richardson extrapolation method to accelerate the recursive method mentioned earlier. Our method is in the spirit of Geske and Johnson (1984), except that we estimate the early exercise boundary first and then calculate the option values and hedge ratios directly. In other words, we calculate option values based on only a few points on the optimal boundary and then extrapolate the correct option value. This extrapolation scheme gains efficiency without sacrificing much accuracy.

For instance, consider a three-point Richardson extrapolation scheme. The Geske and Johnson (1984) formula for extrapolating the American put value P_0 in this case is as follows:

$$\hat{P}_0 = (P_1 - 8P_2 + 9P_3)/2\,, \tag{17}$$

where P_i, $i = 1$, 2, 3, is the price of an i-times exercisable option, and $\hat{P}_0$ denotes an estimate of P_0.

It follows from Eq. (9) that[i]

$$P_1 = p_0\,,$$

$$P_2 = p_0 + \frac{rKT}{2}\, e^{-\frac{rT}{2}}\, N(-d_2(S_0, B_{\frac{T}{2}}, T/2))$$

$$-\frac{\alpha S_0 T}{2}\, e^{-\frac{\alpha T}{2}}\, N(-d_1(S_0, B_{\frac{T}{2}}, T/2))\,,$$

---

[h]The case of a nonconstant exercise price (i.e., an exercise schedule) merits closer attention even for short-term options, since the computational procedure has to adequately capture all the points where there is a change in the exercise price so as to be computationally accurate and efficient.
[i]These approximations are made by replacing the integration in Eq. (9) with a simple sum over exercisable points.

$$P_3 = p_0 + \frac{rKT}{3} [e^{-\frac{rT}{3}} N(-d_2(S_0, B_{\frac{T}{3}}, T/3)) + e^{-\frac{2rT}{3}} N(-d_2(S_0, B_{\frac{2T}{3}}, 2T/3))]$$

$$- \frac{\alpha S_0 T}{3} [e^{-\frac{\alpha T}{3}} N(-d_1(S_0, B_{\frac{T}{3}}, T/3)) + e^{-\frac{2\alpha T}{3}} N(-d_1(S_0, B_{\frac{2T}{3}}, 2T/3))].$$

A special case of the above expressions applies to the case of a non-dividend paying stock, where $\alpha = 0$:

$$P_1 = p_0, \tag{18}$$

$$P_2 = p_0 + \frac{rKT}{2} e^{-\frac{rT}{2}} N(-d_2(S_0, B_{\frac{T}{2}}, T/2)), \tag{19}$$

$$P_3 = p_0 + \frac{rKT}{3} [e^{-\frac{rT}{3}} N(-d_2(S_0, B_{\frac{T}{3}}, T/3))$$

$$+ e^{-\frac{2rT}{3}} N(-d_2(S_0, B_{\frac{2T}{3}}, 2T/3))]. \tag{20}$$

One can see that only three boundary points — $B_{\frac{T}{3}}$, $B_{\frac{T}{2}}$, and $B_{\frac{2T}{3}}$ — are needed to calculate $\{P_i\}$ and consequently $\hat{P}_0$.

Given $B_{\frac{T}{3}}$, $B_{\frac{T}{2}}$, and $B_{\frac{2T}{3}}$, the Richardson extrapolation method can also be used to calculate the option hedge ratios such as the delta, gamma, and vega. For example, the delta value can be extrapolated as follows:

$$\hat{\Delta}_0 = (\Delta_1 - 8\Delta_2 + 9\Delta_3)/2, \tag{21}$$

where, for an option on a non-dividend-paying stock,

$$\Delta_1 = -N(-d_1(S_0, K, T)), \tag{22}$$

$$\Delta_2 = -N(-d_1(S_0, K, T)) - \frac{rK\sqrt{T/2}}{\sigma S_0} e^{-\frac{rT}{2}} n(d_2(S_0, B_{\frac{T}{2}}, T/2)), \tag{23}$$

$$\Delta_3 = -N(-d_1(S_0, K, T)) - \frac{rK\sqrt{T/3}}{\sigma S_0} e^{\frac{-rT}{3}} n(d_2(S_0, B_{\frac{T}{3}}, T/3))$$

$$- \frac{rK\sqrt{2T/3}}{2\sigma S_0} e^{-\frac{2rT}{3}} n(d_2(S_0, B_{\frac{2T}{3}}, 2T/3)). \tag{24}$$

2.3. *Comparison with the Geske and Johnson method*

As mentioned earlier, the accelerated recursive method is similar in some ways to the Geske and Johnson (1984) method. Although both methods employ Richardson extrapolation, they use different formulae for computing $\{P_i, i = 1, \ldots, n\}$.[j] In the

---

[j]Other similar methods are Breen (1991), Bunch and Johnson (1992), Carr and Faguet (1994), and Ho, Stapleton, and Subrahmanyam (1994), which use the binomial method, a maximization scheme, the method of lines, and an exponential approximation, respectively, to calculate $\{P_i\}$, and in turn $\hat{P}_0$.

Geske and Johnson framework, the exact option price is the limit $(n \uparrow \infty)$ of the price of an option exercisable on n dates. As Geske and Johnson showed, the price of such an option, P_n, is a sum over several multivariate cumulative normal density functions. Specifically, for a given n, P_n involves two univariate integrals (from a European option value), two bivariate integrals, two trivariate integrals, ..., and two n-variate integrals [see Eq. (5) of Geske and Johnson (1984)]. It is known that the computation of a multivariate normal integral is time intensive for large n. Consequently, the Geske and Johnson method has been implemented only for $n \leq 4$. On the other hand, as shown in Eq. (9), P_n in the recursive method involves only the *univariate* normal integral. For example, in the case of options on non-dividend-paying stocks, P_n involves n univariate normal integrals, even under the crudest integration scheme, using a step function. As a result, the recursive method can be easily implemented, in principle, for any value of n without too much computational difficulty.

Due to its ease of implementation, the recursive method can improve the convergence of the extrapolation. As Omberg (1987a) pointed out, the arithmetic sequence of $n = \{1, 2, 3, \ldots\}$ used by Geske and Johnson in their extrapolation may not yield uniform convergence in $\hat{P}_0$, whereas a geometric sequence of $n = \{1, 2, 4, \ldots\}$ would give such convergence. However, it is not feasible to implement the Geske and Johnson method using a geometric sequence of n due to the substantial increase in computational cost beyond $n = 3$. On the other hand, one can see from Eq. (9) that the recursive method can be implemented easily using a geometric sequence and can, therefore, improve the convergence in $\hat{P}_0$.

The difference between the two methods in the dimensionality of the integrals involved also indicates that the recursive method can improve the efficiency of the Geske and Johnson method. To illustrate this, consider a three-point (an arithmetic sequence of n) Richardson extrapolation scheme. In this case, computation of $\hat{P}_0$ using the Geske and Johnson method would require calculations of six single integrals, four double integrals, and two triple integrals, whereas using the recursive method would require only five single integrals [compare Eqs. (18) through (21)].[k] Although it is difficult to say precisely how much more time it takes to calculate a multinormal integral than a single one, we can obtain a rough estimate. The most common algorithm to calculate a univariate integral is Hill's (1973), which approximates the integral by a summation of four terms. A commonly used algorithm to calculate a bivariate integral is the one by Drezner (1978), which uses a Gauss quadrature method. Typically, this method approximates the double integral by a sum of 9 to 25 terms. Schervish's (1984) algorithm for a multinormal integral also uses a Gauss quadrature method. In this method, calculation of an n integral $(n \geq 3)$ needs computation of at least two $(n - 1)$ dimensional integrals. It would not, therefore, be unreasonable to conclude from these facts that the recursive method is likely to

---

[k]In the Geske and Johnson method, P_1 involves the calculation of two single integrals, P_2 the calculation of two single and two double integrals, and P_3 the calculation of two single, two double, and two triple integrals.

be much more efficient. Furthermore, the larger the n value, the more efficient is the recursive method compared to the Geske and Johnson method, which involves multinormal integrals.[l]

As far as the accuracy is concerned, the recursive method and the Geske and Johnson method should be similar, given that both use the same extrapolation scheme, *provided* that $\{P_n\}$ values calculated in the two methods are equally accurate. Under a three-point extrapolation scheme, the truncation error of using $\hat{P}_0$ defined in Eq. (17) is $O(h^3)$ where $h = T/3$.[m] However, given the relative ease of computation, the recursive method can be implemented under a higher-order extrapolation, say a five-point extrapolation [whose truncation error is $O(h^5)$], to reduce the truncation error, that is, to improve the accuracy.[n]

3. Applications

In this section, we illustrate the application of the analytical valuation formula in Eq. (9) and its implementation using the accelerated recursive method to two common American option pricing problems: (1) American options on stocks and (2) American quanto options. It should be emphasized that this method can be applied to other American-style contracts, such as futures options, Asian options, barrier options, and look-back options, since a closed-form solution for the corresponding European-style option exists.

3.1. *Stock options*

Assume that the underlying security is a common stock without a dividend payout. This case is chosen for ease of comparison with results in previous papers. Then, the American put price is obtained from Eq. (9) by letting $b \equiv r$:[o]

$$P_0 = p_0 + \int_0^T rK\, e^{-rt} N(-d_2(S_0,\, B_t,\, t))\, dt\,, \tag{25}$$

where p_0 is the Black and Scholes European put price:

$$p_0 = K\, e^{-rT}\, N(-d_2(S_0,\, K,\, T)) - S_0\, N(-d_1(S_0,\, K,\, T))\,. \tag{26}$$

---

[l]Selby and Hodges (1987) present an identity relating a sum of nested multinormal integrals to a single multidimensional integral. Schroder (1989) improves Curnow and Dunnett's (1962) reduction formula for high dimensional normal integrals. However, the Selby and Hodges method still involves computing high dimensional integrals. The Schroder method has a significant advantage over the Curnow and Dunnett formula only when the dimension is $n > 5$, and involves at least 200 univariate integrals for $n \geq 5$. As a result, a four-point extrapolation with these new reduction schemes is still slow, as noted by Bunch and Johnson (1992).

[m]This result is based on the assumption that the estimate $\hat{P}_0$ of the true price P_0 has the following structure: $\hat{P}_0 = P_0 + \sum_{i=1}^m a_i h^i$, where a_i are constants independent of P_0, $m(\geq 1)$ is an integer, and the series does not necessarily converge. Notice that the Richardson extrapolation method is valid only if this assumption is satisfied [see, e.g., Dahlquist and Bjorck (1974)]. However, it is not clear that the Taylor expansion argument used in Geske and Johnson (1984) can indeed justify this assumption. Results from our attempt to answer this question numerically are inconclusive.

[n]An improvement in the accuracy may also be achieved at a lower-order extrapolation using a more sophisticated integration scheme, say Simpson's rule.

[o]This formula was obtained by Kim (1990) and Carr, Jarrow, and Myneni (1992).

The integral equation defining the early exercise boundary B_t is obtained from Eq. (11):

$$K - B_t = p(B_t, t) + rK \int_t^T e^{-r(s-t)} N(-d_2(B_t, B_s, s-t)) \, ds, \qquad (27)$$

with the boundary condition $B_T = K$. The formulae for option hedge parameters can be obtained from Eqs. (12) through (16) by letting $b \equiv r$.

To test the accuracy and speed of the (accelerated) recursive method, we compare the results of option prices and hedge parameters calculated from the recursive method with those from three widely used numerical methods: the finite difference method, the binomial method, and Breen's (1991) accelerated binomial method.[p] We also report the results from Geske and Johnson (1984) as a basis for comparison. The accuracy is measured by the deviation from a benchmark, which is chosen to be results from the binomial method with $N = 10,000$ time steps.[q] The speed is measured by the CPU time required to compute option prices or hedge parameters for a given set of contracts. The set of contracts is chosen to be the one in Table I of Geske and Johnson (1984).

Table 1 reports the results of option prices from the five alternative methods. Column 4 shows the numerical results from the binomial method with $N = 10,000$ time steps (the benchmark). Column 5, reported in Table I of Geske and Johnson (1984), is included here for comparison. Columns 6 through 9 show the results from the binomial method with $N = 150$ time steps, the accelerated binomial method with $N = 150$ time steps, the finite difference method with 200 steps in both time and underlying state variables, and the recursive method with a four-point extrapolation, respectively. The accuracy of a particular method is measured by its root of the mean squared error (RMSE), as shown in the second-to-last row. The CPU time in seconds is shown in the last low. Note that the CPU time in column 5 is absent because the corresponding results are taken directly from Geske and Johnson (1984).

Table 1 indicates that the RMSE of the recursive method has the same order of magnitude as that of the binomial method with $N = 150$, in spite of the fact that the former method is implemented with a step-function discretization.[r] One can also see that the amount of CPU time using the recursive method is significantly less than that using the other three numerical methods listed in the table. In particular, the

---

[p]It would be interesting to include more methods for comparison, for instance, Carr and Faguet (1994) and Broadie and Detemple (1994). However, as mentioned before, the implementation of these two methods involves using some regression coefficients and adjustment factors, whose values are determined from data sets that are different from ours. See these two papers for a comparison of their methods against other methods.

[q]Broadie and Detemple (1994) also use option values from the binomial method with $N = 10,000$ time steps as their benchmark, Carr and Faguet (1994) use the average of values from $N = 1,000$ and $N = 1,001$ as their benchmark.

[r]One alternative measure is the root of the mean squared *relative* error. In fact, the recursive method is more accurate under this measure than the RMSE measure.

Table 1. Values of American put options on stocks using different numerical methods ($S = \$40$; $r = 4.88\%$).

(1) K	(2) σ	(3) T (yr)	(4) Binomial (N = 10,000)	(5) Geske and Johnson	(6) Binomial (N = 150)	(7) Accelerated binomial	(8) Finite difference	(9) Recursive method
35	0.2	0.0833	0.0062	0.0062	0.0061	0.0061	0.0278	0.0062
35	0.2	0.3333	0.2004	0.1999	0.1995	0.1994	0.2382	0.2004
35	0.2	0.5833	0.4328	0.4321	0.4340	0.4331	0.4624	0.4337
40	0.2	0.0833	0.8522	0.8528	0.8512	0.8517	0.9874	0.8543
40	0.2	0.3333	1.5798	1.5807	1.5783	1.5752	1.6244	1.5873
40	0.2	0.5833	1.9904	1.9905	1.9886	1.9856	2.0177	1.9987
45	0.2	0.0833	5.0000	4.9985	5.0000	4.9200	5.0052	5.0020
45	0.2	0.3333	5.0883	5.0951	5.0886	4.9253	5.1327	5.0954
45	0.2	0.5833	5.2670	5.2719	5.2677	5.2844	5.2699	5.2631
35	0.3	0.0833	0.0774	0.0744	0.0775	0.0772	0.1216	0.0775
35	0.3	0.3333	0.6975	0.6969	0.6993	0.6977	0.7300	0.6978
35	0.3	0.5833	1.2198	1.2194	1.2239	1.2218	1.2407	1.2233
40	0.3	0.0833	1.3099	1.3100	1.3083	1.3095	1.3860	1.3116
40	0.3	0.3333	2.4825	2.4817	2.4799	2.4781	2.5068	2.4919
40	0.3	0.5833	3.1696	3.1733	3.1665	3.1622	3.1819	3.1842
45	0.3	0.0833	5.0597	5.0599	5.0600	5.0632	5.1016	5.0604
45	0.3	0.3333	5.7056	5.7012	5.7065	5.6978	5.7193	5.6970
45	0.3	0.5833	6.2436	6.2365	6.2448	6.2395	6.2477	6.2303
35	0.4	0.0833	0.2466	0.2466	0.2454	0.2456	0.2949	0.2467
35	0.4	0.3333	1.3460	1.3450	1.3505	1.3481	1.3696	1.3468
35	0.4	0.5833	2.1549	2.1568	2.1602	2.1569	2.1676	2.1603
40	0.4	0.0833	1.7681	1.7679	1.7658	1.7674	1.8198	1.7694
40	0.4	0.3333	3.3874	3.3632	3.3835	3.3863	3.4011	3.3970
40	0.4	0.5833	4.3526	4.3556	4.3480	4.3426	4.3567	4.3699
45	0.4	0.0833	5.2868	5.2855	5.2875	5.2863	5.3289	5.2853
45	0.4	0.3333	6.5099	6.5093	6.5103	6.5054	6.5147	6.5128
45	0.4	0.5833	7.3830	7.3831	7.3897	7.3785	7.3792	7.3865
RMSE			0.0000	5.3383e-03	2.6380e-03	3.5302e-02	4.1041e-02	6.7047e-03
CPU time (sec)					1.1766e+00	7.8330e-01	6.4664e+00	1.3333e-01

Columns 1–3 represent the values of the parameters, K (strike price), σ (volatility), and T (the time to expiration), respectively. Column 4 shows the numerical results of option values from the binomial method with $N = 10,000$ time steps (the benchmark). Column 5 is as reported in Table 1 of Geske and Johnson (1984). Columns 6–9 show the results from the binomial method with $N = 150$ time steps, the accelerated binomial method with $N = 150$ time steps, the finite difference method with 200 steps in both time and the underlying state variable, and the recursive method with $n = 4$ (a four-point extrapolation), respectively. The second-to-last row shows the root of the mean squared errors (RMSE), a measure of deviation from the benchmark. The CPU time is the time required on a SPARC-10 machine to compute the option values for all 27 contracts listed in the table.

Table 2. Hedge ratios (deltas) of American put options on stocks using different numerical methods ($S = \$40$; $r = 4.88\%$).

(1) K	(2) σ	(3) T (yr)	(4) Binomial (N = 10,000)	(5) Geske and Johnson	(6) Binomial (N = 150)	(7) Accelerated binomial	(8) Finite difference	(9) Recursive method
35	0.2	0.0833	−0.0080	−0.008	−0.0081	−0.0081	−0.0522	−0.0080
35	0.2	0.3333	−0.0901	−0.090	−0.0911	−0.0913	−0.1337	−0.0900
35	0.2	0.5833	−0.1338	−0.134	−0.1355	−0.1352	−0.1692	−0.1351
40	0.2	0.0833	−0.4693	−0.470	−0.4698	−0.4690	−0.5738	−0.4712
40	0.2	0.3333	−0.4435	−0.443	−0.4443	−0.4425	−0.5083	−0.4416
40	0.2	0.5833	−0.4287	−0.427	−0.4299	−0.4297	−0.4821	−0.4228
45	0.2	0.0833	−1.0000	−1.000	−1.0000	−0.9867	−0.9617	−0.9993
45	0.2	0.3333	−0.8812	−0.888	−0.8803	−0.8762	−0.8897	−0.8964
45	0.2	0.5833	−0.7948	−0.805	−0.7944	−0.7439	−0.8355	−0.8112
35	0.3	0.0833	−0.0516	−0.052	−0.0526	−0.0524	−0.1041	−0.0516
35	0.3	0.3333	−0.1741	−0.174	−0.1758	−0.1758	−0.2067	−0.1744
35	0.3	0.5833	−0.2126	−0.213	−0.2144	−0.2140	−0.2372	−0.2145
40	0.3	0.0833	−0.4695	−0.470	−0.4699	−0.4697	−0.5437	−0.4710
40	0.3	0.3333	−0.4420	−0.442	−0.4429	−0.4409	−0.4810	−0.4429
40	0.3	0.5833	−0.4256	−0.425	−0.4268	−0.4253	−0.4544	−0.4243
45	0.3	0.0833	−0.9233	−0.926	−0.9220	−0.8762	−0.9111	−0.9293
45	0.3	0.3333	−0.7266	−0.726	−0.7260	−0.7302	−0.7516	−0.7229
45	0.3	0.5833	−0.6520	−0.651	−0.6520	−0.6576	−0.6776	−0.6459
35	0.4	0.0833	−0.1062	−0.106	−0.1073	−0.1072	−0.1526	−0.1062
35	0.4	0.3333	−0.2260	−0.226	−0.2277	−0.2278	−0.2502	−0.2265
35	0.4	0.5833	−0.2539	−0.254	−0.2557	−0.2553	−0.2714	−0.2558
40	0.4	0.0833	−0.4668	−0.467	−0.4673	−0.4673	−0.5236	−0.4679
40	0.4	0.3333	−0.4360	−0.437	−0.4370	−0.4362	−0.4626	−0.4376
40	0.4	0.5833	−0.4173	−0.418	−0.4186	−0.4162	−0.4360	−0.4177
45	0.4	0.0833	−0.8363	−0.835	−0.8350	−0.8364	−0.8478	−0.8341
45	0.4	0.3333	−0.6475	−0.646	−0.6474	−0.6479	−0.6678	−0.6424
45	0.4	0.5833	−0.5819	−0.580	−0.5819	−0.5833	−0.5982	−0.5753
RMSE			0.0000	2.5289e-03	1.0593e-03	4.2777e-02	4.2216e-02	5.1702e-03
CPU time (sec)				3.5315e+00	3.5315e+00	2.3332e+00	6.4664e+00	1.3333e-01

Columns 1–3 represent the values of the parameters, K (strike price), σ (volatility), and T (the time to expiration), respectively. Column 4 shows the numerical results of option values from the binomial method with $N = 10,000$ time steps (the benchmark). Column 5 is as reported in Table I of Geske and Johnson (1984). Columns 6–9 show the results from the binomial method with $N = 150$ time steps, the accelerated binomial method with 150 time steps, the finite difference method with 200 steps in both time and the underlying state variable, and the recursive method with $n = 4$ (a four-point extrapolation), respectively. The second-to-last row shows the root of the mean squared errors (RMSE), a measure of deviation from the benchmark. The CPU time is the time required on a SPARC-10 machine to compute the delta values for all 27 contracts listed in the table.

Table 3. Convexity parameter (gammas) of American put options on stocks using different numerical methods ($S = \$40$; $r = 4.88\%$).

(1) K	(2) σ	(3) T (yr)	(4) Binomial ($N = 10,000$)	(5) Binomial ($N = 150$)	(6) Accelerated binomial	(7) Finite difference	(8) Recursive method
35	0.2	0.0833	0.0095	0.0096	0.0095	0.0273	0.0095
35	0.2	0.3333	0.0357	0.0358	0.0359	0.0434	0.0358
35	0.2	0.5833	0.0364	0.0364	0.0361	0.0412	0.0374
40	0.2	0.0833	0.1776	0.1772	0.1757	0.1638	0.1759
40	0.2	0.3333	0.0923	0.0922	0.0928	0.0942	0.0870
40	0.2	0.5833	0.0719	0.0718	0.0736	0.0748	0.0669
45	0.2	0.0833	0.0000	0.0000	0.0132	0.0106	0.0010
45	0.2	0.3333	0.0827	0.0829	0.3140	0.0541	0.0804
45	0.2	0.5833	0.0787	0.0785	-0.2757	0.0755	0.0860
35	0.3	0.0833	0.0306	0.0308	0.0307	0.0422	0.0306
35	0.3	0.3333	0.0376	0.0376	0.0376	0.0409	0.0381
35	0.3	0.5833	0.0326	0.0325	0.0322	0.0344	0.0332
40	0.3	0.0833	0.1170	0.1168	0.1160	0.1160	0.1173
40	0.3	0.3333	0.0597	0.0596	0.0592	0.0601	0.0579
40	0.3	0.5833	0.0459	0.0458	0.0461	0.0462	0.0434
45	0.3	0.0833	0.0578	0.0581	-0.1188	0.0393	0.0615
45	0.3	0.3333	0.0572	0.0571	0.0599	0.0519	0.0614
45	0.3	0.5833	0.0485	0.0484	0.0509	0.0471	0.0517
35	0.4	0.0833	0.0398	0.0399	0.0399	0.0469	0.0398
35	0.4	0.3333	0.0330	0.0329	0.0329	0.0345	0.0334
35	0.4	0.5833	0.0269	0.0268	0.0266	0.0276	0.0272
40	0.4	0.0833	0.0872	0.0871	0.0866	0.0869	0.0877
40	0.4	0.3333	0.0441	0.0440	0.0437	0.0441	0.0434
40	0.4	0.5833	0.0336	0.0336	0.0333	0.0337	0.0324
45	0.4	0.0833	0.0590	0.0591	0.0608	0.0459	0.0612
45	0.4	0.3333	0.0439	0.0438	0.0450	0.0417	0.0431
45	0.4	0.5833	0.0355	0.0354	0.0365	0.0346	0.0347
RMSE			0.0000	1.4332e-04	8.8301e-02	9.2263e-03	2.5499e-03
CPU time (sec)				3.5332e+00	2.3332e+00	6.4664e+00	1.3333e-01

Columns 1–3 represent the values of the parameters, K (strike price), σ (volatility), and T (the time to expiration), respectively. Columns 4–8 show the numerical results of gamma values from the binomial method with $N = 10,000$ time steps (the benchmark), with $N = 150$ time steps, the accelerated binomial method with 150 time steps, the finite difference method with 200 steps in both time and the underlying state variable, and the recursive method with $n = 4$ (a four-point extrapolation), respectively. The second-to-last row shows the root of the mean squared errors (RMSE), a measure of deviation from the benchmark. The CPU time is the time required on a SPARC-10 machine to compute the gamma values for all 27 contracts listed in the table.

recursive method is about *five times* faster than the accelerated binomial method, and more accurate as well.

Table 2 reports the results of deltas for the same contracts as shown in Table 1. As before, the results in column 4 are taken as the benchmark. Again, column 5, reported in Table I of Geske and Johnson (1984), is included here for comparison. One can see from the table that the recursive method again produces much smaller RMSE than the accelerated binomial and the finite difference methods do. It is worth noticing that it takes almost the same amount of CPU time for the recursive method (and also the finite difference method) to compute deltas as option prices, whereas it takes *twice* as much CPU time to compute deltas as option prices for both the binomial and the accelerated binomial methods. This indicates that the recursive method is particularly efficient in calculating hedge parameters.

Table 3 reports the results of gammas for the same contracts as shown in Table 1.[s] Again, the RMSE of the recursive method is much smaller than that of the accelerated binomial method. Also, the speed of computation is faster by at least one *order of magnitude* compared with the alternatives. It is apparent that the relative computational merit of the recursive method increases as we get to *higher-order* hedge parameters.[t]

3.2. *American quanto options*

We now consider the application of our method to nonstandard options, in particular, to quanto options. We do this to illustrate the use of the method for any case where a closed-form solution exists for the corresponding European-style option. Quanto options are options based on two (or more) underlying state variables, usually the foreign exchange rate and an asset price. Examples of these options available to U.S. investors are Nikkei index warrants and currency-protected warrants on foreign treasury bonds, warrants on cross-currency rates denominated in dollars, as well as a range of over-the-counter products. Some basic features and valuation results on these instruments are described in, among others, Derman, Karasinski, and Wecker (1990) and Dravid, Richardson, and Sun (1993; henceforth DRS). These papers examine the issues of hedging, sensitivity analysis, and parameter estimations of these contracts. In this section, we focus on the valuation of American-style put warrants using our alternative approach.[u] The analysis of the American-style call warrants can be analogously developed.

Let S_t^* be the yen price at time t of the underlying Nikkei index with an instantaneous dividend payout rate δ_t. Assume that Y_t denotes the spot exchange rate specified in units of dollars (or cents) per yen. Define the dollar price of the Nikkei index at t by $S_t \equiv S_t^* Y_t$. Denote by r_t and r_t^* the instantaneous U.S. and Japanese riskless interest rates, respectively. Following DRS, we have, under the risk neutral probability measure,

$$dS_t = (r_t - \delta_t)S_t\,dt + \sigma_{s^*}\,S_t\,dZ_t^{(s^*)} + \sigma_y\,S_t\,dZ_t^{(y)}, \tag{28}$$

$$dY_t = (r_t - r_t^*)Y_t\,dt + \sigma_y\,Y_t\,dZ_t^{(y)}, \tag{29}$$

$$dS_t^* = r_f\,S_t^*\,dt + \sigma_{s^*}\,S_t^*\,dZ_t^{(s^*)}, \tag{30}$$

---

[s]The Geske and Johnson method is left out in Table 3 because no results about gammas for the same contracts are reported in Geske and Johnson (1984).

[t]Notice that two of the gamma values from the accelerated binomial method in column 6 are negative and that another one with $K = 45$, $\sigma = 0.2$, and $T = 0.3333$, though positive, deviates significantly from the corresponding benchmark value. This may be due to the fact that the accelerated binomial method is an approximation of the Geske and Johnson formula (also an approximation). This calls for caution when one uses the accelerated binomial method to compute hedge parameters.

[u]In this article, we present the valuation framework in terms of the Nikkei index put warrants. The same intuition and valuation methodology can be applied to the valuation of other types of contingent claims on a foreign stock index, as well as the valuation of dollar-denominated cross-currency warrants, such as warrants on the yen/DM cross-currency rate.

where σ_{s^*} and σ_y are the instantaneous volatilities of the Nikkei index and the exchange rate, $\{Z_t^{(s^*)}; t \geq 0\}$ and $\{Z_t^{(y)}; t \geq 0\}$ are two standard one-dimensional Brownian motions with a correlation coefficient ρ, and $r_f \equiv r_t^* - \delta_t - \rho\sigma_{s^*}\sigma_y$.

Consider a put option whose yen payoffs can be converted into U.S. currency at a prespecified exchange rate $\overline{Y}$.[v] With a fixed exchange rate, the European put price depends on the fixed exchange rate and the parameters of the exchange rate dynamics, but is independent of the current level of the exchange rate. This is because the U.S. investor has to continuously rebalance his position in the Nikkei index through the foreign currency market, by converting dollars into yen, and *vice versa*, in order to hedge this option (since the Nikkei index is not directly traded in dollars). Equivalently, it can be said that simply by using the Nikkei index (denoted in yen) and riskless Japanese bonds, the U.S. investor completely hedges the European put option, but is exposed to the foreign exchange risk, which can be hedged over time by continuously rebalancing the positions in the Nikkei index through the foreign currency market. Consequently, by using riskless Japanese bonds and the Nikkei index, the U.S. investor can also hedge/replicate the American put, but in terms of yen. This essentially makes the pricing of fixed exchange rate American Nikkei puts a one-dimensional problem. Namely, the problem is equivalent to pricing an American put option whose underlying asset price process is described by Eq. (30), and hence can be recast in the framework developed in Sec. 1.

Substituting $\alpha = r - r_f$ and $\sigma = \sigma_{s^*}$ into Eqs. (9) through (11), we have that the value of a fixed exchange rate American put (in yen) with a strike price K^* is

$$P_0/\overline{Y} = p_0 + \int_0^T [r\, K^* e^{-rs}\, N(-x_2(S_0^*, B_s, s)) - (r - r_f)S_s^*$$

$$\times e^{-(r-r_f)s}\, N(-x_1(S_0^*, B_s, s))]\, ds\,, \tag{31}$$

and that the value of the corresponding European put option in yen is (see DRS)

$$p_0 = K^* e^{-rT}\, N(-x_2(S_0^*, K^*, T)) - S_0^* e^{-(r-r_f)T}\, N(-x_1(S_0^*, K^*, T))\,, \tag{32}$$

where

$$x_1(x, y, t) \equiv \frac{\ln(x/y) + (r_f + \sigma_{s^*}^2/2)t}{\sigma_{s^*} \sqrt{t}}\,,$$

$$x_2(x, y, t) \equiv x_1(x, y, t) - \sigma_{s^*} \sqrt{t}\,,$$

and B_t is the early exercise boundary at time t, in terms of the Nikkei index. The value of B_t is determined by

$$K^* - B_t = p_1 + \int_t^T [rK^* e^{-r(s-t)}\, N(-x_2(B_t, B_s, s - t))$$

$$- (r - r_f)B_t\, e^{-(r-r_f)(s-t)}\, N(-x_1(B_t, B_s, s - t))]\, ds\,, \tag{33}$$

[v]Our method also applies to some variations on the basic quanto structure analyzed here. An example is the flexible exchange rate case. [See DRS for details.]

where

$$p_1 = K^* e^{-r(T-t)} N(-x_2(B_t^*, K^*, T-t))$$
$$- S_t^* e^{-(r-r_f)(T-t)} N(-x_1(B_t^*, K^*, T-t)),$$

and x_1 and x_2 are as defined in Eq. (32). The formulae for hedge ratios can be obtained in a straightforward manner from Eq. (31), and will not be presented here in the interest of brevity. Once the exercise boundary is computed using methods analogous to those described in Sec. 2, the values and hedge parameters of options exercisable at discrete points can be obtained from Eq. (31). The use of Richardson extrapolation then yields the estimates of the values and hedge parameters of the American-style quanto options.[w]

It should be noted that the pricing formula in Eq. (31) can also be obtained directly using the arguments underlying Eq. (7) and the fact that a closed-form solution exists for the corresponding European-style option. Further, this scheme also applies to those nonstandard American-style pricing problems that cannot be recast in the framework of the general pricing formula in Eq. (9).

The above analysis illustrates the general principles to be applied in implementing the recursive method to nonstandard American-style options. Given that a closed-form solution exists for the corresponding European-style option, one can set up the value-matching condition as in Eq. (33) above. One can also write down the discrete time analog of Eq. (31) for the i-times exercisable option, P_i, $i = 1, 2, 3 \ldots$, and the corresponding ones for the hedge parameters. Richardson extrapolation can then be used to obtain the values of the American-style option and the hedge parameters.

4. Conclusions

This article has presented a method of recursive implementation of analytical formulae, taking the early exercise boundary as given, for the value and hedge parameters of an American-style option. The early exercise boundary is computed by using Richardson extrapolation based on options exercisable on (one, two, three, etc.) discrete dates. The method can be used for *any* American-style option for which a closed-form solution exists for the equivalent European-style option. A unified analytical formula was derived for American-style put options on an underlying security with a cost of carry. This formula can be applied to a range of standard American-style options: options on stocks with dividends, options on futures, options on foreign exchange, and quanto options. The method was implemented for American put options on non-dividend-paying stocks, for ease of comparison with other methods in the literature.

The method presented in this article has three attractive features. First, since its implementation is based on using an analytical formula for both option values and hedge parameters, the latter are computed directly, rather than by perturbation

---

[w]Numerical results are available from the authors upon request.

of the option pricing formula. Second, as a result, the computation is both efficient and accurate, since the analytical formula involves only univariate integrals. The improvement in efficiency over competing methods is especially advantageous in the management of options books in practice, since the implementation of dynamic trading strategies involves the computation of values and hedge ratios for large numbers of options on a continuous basis. Third, the analytical formulae presented here permit an intuitive decomposition of the value and the hedge ratios into three terms: that of a corresponding European-style option, the gain in the time value of money, and the loss in the insurance value from early exercise.

Several modifications and extensions of the methods proposed in this article can be explored in further research. First, the accuracy and efficiency of the method can be improved, particularly for long-term options, by exploring the possibility of unequal spacing of the discrete points or by explicitly taking into account the slope of the exercise boundary between the discrete points. Second, the method can be applied to the valuation of other American options such as those with two or more sources of uncertainty, for instance, mortgage-backed securities, options on bonds, foreign exchange, or real assets in capital budgeting applications. In particular, the applications to exotic options in these situations can be studied in greater detail. Third, the valuation of American options can be studied in the context of alternative stochastic processes for the underlying asset. We leave these directions of enquiry for later work.

References

G. Barone-Adesi and R. Whaley, "Efficient Analytic Approximation of American Option Values", *Journal of Finance* **42** (1987) 301–320.

F. Black, "The Pricing of Commodity Contracts", *Journal of Financial Economics* **3** (1976) 167–179.

F. Black and M. Scholes, "The Pricing of Options and Corporate Liabilities", *Journal of Political Economy* **81** (1973) 637–659.

R. Breen, "The Accelerated Binomial Option Pricing Model", *Journal of Finanacial and Quantitative Analysis* **26** (1991) 153–164.

M. Brennan and E. Schwartz, "The Valuation of American Put Options", *Journal of Finance* **32** (1977) 449–462.

M. Broadie and J. Detemple, "American Option Valuation, New Bounds, Approximations, and a Comparison of Existing Methods", forthcoming, *Review of Financial Studies* (1994).

D. S. Bunch and H. Johnson, "A Simple and Numerically Efficient Valuation Method for American Puts Using a Modified Geske-Johnson Approach", *Journal of Finance* **47** (1992) 809–816.

P. Carr and D. Faguet, "Fast Accurate Valuation of American Options", working paper, Cornell University (1994).

P. Carr, R. Jarrow and R. Myneni, "Alternative Characterizations of American Put Options", *Mathematical Finance* **2** (1992) 87–106.

G. R. Courtadon, "A More Accurate Finite Difference Approximation for the Valuation of Option", *Journal of Financial and Quantitative, Analysis* **17** (1982) 697–703.

J. C. Cox and S. A. Ross, "The Valuation of Options for Alternative Stochastic Processes", *Journal of Financial Economics* **3** (1976) 145–166.

J. C. Cox, S. A. Ross and M. Rubinstein, "Option Pricing: A Simplified Approach", *Journal of Financial Economics* **7** (1979) 229–264.

R. N. Curnow and C. W. Dunnett, "The Numerical of Certain Multivariate Normal Integrals", *Annals of Mathematical Statistics* **33** (1962) 571–579.

G. Dahlquist and A. Bjorck, *Numerical Methods*, Prentice-Hall, Englewood Cliffs, N.J. (1974).

E. Derman, P. Karasinski and J. S. Wecker, "Understanding Guaranteed Exchange-Rate Contracts in Foreign Stock Investments", Working paper, Goldman, Sachs & Co. (1990).

A. Dravid, M. Richardson and T. S. Sun, "Pricing Foreign Index Contingent Claims: An Application to Nikkei Index Warrants", *Journal of Derivatives* **1** (1993) 33–51.

Z. Drezner, "Computation of the Bivariate Normal Integral", *Mathematics of Computations* **32** (1978) 277–279.

M. Garman and S. Kohlhagen, "Foreign Currency Option Values", *Journal of International Money and Finance* **2** (1983) 231–237.

R. Geske and H. E. Johnson, "The American Put Option Valued Analytically", *Journal of Finance* **39** (1984) 1511–1524.

J. O. Grabbe, "The Pricing of Call and Put Options on Foreign Exchange", *Journal of International Money and Finance* **22** (1983) 239–254.

J. M. Harrison and D. M. Kreps, "Martingale and Arbitrage in Multiperiod Securities Markets", *Journal of Economic Theory* **20** (1979) 381–408.

I. D. Hill, "The Normal Integral", *Applied Statistics* **22** (1973) 425–427.

T. S. Ho, R. C. Stapleton and M. G. Subrahmanyam, "A Simple Technique for the Valuation and Hedging of American Options", *Journal of Derivatives* **2** (1994) 52–66.

S. D. Jacka, "Optimal Stopping and the American Put", *Mathematical Finance* **1** (1991) 1–14.

I. J. Kim, "The Analytic Valuation of American Options", *Review of Financial Studies* **3** (1990) 547–572.

I. J. Kim and G. Yu, "A Simplified Approach to the Valuation of American Options and its Applications", working paper, New York University (1993).

L. W. MacMillan, "An Analytic Approximation for the American Put Price", *Advances in Futures and Options Research* **1** (1986) 119–139.

H. P. McKean, Jr., "Appendix: A Free Boundary Problem for the Heating Function Arising from a Problem in Mathematical Economics", *Industrial Management Review* **6** (1965) 32–39.

R. C. Merton, "Theory of Rational Option Pricing", *Bell Journal of Economics and Management Science* **4** (1973) 141–183.

E. Omberg, "A Note on the Convergence of Binomial Pricing and Compound Option Models", *Journal of Finance* **42** (1987a) 463–470.

E. Omberg, "The Valuation of American Put Options with Exponential Exercise Policies", *Advances in Futures and Options Research* **2** (1987b) 117–142.

M. J. Schervish, "Multivariate Normal Probabilities with Error Bound", *Applied Statistics* **33** (1984) 81–94.

M. Schroder, "A Reduction Method Applicable to Compound Option Formulas", *Management Science* **35** (1989) 823–827.

E. Schwartz, "The Valuation of Warrants: Implementing a New Approach", *Journal of Financial Economics* **4** (1977) 79–93.

M. J. P. Selby and S. D. Hodges, "On the Evaluation of Compound Options", *Management Science* **33** (1987) 347–355.

Reprinted from Ann. Stat. **25**(6) (1997) 2592–2606
© Inst. Math. Stat.

PIECEWISE CONVEX FUNCTION ESTIMATION:
PILOT ESTIMATORS*

KURT S. RIEDEL

Courant Institute of Mathematical Sciences,
New York University, New York, New York 10012–1185, USA
E-mail: riedel@cims.nyu.edu

Received June 1995
Revised October 1996

Given noisy data, function estimation is considered when the unknown function is known *a priori* to consist convex or concave on each of a small number of regions where the function. When the number of regions is unknown, the model selection problem is to determine the number of convexity change points. For kernel estimates in Gaussian noise, the number of false change points is evaluated as a function of the smoothing parameter. To insure that the number of false convexity change points tends to zero, the smoothing level must be larger than is generically optimal for minimizing the mean integrated square error (MISE). A two-stage estimator is proposed and shown to achieve the optimal rate of convergence of the MISE. In the first stage, the number and location of the change points is estimated using strong smoothing. In the second stage, a constrained smoothing spline fit is performed with the smoothing level chosen to minimize the MISE. The imposed constraint is that a single change point occur in a region about each empirical change point from the first-stage estimate. This constraint is equivalent to the requirement that the third derivative of the second-stage estimate has a single sign in a small neighborhood about each first-stage change point. The change points from the second stage are near the first-stage change points, but need not be at the identical locations.

Keywords and phrases: Pilot estimators, shape constraints, nonparametric estimation, convexity.

AMS 1991 subject classification: 62G07, 65D10, 65D07, 60G35.

1. Introduction

Our basic tenet is: "Most real world functions are piecewise l-convex with a small number of change points of convexity." Given N measurements of the unknown function, $f(t)$, contaminated with random noise, we seek to estimate $f(t)$ while preserving the geometric fidelity of the estimate, $\hat{f}(t)$, with respect to the true function. In other words, the number and location of the change points of convexity of $\hat{f}(t)$ should approximate those of $f(t)$. We say that $f(t)$ has K change points of

*Supported by U.S. Department of Energy Grant DE-FG02-86ER53223.

240

l-convexity with change points $x_1 \leq x_2 \ldots \leq x_K$ if $(-1)^{k-1} f^{(l)}(t) \geq 0$ or ≤ 0 for $x_k \leq t \leq x_{k+1}$.

The idea of constraining the function fit to preserve *prescribed l*-convexity properties has been considered by a number of authors [6, 10, 18, 19]. The more difficult problems of determining the number and location of the l-convexity breakpoints will be a focus of this article. Historical perspectives to the problem may be found in [5, 7]. "Bump hunting" dates back at least to [3]. Silverman [16], Mammen [8] and Mammen, Marron and Fisher [9] formulated the problem as a sequential hypothesis testing problem. We refer to the estimation of the number of change points as the "model selection problem" because it resembles model selection in an infinite family of parametric models.

An interesting result of [8, 9, 15] is that kernel smoothers will often produce too many inflection points or wiggles. If the amount of smoothing is chosen to minimize the mean integrated square error (MISE), then with nonvanishing asymptotic probability, the estimate will have multiple inflection points in a neighborhood of an actual one. For many applications, estimating the correct shape is more important than minimizing the MISE.

In this article, we propose a class of two-stage estimators which estimate the l-change points in the first stage and then perform a constrained regression in the second stage. In the first stage, the function is strongly smoothed while the smoothing in the constrained second stage is optimized for the minimal mean square error. When the change points are correctly specified, the constrained spline estimate has a smaller square error (as measured in a particular norm) than the unconstrained estimate.

Our second-stage estimate achieves the asymptotically optimal MISE convergence rate while suppressing artificial change points that can occur with the unconstrained method. Our proof does not exclude the possibility that the second-stage estimate has spurious inflection points far from the first-stage inflection points. We believe that our estimator has the same *relative* convergence rate as standard nonparametric methods. Thus our estimator suppresses artificial wiggles at nontrivial computational costs but no loss of MISE.

In Sec. 2, we show that the constrained smoothing spline estimate achieves the optimal rate of convergence for the expected square error even when the constraints are occasionally misspecified, provided that the misspecification rate is sufficiently small.

In Sec. 3, we evaluate the expected number of false (extra) empirical change points [9] for kernel smoothers when the errors are Gaussian. By adjusting the smoothing parameter, we can guarantee an asymptotically small probability of an error in our imposed constraints.

In Sec. 4, we propose two-stage estimators which estimate the number and location of the l-change points in the first stage. In the second stage, we impose that $\hat{f}^{(l+1)}$ be positive/negative in a small region about each of the empirical l-change points. The main advantage of the two-stage procedure is that less smoothing is required in the second-stage than in the first stage while preserving the geometric fidelity.

Section 5 discusses potential extensions of the method. Section 6 describes a global shape optimization that is *heuristically* designed to be efficient in the amount of smoothing subject to determining the number of change points consistently.

2. Expected Error under Inexact Convexity Constraints

We are given N measurements of the unknown function, $f(t)$:

$$y_i = f(t_i) + \varepsilon_i. \tag{1}$$

The mean integrated squared error (MISE) for the estimate, $\hat{f}^{(j)}$, of the jth derivative is defined as $\mathbf{E}[\int |\hat{f}^{(j)}_{N,\lambda}(t) - f^{(j)}(t)|^2]$. We consider the MISE for constrained estimation as the number of measurements, N, tends to infinity. In describing the large N asymptotics, we consider a sequence of measurement problems. For each N, the measurements occur at $\{t_i^N, i = 1, \ldots, N\}$, we suppress the superscript, N, on the measurement locations $t_i \equiv t_i^N$. We define the empirical distribution of measurements, $F_N(t) = \sum_{t_i \leq t} 1/N$, and let $F(t)$ be its limiting distribution.

Assumption A. Consider the sequence of estimation problems: $y_i^N = f(t_i^N) + \varepsilon_i^N$, where the ε_i^N are zero mean random variables and $\mathbf{Cov}[\varepsilon_i^N, \varepsilon_j^N] = \sigma^2 \delta_{i,j}$. Assume that the distribution of measurement locations, converges in the sup norm: $|F_N(t) - F(t)| \to 0$, where $F(t)$, is $C^\infty[0,1]$ and $0 < c_F < F'(t) < C_F$.

We measure the convergence of a set of measurement times to the continuum limit in terms of the discrepancy of the point set:

Definition. The star discrepancy of $\{t_1 \ldots t_N\}$ with respect to the continuous distribution $F(t)$ is $D_N^* \equiv \sup_t \{F_N(t) - F(t)\}$.

Equivalently, $D_N^* \sim 1/2N + \max_{1 \leq i \leq N} |F(t_i) - (i - 1/2)/N|$. For regularly spaced points, $D_N^* \sim 1/N$, while for randomly spaced points, $D_N^* \sim \sqrt{\ln[\ln[N]]/N}$ by the Glivenko–Cantelli theorem.

A popular linear estimator is the smoothing spline [20]: $\hat{f} = \arg\min \mathrm{VP}[f]$, where

$$\mathrm{VP}[f] \equiv \frac{\lambda}{2} \int |f^{(m)}(s)|^2 ds + \frac{1}{N} \sum_{i=1}^{N} \frac{|y_i - f(t_i)|^2}{\sigma^2}. \tag{2}$$

We denote the standard Sobolev space of functions with square integrable derivatives by $W_{m,2}$ [20]. In general, the smoothing parameter will be decreased as N increases: $\lambda_N \to 0$. For smoothing splines, we add the stronger requirements:

Assumption A* [1]. Let Assumption A hold with $f \in W_{m,2}$ and $m > 3/2$. Consider the sequence of smoothing spline minimizers of (2). Let the smoothing parameter, λ_N, satisfy $\lambda_N \to 0$ and $D_N^* \lambda_N^{-5/(4m)} \to 0$ as $N \to \infty$.

The constraint that $D_N^* \lambda_N^{-5/(4m)} \to 0$ is very weak since the optimal value of λ_N satisfies $\lambda_N \sim N^{-2m/(2m+1)}$. The assumption that $F(t)$ is $C^\infty[0,1]$ can be weakened.

For unconstrained smoothing spline estimates, the MISE has the following upper bound:

Theorem 2.1 [13, 1]. *Let Assumption A^* hold, and denote the smoothing spline estimate from (2) by $\hat{f}_{N,\lambda}(t)$. As $N \to \infty$,*

$$\mathbf{E}\left[\int |\hat{f}_{N,\lambda}^{(j)}(t) - f^{(j)}(t)|^2\right] \le \alpha_j \lambda^{(m-j)/m}\|f\|_m^2 + \frac{\beta_j \sigma^2}{N\lambda^{(2j+1)/(2m)}}, \quad j \le m, \quad (3)$$

where α_j and β_j are positive constants. The MISE is minimized by $\lambda = O(N^{-2m/(2m+1)})$ and

$$E\left[\int |\hat{f}_{N,\lambda}^{(j)}(t) - f^{(j)}(t)|^2 dt\right] = O\left(N^{-2(m-j)/(2m+1)}\right). \quad (4)$$

For uniformly spaced measurement points, this result is in [13], while [1] generalizes the result to these more general conditions on the measurement points. Cox imposes the condition that $j < m$ while his proof applies to $j = m$ as well.

Given change points, $\{x_1, x_2, \ldots, x_K\}$, we define the closed convex cone:

$$V_{m,2}^{K,l}[x_1, \ldots, x_K] = \{f \in W_{m,p}|(-1)^{k-1}f^{(l)}(t) \ge 0 \text{ for } x_{k-1} \le t \le x_k\}, \quad (5)$$

where $x_0 \equiv 0$ and $x_{K+1} \equiv 1$. When the change points are unknown, we need to consider $V_{m,2}^{K,l} = \bigcup_{\mathbf{x}\in[0,1]^K} V_{m,2}^{K,l}[\mathbf{x}] \cup (-V_{m,2}^{K,l}[\mathbf{x}])$, where $\mathbf{x} \equiv (x_1, x_2, \ldots, x_K)$.

If we know the change point locations $\mathbf{x}$, a natural estimator is $\hat{f} = \arg\min \text{VP}$ $[f]$ subject to the convex constraints that $\hat{f} \in V_{m,2}^{K,l}[\mathbf{x}]$. This constrained spline estimate is the basis of our analysis. Detailed representation and duality results are in [14]. If we correctly impose the constraint that $f \in V_{m,2}^{K,l}[\mathbf{x}]$, the constrained spline fit always outperforms the nonconstrained fit as the following theorem indicates:

Theorem 2.2 [18]. *Let f be in a closed convex cone, $C \subseteq W_{m,2}$. Let $\hat{f}_u$ be the unconstrained minimizer of (2) given y_i, and $\hat{f}_c$ be the constrained minimizer. Then $\|f - \hat{f}_c\|_V \le \|f - \hat{f}_u\|_V$, where*

$$\|f\|_V^2 \equiv \frac{\lambda}{2}\int |f^{(m)}(s)|^2 ds + \frac{1}{N\sigma^2}\sum_{i=1}^N |f(t_i)|^2. \quad (6)$$

Theorem 2.2 applies to any set of y_i and does not use $y_i = f(t_i) + \varepsilon_i$. Theorem 2.2 shows that if one is certain that f is in a particular closed convex cone, the constrained estimate is always better than the unconstrained one. Unfortunately, Theorem 2.2 does not generalize to unions of convex cones, and thus does not apply to $V_{m,2}^{K,l}$.

Theorem 2.3. *Consider $f \in W_{m,2}$ and let Assumption A^* hold. Consider a sequence of estimates, $\hat{f}_{u,N}$, which satisfy the error bound given by (3), and a second sequence of estimates, $\hat{f}_{c,N}$, which satisfy the error bound:* $\|f - \hat{f}_c\|_V \leq \|f - \hat{f}_u\|_V$. *The asymptotic error bound given by (3) holds for $\hat{f}_{c,N}$ with different constants, α'_j and β'_j.*

Proof. For uniform sampling, this theorem is proved in [18] using interpolation inequalities. To generalize Utreras's result to our sampling hypotheses, we replace his Lemma 4.3 with (15) and (16) in Appendix A (with $\delta = 0.05$) and substitute $D_N^{*^{0.45}}$ everywhere $1/n$ appears. □

This generalization of Utreras's result does not require the ratio of $\overline{\Delta}_N \equiv \sup_{i<N}(t_{i+1} - t_i)$ to $\underline{\Delta}_N \equiv \inf_{i<N}(t_{i+1} - t_i)$ to be bounded. In practice, we choose our constraints empirically and sometimes impose an incorrect constraint. We now show that occasionally imposing the wrong constraint does not degrade the asymptotic rate of convergence provided that the probability of an incorrect constraint is sufficiently small. The following theorem is a basis for our data adaptive estimators in Sec. 4.

Theorem 2.4 (Occasional Misspecification). *Consider a two-stage estimator that with probability $1 - \mathcal{O}(p_N)$, correctly chooses a closed convex cone C, with $f \in C$, in the first stage and then performs a constrained regression as in (2). Under Assumption A^*, the estimate $\hat{f}^{(j)}$ satisfies the asymptotic bound (3) (with different constants, α'_j and β'_j) provided that, as $N \to \infty$, p_N vanishes rapidly enough: $P_N/\lambda_N \to 0$ and $P_N N \lambda^{1/(2m)} \to 0$.*

The proof is given in Appendix B.

3. Convergence of Kernel Smoothers

In this section, we examine the expected number of zeros of a kernel smoother as a function of the halfwidth parameter. The results in this section are proven in [15]. We begin by presenting convergence results for kernel estimators $\hat{f}_N^{(l)}(t)$, of $f^{(l)}(t)$ as $N \to \infty$. We define

$$\sigma_N^2(t) = \mathbf{Var}\left[\hat{f}_N^{(l)}(t)\right], \quad \xi_N^2(t) = \mathbf{Var}\left[\hat{f}_N^{(l+1)}(t)\right],$$

$$\mu_N(t) = \mathbf{Corr}\left[\hat{f}_N^{(l)}(t), \hat{f}_N^{(l+1)}(t)\right].$$

We use the notation $\mathcal{O}_R(\cdot)$ to denote a size of $\mathcal{O}(\cdot)$ relative to the main term: $\mathcal{O}_R(\cdot) = \times [1 + \mathcal{O}(\cdot)]$ and $\mathcal{O}_R = \times [1 + \mathcal{O}(\cdot)]$. We define $\|f\|_{bv}$ to be the sum of the L_∞ and total variation norms of f.

Lemma 3.1 (Generalized Gasser–Müller [2, 15]). *Let $f(t) \in C^{l+1}[0,1] \cap TV[-1,1]$ and consider a sequence of estimation problems satisfying Assumption A.*

Let $\hat{f}_N^{(l)}(t)$ be a kernel smoother estimate of the form

$$\hat{f}^{(l)}(t) = \frac{1}{Nh_N^{l+1}} \sum_i^N \frac{y_i w_i}{F'(t_i)} \kappa^{(l)}\left(\frac{t - t_i}{h_N}\right), \tag{7}$$

where h_N is the kernel halfwidth and the weights w_i satisfy $|w_i - 1| \sim \mathcal{O}(D_N^*/h_N)$. Let the kernel, $\kappa^{(l+1)} \in TV[-1,1] \cap C[-1,1]$, satisfy the moment condition $\int_{-1}^1 \kappa(s) \, ds = 1$, and the boundary conditions $\kappa^{(j)}(-1) = \kappa^{(j)}(1) = 0$ for $0 \le j \le l$. Choose the kernel halfwidths such that $h_N \to 0$ and $D_N^*/h_N^{l+2} \to 0$. Then

$$\mathbf{E}\,[\hat{f}_N^{(l)}](t) \to f^{(l)}(t) + \mathcal{O}_{\mathcal{R}}(h_N + D_N^*/h_N^{l+1}),$$

$$\mathbf{E}\,[\hat{f}_N^{(l+1)}](t) = \int_{-1}^1 f^{(l)}(t + hs)\kappa(-s)ds + \mathcal{O}(\|f\kappa^{(l+1)}\|_{bv} D_N^*/h_N^{l+2}),$$

$$\sigma_N^2(t) \to \sigma^2\|\kappa^{(l)}\|^2/(NF'(t)h_N^{2l+1}) + \mathcal{O}_{\mathcal{R}}(h_N + D_N^*/h_N),$$

$$\xi_N^2(s) \to \sigma^2\|\kappa^{(l+1)}\|^2/(NF'(s)h_N^{2l+3}) + \mathcal{O}_{\mathcal{R}}(h_N + D_N^*/h_N)$$

and

$$\mu_N^2(t) \to \mathcal{O}(h_N + D_N^*/h_N) \text{ uniformly in the interval } [h_N, 1 - h_N].$$

Lemma 3.1 applies to all of the common kernel smoother weightings [2] such as Priestley–Chao and Gasser–Müller. This result is slightly stronger than previous theorems on kernel smoothers [2]. Our hypotheses are stated in terms of the star discrepancy while previous convergence theorems [2] place restrictions on both $\overline{\Delta}_N \equiv \sup_{i<N}\{t_{i+1} - t_{i-1}\}$ and $\varepsilon_N \equiv \sup_{i<N}\{|1 - (t_{i+1} - t_i)NF'(t_i)\}$.

We now evaluate the expected number of false l-change points for a sequence of kernel estimates of $f^{(l)}(t)$. We restrict to independent *Gaussian* errors: $\varepsilon_i \sim N(0, \sigma^2)$. Thus, $\hat{f}_N^{(l)}(t)$ a Gaussian process. The following assumption rules out nongeneric cases:

Assumption B. Let $f(t) \in C^{l+1}[0,1]$ have K l-change points $\{x_1, \ldots, x_K\}$, with $f^{(l+1)}(x_k) \neq 0$, $f^{(l)}(0) \neq 0$ and $f^{(l)}(1) \neq 0$. Consider a sequence of estimation problems with independent, normally distributed measurement errors ε_i^N with variance σ^2. Let $\hat{f}_N^{(l)}(t)$ be a sequence of kernel estimates of kernel estimates of $f^{(l)}$ on the sequence of intervals $[\delta_N, 1 - \delta_N]$.

To neglect boundary effects, we take $\delta_N = h_N$ for kernel estimators and $\delta_N = \delta$ for splines. For each change point x_k, we define the change point variance [2]: $\sigma_{if}^2(x_k) \equiv \mathbf{Var}[\hat{f}_N^{(l)}(x_k)]/f^{(l+1)}(x_k)|^2$. The following theorem bounds the probability of a false estimate of a change point far away from a true change point.

Theorem 3.2. *Let Assumption B hold and consider a sequence of kernel estimators $\hat{f}_N^{(l)}(t)$ that satisfy the hypotheses of Lemma 3.1. Choose kernel halfwidths,*

h_N, and uncertainty intervals w_N such that $h_N/w_N \to 0$, $w_N \to 0$, $w_{N,k}^2 N h_N^{2l+1} \geq$ 1. The probability, $p_N(w_N)$, that $\hat{f}_N^{(l)}$ has a false change point outside of a width of w_N from the actual $(l+1)$-change points satisfies

$$p_N(w_N) \leq \sum_{k=1}^{K} \mathcal{O}\left(\frac{\sigma_{if}(x_k)}{h_N} \exp\left(\frac{-w_N^2}{2\sigma_{if}^2(x_k)}\right)\right), \tag{8}$$

where $\sigma_{if}^2(x_k) \to \sigma^2 \|\kappa^{(l)}\|^2 / |f^{(l+1)}(x_k)|^2 N F'(x_k) h_N^{2l+1}$ on the interval $[h_N, 1 - h_N]$.

In [8, 9], Mammen and co-workers derive the number of false change points for kernel estimation of a probability density. We present the analogous result for regression function estimation.

Theorem 3.3 (Analog of [8, 9]). *Let Assumption B hold. Consider a sequence of kernel smoother estimates $\hat{f}_N$ which satisfy the hypotheses of Lemma 3.1 with $\int_{-1}^{1} s\kappa(s)ds = 0$. Let the sequence of kernel halfwidths h_N satisfy $D_N^* N^{1/2} h_N^{\frac{1}{2}} \to 0$ and $0 < \lim\inf_N h_N N^{1/(2l+3)} \leq \lim\sup_N h_N N^{1/(2l+3)} < \infty$. The expected number of l-change points of $\hat{f}_N$ in the estimation region $[h_N, \, 1 - h_N]$ is asymptotically*

$$\mathbf{E}[\hat{K}] - K = 2\sum_{k=1}^{K} H\left(\sqrt{\frac{|f^{(l+1)}(x_k)|^2 N F'(x_k) h^{2l+3}}{\sigma^2 \|\kappa^{(l+1)}\|^2}}\right) + o_{\mathcal{R}}(1) \tag{9}$$

where $H(z) \equiv \phi(z)/z + \Phi(z) - 1$ with ϕ and Φ being the Gaussian density and cdf respectively. If $f^{(l+1)}(t)$ has Hölder smoothness of order ν for some $0 < \nu < 1$, and $h_N N^{1/(2l+3)} \to 0$, then (9) remains valid provided that $h_N N^{1/(2l+3+2\nu)} \to 0$.

In [8, 9], the correction in (9) is shown to be $o(1)$ when $\lim\sup_N h_N N^{1/(2l+3)} < \infty$. We strengthen this result by showing that (9) continues to represent the leading order asymptotics even when $h_N N^{1/(2l+3)} \to \infty$.

For each change point, x_k, we define the α uncertainty interval by

$$[x_k - z_\alpha \sigma_{if}(x_k), x_k + z_\alpha \sigma_{if}(x_k)],$$

where z_α is the two-sided $\alpha[1 + 2H(h_N\|\kappa^{(l)}\|/\sigma_{if}(x_k)\,\|\kappa^{(l+1)}\|))]$-critical value for a normal distribution. The probability that an empirical change point is more than $z_\alpha \sigma_{\mathrm{if}}(x_k)$ away from the kth actual change point is less than α. We consider two change points well resolved if the two uncertainty intervals do not overlap.

A similar variance for change point estimation is given in [11], where Müller shows that the left-most change point is asymptotically normally distributed with variance $\sigma_{if}^2(x_k)$. For his result, Müller imposes stricter requirements on $f^{(l+1)}$ and $\hat{f}^{(l+1)}$, and does not obtain results pertaining to the expected number of false change points.

When $f \in C^m$, the halfwidth which minimizes the MISE scales as $h_N \sim N^{-1/(2m+1)}$. Other schemes for piecewise convex fitting [5] choose the kernel halfwidth or smoothing parameter to be the smallest value, $h_{cr,K}$, that yields only K change points in an unconstrained fit. Theorem 3.3 shows that $h_{cr,K}$ is asymptotically larger than the halfwidth which minimizes the MISE for $l = m, m-1$. As a result, these schemes oversmooth.

4. Data-Based Pilot Estimators with Geometric Fidelity

We consider two-stage estimators that begin by estimating $f^{(l)}$ and $f^{(l+1)}$ using an unconstrained estimate with $h_N \gtrsim \log(N)N^{1/(2l+3)}$. From the pilot estimate, we determine the number, $\hat{K}$, and approximate locations of the change points. In the second stage, we perform a constrained fit; requiring that $\hat{f}^{(l)}$ be monotone in small regions about each empirical change point. Since spurious change points asymptotically occur only in a neighborhood of an actual change point, the second-stage fit need only be constrained in a vanishingly small portion of the domain asymptotically.

Theorem 4.1 (Asymptotic MISE for pilot estimation). *Let $f(t)$ satisfy Assumption B and consider a sequence of two-stage estimators. In the first stage, let the hypotheses of Theorem 3.2 be fulfilled. From the first-stage estimate, denote the empirical l-change points by $\hat{x}_k, k = 1, \ldots, \hat{K}$. Choose widths $w_{N,k}$ such that $w_{N,k} \to 0$, $h_N/w_{N,k} \to 0$ and $w_{N,k}^2 N h_N^{2l+1}/\ln(N) \to \infty$, where h_N is the first-stage halfwidth. In the second-stage, perform a constrained regression as in (2), where the second-stage smoothing parameter, λ_N, satisfies $\lambda_N \to 0$ and $D_N^* \lambda_N^{-5/4m} \to 0$. In the second-stage regression, impose the constraints that the second-stage $\hat{f}^{(l+1)}$ has a single sign in the regions $[\hat{x}_k - w_{N,k}, \hat{x}_k + w_{N,k}]$ [which matches the sign of $\hat{f}^{(l)}(\hat{x}_k + w_{N,k} - \hat{f}^{(l)}(\hat{x}_k - w_{N,k}))$]. For $f \in W_{m,2}$, the second-stage estimate $\hat{f}$ satisfies the expected error bounds of (3) (with different constants).*

Proof. Theorem 3.2 shows that $\hat{x}_k$ lie within $w_N/2$ of the x_k with probability $1 - p_N$, where $p_N = \mathcal{O}(\exp(-c_0^2 w_N^2 N h_N^{2l+1}/8\sigma^2))$ and $c_o = \inf_k\{|f^{(l+1)}(x_k)|\}$. In the remainder of the proof, we implicitly neglect this set of measure p_N and use arguments that are valid for large N. By Assumption B, there are no zeros of $f^{(l+1)}(x_k)$ in $[\hat{x}_k - w_{N,k}, \hat{x}_k + w_{N,k}]$ and thus $|f^{(l)}(\hat{x}_k + w_{N,k}) - f^{(l)}(\hat{x}_k - w_{N,k})| > c_o w_{N,k}$. Note that $\hat{f}^{(l)}(\hat{x}_k + w_{N,k}) - \hat{f}^{(l)}(\hat{x}_k - w_{N,k})$ has a Gaussian distribution with variance $2\sigma_N^2(t)$ and a bias error bounded by $\mathcal{O}(h_N)$ [2]. Thus, the sign of $\hat{f}^{(l)}(\hat{x}_k + w_{N,k}) - \hat{f}^{(l)}(\hat{x}_k - w_{N,k})$ is determined correctly with a probability of $1 - p_N$. The result follows from Theorem 2.4. $\square$

The trick of Theorem 4.1 is to constrain $\hat{f}^{(l+1)}$ to be positive (or negative) in the uncertainty interval of the estimated change points (linear constraints) rather than constraining $\hat{f}^{(l)}$ to have a single zero around $\hat{x}_j$ (nonlinear constraints). Theorem 4.1 implies that the second-stage estimate has no false l change points within $\pm w_{N,k}$ of $\hat{x}_k$ with high probability. It does not exclude the possibility that false change points occur outside of $[\hat{x}_k - w_{N,k}, \hat{x}_k + w_{N,k}]$, but we believe that such false change points seldom occur in practice. We believe that it is adequate to eliminate false inflection points in the regions where they occur in the unconstrained nonparametric estimates.

Asymptotically, the zeros of $\hat{f}^{(l+1)}$ will occur in clusters with an odd number of zeros. If a cluster with an even number of zeros occurs in the first stage, it is spurious (with high asymptotic probability). We recommend imposing the constraint that the second-stage $\hat{f}^{(l)}$ (not $\hat{f}^{(l+1)}$) has a single sign in each neighborhood where an even number of change points of the first stage occur.

For data adaptive methods, we modify Theorem 4.1 slightly:

Corollary 4.2. *The hypotheses of Theorem 4.1 on the first-stage estimate [such as $D_N^*/h_N^{l+2} \to 0$, $w_{N,k}^2 N h_N^{2l+1}/\ln(N) \to \infty$, $w_{N,k} \to 0$ and $h_N/w_{N,k} \to 0$] need only be true with probability $(1 - p_N)$, for the conclusion of Theorem 4.1 to be valid, where p_N satisfies $p_N/\lambda_N \to 0$ and $p_N N \lambda^{1/(2m)} \to 0$.*

Let h_{GCV} denote the smoothing bandwidth chosen by generalized cross-validation. Under certain conditions [4], it can be shown that h_{GCV} has an asymptotically normal distribution with mean $cN^{-\beta}$, where c and β depend on f and the kernel shape. For the first-stage halfwidth, we propose using $h_N = \iota(N)h_{GCV}$, where $\iota(N)$ is chosen such that $\iota(N)h_{GCV}$ satisfies Corollary 4.2. If $h_{GCV} \to cN^{-1/(2l+1)}$, we recommend choosing $\iota(N) \approx log(N)N^\alpha$ with $\alpha = 1/(2l + 1) - 1/(2l + 3)$. This scaling corresponds to a uniformly consistent estimate of $f^{(l+1)}$ [2]. The overall moral is: the smoothing level chosen by GCV is asymptotically optimal for estimating functions, but derivative estimation requires more smoothing.

Numerical implementations [19] of constrained least squares usually apply the active set method of quadratic programming. The constrained smoothing spline regression reduces to a finite dimensional minimization when the constraints are on the mth derivative. Since we constrain $\hat{f}_N^{(l+1)}(t)$ in each neighborhood of an estimated l-change point, our algorithm is most readily implemented for $l = m - 1$ using the duality result of [14].

To reduce the number of constraints, we seek to minimize the length of the constraint intervals, $w_{N,k}$. When the hypotheses of Theorem 3.3 are satisfied with $h_N N^{1/2l+3} \to \infty$, we can estimate the uncertainty interval for inflection points by $\hat{\sigma}_{if}^2(\hat{x}_k) \equiv \sigma^2 \|\kappa^{(l)}\|^2 / [|\hat{f}^{(l+1)}(\hat{x}_k)|^2 N F'(\hat{x}_k) h_N^{2l+1}]$. We recommend choosing the constraint width such that $w_{N,k} \gg \hat{\sigma}_{if}^2(\hat{x}_k)$. (For smaller constraint widths, Theorem 4.1 is true, but false inflection points can still occur near the actual inflection point.)

5. Potential Extensions

1. At present, we cannot exclude the posibility of false change points arising in the second-stage estimate away from the constraint intervals. We believe that this is a technical gap in our analysis and not a practical difficulty. The risk of

false change points can be reduced by choosing larger intervals to impose the constraints on $\hat{f}^{(l+1)}$. Let $\{\hat{u}_j\}$ denote the zeros of $\hat{f}^{(l+1)}$ in the first stage. Let u_j and u_{j+1} be the two closest zeros of $\hat{f}^{(l+1)}$ to x_k with $u_j < x_k < u_{j+1}$. A judicious choice of constraint intervals is $[(x_k - u_j)/2, (x_k + u_{j+1})/2]$, which gives large constraint intervals with only a small chance of imposing an incorrect constraint on the second-stage $\hat{f}^{(l+1)}$.

2. The pilot method suppresses false zeros of $f^{(l)}(t)$, but does not suppress false zeros of $f^{(l)}(t) - c$ where c is a nonzero real number. It may be more desirable to apply the pilot estimator to $f^{(l)}(t) - q(t)$, where $q(t)$ is a prescribed function possibly involving a small number of empirically estimated free parameters. The constraints in the second stage will virtually never need to be imposed if $f^{(l)}(t)$ is always positive or negative. Thus, we suggest centering $f^{(l)}(t)$ about zero by subtracting off a polynomial fit of order l (and thereby centering $f^{(l)}(t)$ about zero) prior to applying our two-stage estimator.

3. Asymptotically, smoothing splines are equivalent to kernel smoothers [1, 17]. Using this convergence, analogous results to Theorems 3.2, 3.3 and 4.1 can be proved for when smoothing splines are used in the first-stage estimate.

4. It is tempting to try the pilot estimation procedure using local polynomial regression (LPR) in the second stage. Unforunately, there is a difficulty with shape constrained LPR. Let $\hat{f}(x) \sim a_0(t) + a_1(t)[x - t] + a_2(t)[x - t]^2/2$ for $|x - t| < h$. If $a_2(t)$ is constrained to be nonnegative, $a_0(t)$ need not be convex because LPR does not require $a_0''(t) = a_2(t)$.

5. Our data adaptive convergence results in Secs. 3 and 4 are for quadratic estimation with Gaussian errors. The results should be extendable to non-Gaussian errors using the central limit theorem and Brownian bridges as in [9].

6. Piecewise Convex Information Criterion

Instead of the two-stage pilot estimator, we now propose a second class of estimators which penalize both smoothness and the number of change points. Information or discrepancy criteria are used to measure whether the improvement in the goodness of fit is sufficient to justify using additional free parameters. Both the number of free parameters and their values are optimized with respect to the discrepancy criterion, $d(\hat{f}, \{y_i\})$. Let $\hat{\sigma}^2$ be a measure of the average residual error:

$$\hat{\sigma}^2(\hat{f}, \{y_i\}) = \frac{1}{N\sigma^2} \sum_{i=1}^{N} [y_i - \hat{f}(t_i)]^2 \,,$$

or its L_1 analog. Typical discrepancy functions are

$$d^I(\hat{f}, \{y_i\}) = \hat{\sigma}^2/[1 - (\gamma_1 p/N)]^2 \tag{10}$$

and

$$d^B(\hat{f}, \{y_i\}) = \hat{\sigma}^2[1 + (\gamma_2 p \ln(N)/N)] \,, \tag{11}$$

where p is the effective number of free parameters in the smoothing spline fit. For $\gamma_1 = 1$, $d^I(\hat{f}, \{y_i\})$ is generalized cross-validation (GCV) which has the same asymptotic behavior as the Akaike information criterion. For $\gamma_2 = 1$, d^B is the Bayesian or Schwartz information criterion. For a nested family of models, $\gamma_2 = 1$ is appropriate while $\gamma_2 = 2$ corresponds to a nonnested family with $2\binom{N}{K}$ candidate models at the kth level. In very specialized settings in regression theory and time series, it has been shown that functions like d^I are asymptotically efficient while those like d^B are asymptotically consistent. In other words, using d^I-like criteria will asymptotically minimize the expected error at the cost of not always yielding the correct model. In contrast, the Bayesian criteria will asymptotically yield the correct model at the cost of having a larger expected error.

Our goal is to consistently select the number of convexity change points and efficiently estimate the model subject to the change point restrictions. Therefore, we propose the following *new* discrepancy criterion:

$$\mathbf{PCIC} = \sigma^2(\hat{f}, \{y_i\}) \left[\frac{1 + \gamma_2 K \ln(N)/N}{(1 - \gamma_1 p/N)^2} \right] , \qquad (12)$$

where K is the number of convexity change points and p is the number of free parameters. PCIC stands for Piecewise Convex Information Criterion. In selecting the positions of the K change points, there are essentially $2\binom{N}{K}$ possible combinations of change point locations if we categorize the change points by the nearest measurement location. Thus, our default values are $\gamma_1 = 1$ and $\gamma_2 = 2$.

We motivate PCIC: to add a change point requires an improvement in the residual square error of $O(\sigma^2 \ln(N))$, which corresponds to an asymptotically consistent estimate. If the additional knot does not increase the number of change points, it will be added if the residual error decreases by $\gamma_1 \sigma^2$. Presently, PCIC is purely a heuristic principle. We conjecture that it consistently selects the number of change points and is asymptotically efficient within the class of methods that are asymptotically consistent with regards to convexity change points.

7. Summary

Theorem 3.3 shows that for $l = m$ and $l = m - 1$, the amount of smoothing necessary for geometric fidelity is larger than the optimal value for minimizing the mean integrated square error. Therefore, we have considered two-stage estimators which estimate the l-change points and their uncertainty intervals in the first stage. In the second stage, a constrained smoothing spline fit is applied using a data-adaptive estimate of the smoothing parameter.

Our main result is that such two-stage schemes achieve the same asymptotic rate of convergence as standard methods such as GCV that do not guarantee geometric fidelity. We prove this result incrementally. Theorem 2.4 evaluates an acceptable rate of failure for imposing the wrong constraints. Theorem 4.1 proves the asymptotic MISE result when the change points are estimated in the first stage while the widths of the constraint intervals and the smoothing parameter in the first stage satisfy certain scaling bounds. The second-stage estimates have no false change points in the regions where the unconstrained estimators have all of their false change points with probability approaching unity.

Linear constraints are necessary only in small neighborhoods about each l-change point asymptotically. This suggest that the ratio of the MISE from our two-stage estimate to that of kernel smoothers or spline tends to 1. Our estimators should be useful in situations where obtaining the correct shape is important and computational costs are not an issue. Piecewise convex fitting may offer larger potential gains in the MISE in small sample situations because the *a priori* knowledge that there are only a small number of inflections points should be of more value when less data is available. Numerical simulations are underway and will be discussed elsewhere.

Appendix A

Interpolation Inequalities

We measure the distance of an arbitrary set of measurement times to an equispaced set of points in terms of the *discrepancy* as defined in Sec. 2. The discrepancy is useful because it describes how closely a discrete sum over an arbitrarily placed set of points approximates an integral. In this appendix, we summarize these results and present a new interpolation identity for discrete sums. An useful condition is the following:

Assumption 0. Assume that the limiting distribution of the measurement locations, $F(t)$, is $C^1[0,1]$ and $0 < c_F < F'(t) < C_F$.

We denote the set of functions of bounded variation by $TV[0,1]$ and the corresponding norm by $\|\cdot\|_{TV}$. The discrepancy is useful in evaluating the approximation accuracy of a discrete sum of arbitrarily placed points to the corresponding integral:

Theorem A.1 (Generalized Koksma [15]). *Let g be a bounded function of bounded variation, $\|g\|_{TV}$, on $[0,1] : g \in TV[0,1] \cap L_\infty[0,1]$. Let the star discrepancy be measured by a distribution, $F(t)$, which satisfies Assumption 0. If the discrete sum weights, $\{w_i, i = 1, \ldots, N\}$ satisfy $|w_i - 1| \leq CD_N^*$, then*

$$\left| \int_0^1 g(t)dF(t) - \frac{1}{N}\sum_{i=1}^N g(t_i)w_i \right| \leq [\|g\|_{TV} + C\|g\|_\infty] D_N^*. \qquad (13)$$

In our version of Koksma's theorem, we have added two new effects: a nonuniform weighting, $\{w_i, i = 1, \ldots, N\}$, and a nonuniform distribution of points, dF. The total variation of $g(t(F))$ with respect to dF is equal to the total variation of $g(t)$ with respect to dt. Theorem A.1 follows from Koksma's theorem by a change of variables.

In the continuous case, the following Sobolev interpolation result [13] is well known:

Lemma A.2. *There exists constants c_j depending only on m such that, for all $g \in W_{m,2}[0,1]$ and $\theta \in [0,1]$ $0 \le j \le m$,*

$$\theta^{2j} \int_0^1 |g^{(j)}(s)|^2 ds \le c_j \left[\int_0^1 |g(s)|^2 ds + \theta^{2m} \int_0^1 |g^{(m)}(s)|^2 ds \right]. \tag{14}$$

Using Koksma's theorem and Lemma A.2, we can arrive at the following inequalities:

Corollary A.3. *Let g be in $W_{m,2}[0,1]$ and assume the star discrepancy satisfies Assumption 0 with $m < N$. The following interpolation bounds hold:*

$$\frac{1}{N} \sum_{i=1}^N g^2(t_i) \le C_1 \int_0^1 g^2(t) dt + c_1 D_N^{*^m} \int_0^1 |g^{(m)}(s)|^2 ds, \tag{15}$$

where $C_1 = C_F + c_1 + D_N^*$, and

$$\left[c_F - c_1 D_N^{*^\delta} - D_N^* \right] \int_0^1 |g(s)|^2 \, ds \le \frac{1}{N} \sum_{i=1}^N g(t_i)^2 + c_1 D_N^{*^{m(1-2\delta)}} \int_0^1 |g^{(m)}(s)|^2 ds, \tag{16}$$

for all δ in $(0, 1/2)$ such that $c_F > c_1 D_N^{*^\delta} + D_N^*$.

Proof. For $g \in W_{1,2}$, Koksma's theorem implies

$$\left| \int_0^1 g^2(t) dF(t) - \frac{1}{N} \sum_{i=1}^N g^2(t_i) \right| \le \|g^2\|_{TV} D_N^* \le \left(\|g\|_{0,2}^2 + \|g\|_{1,2}^2 \right) D_N^*.$$

We then apply (14) to $\|g\|_{1,2}^2$ with $\theta = |D_N^*|^{1/2+\delta}$ for arbitrarily small δ.

$$\left| \int_0^1 g^2(t) dF(t) - \frac{1}{N} \sum_{i=1}^N g(t_i)^2 \right| \le c_1 D_N^{*^\delta} \left[\|g\|_{0,2}^2 + D_N^{*^{m(1-2\delta)}} \|g\|_{m,2}^2 \right] + \|g\|_{0,2}^2 D_N^*,$$

yielding the bound (16). □

Appendix B

Proof of Theorem 2.4. If the constraints are correct, Theorem 2.3 yields the asymptotic error bound. We need to show that misspecified models do not contribute significantly to the error. For any realization of the $\{y_i\}$, we have the bound

$$\|\hat{f} - f\|_V^2 \le \lambda(\|f\|_m^2 + \|\hat{f}\|_m^2) + \frac{1}{N\sigma^2} \sum_i^N |\hat{f}(t_i) - f(t) - \varepsilon_i|^2 + \varepsilon_i^2$$

$$\le \lambda \|f\|_m^2 + \frac{1}{N\sigma^2} \sum_i (y_i^2 + \varepsilon_i^2) \le \|f\|_V^2 + \frac{1}{N} \sum_i \frac{\varepsilon_i^2}{\sigma^2}. \tag{17}$$

This paragraph is devoted to bounding the expectation of $\sum_i \varepsilon_i^2$ over the worst possible set with probability p_N. Note $\sum_i \varepsilon_i^2$ has a χ_N^2 distribution with density $p_{\chi_N^2}(w) = w^{N/2-1} \exp(-w/2)/2^{N/2}\Gamma(N/2)$. We seek to bound $I_1 \equiv \int_{\chi_0^2}^{\infty} w\, dp_{\chi_N^2}(w)$, where $\chi_0^2(p_N)$ is defined by $\int_{\chi_0^2}^{\infty} dp_{\chi_N^2}(w) = p_N$. We claim that $I_1 \leq 2Np_N$. We assume that $\chi_0^2(p_N) > 1.5N$. (If $\chi_0^2(p_N) < 1.5N$, we split the integral into $w \leq 1.5N$ and $w > 1.5N$.) We define $\tilde{w} = w/N$ and use Sterling's formula to find $p_{\chi_N^2}(N\tilde{w}) \approx \tilde{w}^{N/2-1} \exp(-N(\tilde{w}-1)/2)/\sqrt{4\pi N}$. Evaluating the integrals by Laplace's method under the assumption that $p_N \ll 1$ and $\chi_0^2 > N$ yields $p_N \approx 2p_{\chi_N^2}(\chi_0^2)\chi_0^2/(\chi_0^2 - N + 1/2)$, and $I_1 \approx 2Np_{\chi_N^2}(\chi_0^2)\chi_0^2/(\chi_0^2 - N)$. For $\chi_0^2 > 1.5N$, we have $I_1 \approx p_N N(\chi_0^2 - N + 1/2)/(\chi_0^2 - N) \ll 0.5Np_N$.

Taking the expectation conditional on $f \notin V$ yields $\mathbf{E}_{f\notin V}\|\hat{f} - f\|_V^2 \leq \|f\|_V^2 + 2$. Since $(1/N)\sum_i |f(t_i)|^2 \to \int_0^1 f(s)^2 dF(s)$, we select N large enough that $\lambda_N\|f\|_m^2 > \frac{p_N}{N\sigma^2}\sum_i |f(t_i)|^2$. We now bound the contribution to $\mathbf{E}[\|\hat{f} - f\|_m^2]$ and $\mathbf{E}[\|\hat{f} - f\|_0^2]$ from $f \notin V$:

$$p_N \mathbf{E}_{f\notin V}\|\hat{f} - f\|_m^2 \leq \frac{p_N}{\lambda_N}\left(2 + \lambda_N\|f\|_m^2 + \frac{1}{N\sigma^2}\sum_i^N f(t_i)^2\right)$$

$$\leq \frac{2\sigma^2}{N\lambda^{(2m+1)/(2m)}} + [1 + \mathcal{O}(p_N)]\,\|f\|_m^2, \qquad (18)$$

by assumption on p_N. To bound $\mathbf{E}_{f\notin V}\|\hat{f} - f\|_0^2$, we apply Lemma A.2 with $\delta = 0.05$ in (16) and D_N^* large enough that $1/(c_F - c_1 D_N^{*\delta} + D_N^*)$ is bounded by a constant, γ:

$$p_N \mathbf{E}_{f\notin V}\|\hat{f} - f\|_0^2 \leq \gamma p_N \mathbf{E}_{f\notin V}\left[\frac{1}{N}\sum_{i=1}^N |\hat{f}(t_i) - f(t_i)|^2 + c_1 D_N^{*0.9m}\|\hat{f} - f\|_m^2\right]$$

$$\leq \gamma p_N \left(\sigma^2 + \frac{D_N^{*0.9m}}{\lambda}\right) \mathbf{E}_{f\notin V}\left[\|\hat{f} - f\|_V^2\right]$$

$$\leq \gamma p_N \left(\sigma^2 + \frac{D_N^{*0.9m}}{\lambda}\right)\left(2 + \lambda_N\|f\|_m^2 + \frac{1}{N\sigma^2}\sum_{i=1}^N f(t_i)^2\right)$$

$$\leq \gamma\left(\sigma^2 + 1\right)\left(\frac{\sigma^2}{N\lambda_N^{1/(2m)}} + \lambda_N[1 + \mathcal{O}(p_N)]\,\|f\|_m^2\right). \qquad (19)$$

Equations (18) and (19) are in the form required by Theorem 4.4 of [18]. Using Lemma 14 and duplicating the proof of Theorem 4.6 of [18] yields the result.   □

Acknowledgements

We thank the referee and the Editors for their help.

References

[1] D. D. Cox, *Multivariate smoothing splines functions*, SIAM J. Numer. Anal. **21** (1984) 789–813.

[2] Th. Gasser and H. Müller, *Estimating regression functions and their derivatives by the kernel method*, Scand. J. Statist. **11** (1984) 171–185.

[3] I. J. Good and R. A. Gaskins, *Density estimation and bump-hunting by the penalized likelihood method exemplified by scattering data*, J. Amer. Statist. Assoc. **75** (1980) 43–73.

[4] W. Härdle, P. Hall and J. S. Marron, *How far are automatically chosen regression smoothing parameters from their optimum?* J. Amer. Statist. Assoc. **83** (1988) 86–101.

[5] W. Li, D. Naik and J. Swetits, *A data smoothing technique for piecewise convex/concave curves*, SIAM J. Sci. Comput. **17** (1996) 517–537.

[6] M. Mächler, *Variational Solution of Penalized Likelihood Problems and Smooth Curve Estimation*, Ann. Statist. **23** (1995) 1496–1517.

[7] E. Mammen, *Nonparametric regression under qualitative smoothness assumptions*, Ann. Statist. **19** (1991) 741–759.

[8] E. Mammen, *On qualitative smoothness of kernel density estimates*, Statistics **26** (1995) 253–267.

[9] E. Mammen, J. S. Marron and N. J. Fisher, *Some asymptotics for multimodal tests based on kernel density estimates*, Probab. Theory Related Fields **91** (1992) 115–132.

[10] C. A. Michelli and F. Utreras, *Smoothing and interpolation in a convex set of Hilbert space*, SIAM J. Statist. Sci. Comput. **9** (1985) 728–746.

[11] H. G. Müller, *Kernel estimators of zeros and of the location and size of extrema of regression functions*, Scand. J. Statist. **12** (1985) 221–232.

[12] H. Niederrieter, *Random Number Generators and Quasi-Monte Carlo Methods*, SIAM, Philadelphia (1992).

[13] D. L. Ragozin, *Error bounds for derivative estimation based on spline smoothing of exact or noisy data*, J. Approx. Theory. **37** (1983) 335–355.

[14] K. S. Riedel, *Piecewise convex function estimation and model selection*, Proceedings of Approximation Theory VIII (C. K. Chui and L. L. Schumaker, eds.) (1995) 467–475, World Scientific, Singapore.

[15] K. S. Riedel, *Improved asymptotics for zeros of kernel estimates via a reformulation of the Leadbetter–Cryer integral*, Statist. Probab. Lett. **32** (1997) 351–356.

[16] B. W. Silverman, *Some properties of a test for multimodality based on kernel density estimates*, in Probability, Statistics and Analysis (J. F. C. Kingman and G. E. H. Reuter, eds.) (1983) 248–259, Cambridge Univ. Press.

[17] B. W. Silverman, *Spline smoothing: the equivalent variable kernel method*, Ann. Statist. **12** (1984) 898–916.

[18] F. Utreras, *Smoothing noisy data under monotonicity constraints — existence, characterization and convergence rates*, Numer. Math. **47** (1985) 611–625.

[19] M. Villalobos and G. Wahba, *Inequality-constrained multivariate smoothing splines with application to the estimation of posterior probabilities*, J. Amer. Statist. Assoc. **82** (1987) 239–248.

[20] G. Wahba, *Spline Models for Observational Data.* SIAM (1991), Philadelphia.

Author's present address: S.A.C., Capital Management, LLC 540 Madison Ave, New York, NY 10022, USA.

Reprinted from Computers in Phys.
8(4) (1994) 402–410

FUNCTION ESTIMATION USING DATA-ADAPTIVE KERNEL
SMOOTHERS — HOW MUCH SMOOTHING?

K. S. RIEDEL and A. SIDORENKO

Courant Institute, New York University, 251 Mercer St.,
New York, NY 10012-1185, USA
E-mail: riedel@cims.nyu.edu
E-mail: sidorenk@cims.nyu.edu

We consider a common problem in physics: how to estimate a smooth function given noisy measurements. We assume that the unknown signal is measured at N different times, $\{t_i\colon i = 1, \ldots, N\}$ and that the measurements, $\{y_i\}$, have been contaminated by additive noise. Thus, the measurements satisfy $y_i = g(t_i) + \varepsilon_i$, where $g(t)$ is the unknown signal and ε_i represents random errors. For simplicity, we assume that the errors are independent and have zero mean and uniform variance σ^2.

As an example, we consider a chirp signal: $g(t) = \sin(4\pi t^2)$. This signal is called a chirp because its "frequency" is growing linearly: $(d/dt)\{\text{phase}\} = 8\pi t$, which corresponds to the changing pitch in a bird's chirp. Figure 1 plots the chirp over two periods. Superimposed on the chirp is a point random realization of the noisy signal with $\sigma = 0.5$.

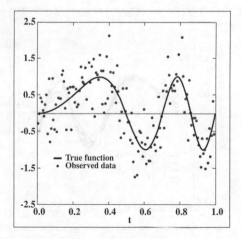

Fig. 1. Chirp signal: $g(t) = \sin(4\pi t^2)$. A random realization of 150 data points is superimposed with $\sigma = 0.5$.

A simple estimator of the unknown signal is a local average:

$$\hat{g}(t_i) = \frac{1}{2L+1} \sum_{j=-L}^{L} y_{i+j}\,, \tag{1}$$

where we assume that the sampling times are uniformly spaced. We denote the sampling rate, $t_{i+1} - t_i$, by Δ, and define the normalized kernel halfwidth $h \equiv (L\Delta)$. The ^ notation denotes the estimate of the unknown function.

When $L = 0$, the estimate is simply the point value: $\hat{g}(t_i) = y_i$, which has variance σ^2. By averaging over $2L + 1$ independent measurements, the variance of the estimate is reduced by a factor of $1/(2L+1)$ to $\sigma^2/(2L+1)$.

As we increase the averaging halfwidth, L, the variance of the estimate will decrease. However, the local average in Eq. (1) includes other data points that systematically differ from $g(t_i)$. As a result, the local average has a systematic bias error in estimating $\hat{g}(t_i)$:

$$\mathbf{E}[\hat{g}(t_i)] = \frac{1}{2L+1} \sum_{j=-L}^{L} g(t_{i+j}) \neq g(t_i)\,,$$

where $\mathbf{E}$ is the expectation of the estimate. As L increases, the variance decreases while the systematic error normally increases. This is a typical example of "bias-versus-variance trade-off" in data analysis.

Figure 2 plots the local average estimate of Eq. (1) for three different values of the halfwidth L: 4, 9, and 14. The averaged curves change discontinuously as measurements are added and deleted from the average. As a result, the averaged curve is harsh and unappealing. The simple average using a rectangular weighting has other disadvantages, such as making less-accurate estimate. Thus, we consider

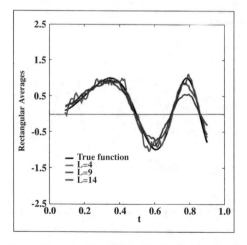

Fig. 2. Three kernel-smoother estimates of the randomized chirp signal using a rectangular window. The curves are actually discontinuous at the points where measurements are added and deleted.

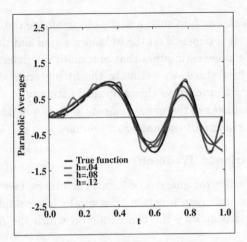

Fig. 3. Smoothed estimates of the randomized chirp signal using a parabolic kernel with three different values of the halfwidth h: 0.04, 0.08, 0.12, corresponding to L: 5, 11, 17. The parabolic weighting makes the curves continuous and more aesthetically pleasing than those of Fig. 2. The $h = 0.04$ average still has a lot of random jitter, indicating that the variance of the estimate is still large. The heights of the last three extrema have been reduced in the $h = 0.08$ and the $h = 0.12$ averages. The $h = 0.12$ average misses the last minimum entirely.

more-general kernel estimates by allowing an arbitrary weighting of the measurements:

$$\hat{g}(t_i) = \sum_{j=-L}^{L} w_j(t_i) y_{i+j}, \tag{2}$$

where the w_j are weights. Typically, the weights are given by a scale function, $\kappa(t)$: $w_j = c\kappa[(t_i - t_j)/h]$. The constant c is chosen such that $\sum_j w_j = 1$. We use a parabolic weighting, $\kappa(t) = (3/4)(1 - t^2)$, for $|t| \leq 1$ and zero otherwise. Near the ends of the data, we modify the kernel [see Eq. (A3)]. The normalized halfwidth h is still a free parameter that determines the strength of the smoothing, and h can depend on the estimation point t_i.

Figure 3 plots the local average estimate using the parabolic weighting for three different values of the halfwidth h: 0.04, 0.08, and 0.12, corresponding to L: 5, 11, and 17. The smoothed curves are continuous and are more aesthetic than those of Fig. 2. The $h = 0.04$ average has a lot of random jitter, indicating that the variance of the estimate is still large. The height of the second maximum and the depth of the two minima have been appreciably reduced in the $h = 0.08$ and the $h = 0.12$ averages. In fact, the $h = 0.12$ average misses the second minimum at $t = 0.935$ entirely. Because the estimated curve significantly depends on h, Fig. 3 shows the need for care in choosing the smoothing parameter.

One of the main issues discussed in this tutorial is how to pick h. In practice, to minimize artifical wiggles and nonphysical aspects, most scientists adjust the smoothing parameter according to their physical intuition. Using intuition makes the analysis subjective and often leads to suboptimal fits.

We would like to choose h in such a way as to minimize the fitting error. Unfortunately, the fitting error depends on the unknown signal and therefore is unknown. We consider multiple stage estimators that automatically determine the smoothing parameter. In the first stage, we estimate the fitting error and then choose the smoothing parameter to minimize this empirical estimate of the error. The combined estimate is nonlinear in the measurements and automatically adapts the local smoothing to the curvature of the unknown function.

1. Bias-Versus-Variance Trade-off

We now give a local-error analysis of kernel smoothers, based on a Taylor-series approximation of the unknown function. Essentially, the sampling rate is required to be rapid in comparison with the time scale on which the unknown function is varying.

The advantage of the weighted local average is that the variance of the estimate is reduced. To see this variance reduction, we rewrite Eq. (2) as

$$\hat{g}(t_i) = \sum_{j=-L}^{L} w_j [(g(t_{i+j}) + \varepsilon_{i+j}] . \tag{3}$$

Because the errors are independent, the resulting variance is

$$\mathbf{Var}[\hat{g}(t_i)] = \sum_{j=-L}^{L} w_j^2 \mathbf{Var}[\varepsilon_{i+j}] = \sigma^2 \sum_{j=-L}^{L} w_j^2 . \tag{4}$$

For the uniform weighting of Eq. (1), $w_j = 1/(2L+1)$ and the variance of the estimate is $\sigma^2/(2L+1)$. This same scaling of the variance holds for the more general class of scale function kernels: $\kappa(t)$: $w_j = c\kappa[(t_i - t_j)/h]$. The disadvantage is that averaging causes systematic error. We define the bias error to be

$$\text{Bias}[\hat{g}(t_i)] = \mathbf{E}[\hat{g}(t_i)] - g(t_i) = \sum_{j=-L}^{L} w_j g(t_{i+j}) - g(t_i) , \tag{5}$$

where $\mathbf{E}$ denotes the expectation. As the averaging halfwidth L increases, $|g(t_{i+L}) - g(t_i)|$ will normally increase, so the bias error will generally grow with increasing amounts of averaging. The expected square error (ESE) is

$$\text{ESE}(\hat{g}) = [\text{Bias}]^2 + \text{Variance} . \tag{6}$$

Figure 4 plots the ESEs of the three parabolic kernel averages. The smallest halfwidth has the largest ESE for $t \leq 0.5$. As time progresses, $g(t)$ oscillates more frequently. As a result, the bias error and the corresponding ESE for all three estimates oscillates more rapidly and increases. For $0.75 \leq t \leq 1.0$, the smallest halfwidth is the most reasonable, illustrates that the halfwidth of the kernel smoother should decrease when the unknown function varies more rapidly.

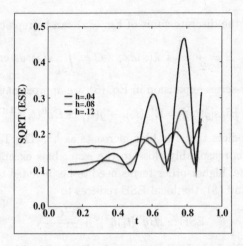

Fig. 4. Expected square error (ESE) of the parabolic-kernel smoothers of Fig. 3. As time progresses, $g(t)$ varies more rapidly, and the bias error grows for all three estimates. As a result, the ESE for all three estimates grows and oscillates proportionally to $|g''(t)|^2$. Each of the halfwidths performs best in a different time interval. The smallest halfwidth has the largest ESE for $t \leq 0.5$. For $0.75 \leq t \leq 1.0$, the smallest halfwidth is the most reasonable.

Our goal in this tutorial is to minimize the ESE of the kernel estimate of $\hat{g}(t)$. Since the ESE is unknown, we estimate the ESE and then optimize the kernel halfwidth with respect to the estimated ESE.

2. Local Error and Optimal Kernels

To understand the systematic error from smoothings, we make a Taylor series expansion of $g(t)$ about t_i: $g(t) = g(t_i) + g'(t_i)(t - t_i) + g''(t_i)[(t - t_i)^2/2] \ldots$. We make this Taylor-series expansion over the kernel halfwidth: $[t_{i-L}, t_{i+L}]$. For this expansion to be valid, the signal $g(t)$ must evolve slowly with respect to this averaging time, $t_{i+L} - t_{i-L}$.

We assume that the kernel weights satisfy the moment conditions:

$$\sum_{j=-L}^{L} w_j = 1, \quad \sum_{j=-L}^{L} (t_{i+j} - t_i)w_j = 0. \tag{7}$$

We define the following moments:

$$\cdot \ B_L(t_i) = \frac{1}{2h^2} \sum_{j=-L}^{L} (t_{i+j} - t_i)^2 w_j, \quad C_L = L \sum_{j=-L}^{L} w_j^2. \tag{8a}$$

We now consider the sampling limit where the number of measurements N in a fixed time interval tends to infinity and the sampling time Δ tends to zero. In this fast-sampling limit, when the weights are given by a scale function, $w_j(t) =$

$(1/L)\kappa[(t - t_j)/h)]$, the discrete sums of Eq. (8a) can be replaced by the integrals:

$$B = \frac{1}{2} \int_{-1}^{1} s^2 \kappa(s) ds \,, \quad C = \int_{-1}^{1} \kappa(s)^2 ds \,. \tag{8b}$$

Making a Taylor-series expansion in Eq. (5), we approximate the local bias as

$$\text{Bias}[\hat{g}(t_i)] = \mathbf{E}[\hat{g}(t_i)] - g(t_i) \simeq Bg''(t_i)h^2 \,. \tag{9}$$

Equation (9) predicts that the bias increases as h^2. The Taylor-series approximation of the bias is reasonably close to the exact bias except when $g''(t)$ nearly vanishes. In this case, higher-order terms need to be included to evaluate the ESE. Using Eqs. (4) and (8), the local ESE reduces to

$$\text{ESE} \simeq [Bg''(t_i)h^2]^2 + \frac{\sigma^2 C \Delta}{h} \,. \tag{10}$$

If we minimize the local ESE [as given by Eq. (10)] with respect to the kernel halfwidth h, the optimal halfwidth is

$$h_{as}(t) = \left[\frac{\sigma^2 C \Delta}{4B^2 |g''(t_j)|^2} \right]^{1/5} \,. \tag{11}$$

For this choice of kernel width h_{as}, the total expected squared error of Eq. (10) is

$$\text{ESE} \simeq 1.65 |Bg''(t)|^{2/5} (\sigma^2 C \Delta)^{4/5} \,. \tag{12}$$

Thus, the optimal h is proportional to $\Delta^{1/5}$, and the total squared error is proportional to $\Delta^{4/5}$. [Equation (12) can be used to give error bars for kernel-smoother estimates.]

Equation (12) is an asymptotic formula and may or may not be a good approximation for a particular signal and data set. In Fig. 5, we compare the local

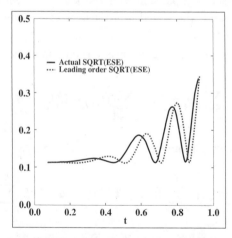

Fig. 5. Comparison of the leading-order ESE given by Eq. (10) with the exact ESE for a kernel-smoother halfwidth of $h = 0.08$. The shape and magnitude of the two curvse are very similar, but there is a phase shift between then due to higher-order Taylor-series terms.

ESE approximation with the actual ESE for the noisy chirp signal. The solid line is the actual value of the ESE for $h = 0.08$, while the dotted line gives the local approximation of the ESE. For $t \leq 0.5$, the two curves agree. For larger times, the shapes and magnitudes of the two curves are very similar, but there is a phase shift between them due to higher-order Taylor-series terms.

Equation (11) drives data-adaptive kernel estimation. In essence, Eq. (11) shows that when $g(t)$ is rapidly varying ($|g''(t)|$ is large), the kernel halfwidth should be decreased. Equation (11) gives an explicit solution for the halfwidth that minimizes the local bias-versus-variance trade-off.

Equation (11) has two major difficulties. First, $g''(t)$ is unknown, and thus Eq. (11) cannot be used directly. Estimating $h_{as}(t)$ is considered in the next section. Second, Eqs. (10)–(12) are based on a Taylor-series expansion, and the expansion parameter is $h_{as} \sim \Delta^{1/5}$. Even when Δ is small, corresponding to fast sampling, $\Delta^{1/5}$ may not be so small.

Figure 6 compares the halfwidth of Eq. (11) with the halfwidth that minimizes the actual ESE. [We calculate the exact ESE by evaluating Eq. (5) for the bias contribution to Eq. (6).] At the four extrema of $|g''(t)|$, $h_{as}(t)$ becomes infinite. Similarly, the actual optimal halfwidth becomes large at four nearby points where the exact bias nearly vanishes. The phase shift between the exact and approximate halfwidths is also apparent.

Figures 5 and 6 illustrate that the local approximation is valuable but can be fallible. Furthermore, the $|g''(t)|^{-2/5}$ dependence of the local optimal halfwidth makes $h_{as}(t)$ depend sensitively on the second derivative of $g(t)$.

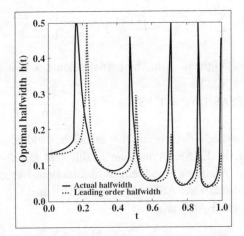

Fig. 6. Optimal halfwidth of kernel smoother as a function of time. The solid line gives the halfwidth that minimizes the exact ESE, while the dashed line gives the halfwidth based on a Taylor-series approximation. At the four extrema of $|g''(t)|$, $h_{as}(t)$ becomes infinite. The phase shift between the extrema of the exact optimal halfwidth and the Taylor-series optimal halfwidth is apparent.

3. How to Select the Halfwidth

There are two main approaches to selecting the smoothing parameter h from the data: goodness-of-fit estimators and plug-in derivative estimators. Figure 6 shows that the kernel halfwidth should be adaptively adjusted as the estimation point t varies. In practice, most codes use a fixed halfwidth for simplicity and stability. In this section, we describe two goodness-of-fit estimators. In the next section, we present a plug-in-derivative scheme with variable halfwidth.

Penalized Goodness-of-Fit Halfwidth Selection.   When the halfwidth is constrained to be constant over the entire interval, we want to choose a constant value of h to minimize the ESE averaged over all sample points, which we denote by EASE for "expected average square error." Since the EASE is unknown, a number of different methods have been developed to estimate it. In the goodness-of-fit methods, the average square residual error (ASR) is evaluated as a function of the kernel halfwidth:

$$\text{ASR}(h) = \frac{1}{N} \sum_i |y_i - \hat{g}(t_i|h)|^2 , \tag{13}$$

where $\hat{g}(t_i|h)$ is the kernel-smoother estimate using the halfwidth h. The ASR systematically underestimates the actual loss, because y_i is used in the estimate $\hat{g}(t_i|h)$ of y_i. The variance term in the EASE is

$$V(h) \equiv \sigma^2 \sum_{j=-L}^{L} w_j^2 \approx \frac{\Delta}{h} \sigma^2 C ,$$

where C is given in Eq. (8b). Correspondingly, the variance term in the ASR is

$$\hat{V}(h) = \sum_{j=-L}^{L} (w_j - \delta_{0,j})^2 \approx V(h) + \sigma^2 \left[1 - \frac{2\Delta}{h} \kappa(0) \right] ,$$

where $\kappa(0)\Delta/h$ is the weight of y_i in the kernel smoother. Thus,

$$\mathbf{E}[\text{ASR}(h)] = \text{EASE} + \sigma^2 \left(1 - \frac{2\kappa(0)\Delta}{h} \right) . \tag{14a}$$

Goodness-of-fit methods determine the kernel halfwidth, h by minimizing an empirical estimate of the EASE. A number of different goodness-of-fit functionals have been proposed. A simple goodness-of-fit functional is given by "inverting" Eq. (14a) under the ansatz that $\text{ASR}(h) \simeq \mathbf{E}[ASR(h)]$ and that $\text{ASR}(h) \simeq \sigma^2$. The resulting estimate of the EASE is

$$\widehat{\text{EASE}} = \text{ASR}(h) \left(1 + \frac{2\kappa(0)\Delta}{h} \right) - \sigma^2 . \tag{14b}$$

Monte Carlo test have shown that minimizing Eq. (14b) with respect to h tends to underestimate the value of the optimal halfwidth. As a result, a number of

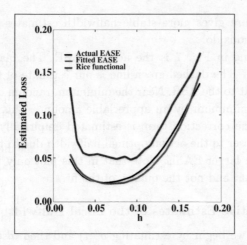

Fig. 7. Empirical fits to the expected average square error. The solid red curve is the actual EASE as a function of kernel halfwidth. The solid black curve is the Rice functional minus σ^2. The dotted green curve is the estimated EASE using the two-parameter fit to the ASR. In this plot, only the locations of the minima are important and not the quality of the fit. Random oscillations can shift the location of the minimum of the corrected ASR appreciably. The minimum of the fitted ASR criterion is less sensitive.

alternative estimates have been proposed. We consider one of the more popular alternatives, the Rice criterion, which selects the kernel halfwidth h_{Rice} by mini- mizing

$$C_R(h) = \text{ASR}(h) \left(1 - \frac{2\kappa(0)\Delta}{h} \right)^{-1}. \tag{14c}$$

Unfortunately, $C_R(h)$, like most of the goodness-of-fit functionals, is often very flat, and its actual minimum can be very sensitive to noise (see Fig. 7). As a result, the halfwidth given by the Rice criterion tends to vary appreciably even when the noise is weak.

Fitted Mean-Square-Error Halfwidths. We now consider methods that produce less-sensitive estimates of the best halfwidth. To reduce the random errors in the h optimization, we fit the ASR(h) with the parametric model:

$$\text{ASR}(h) \sim a\hat{V}(h) + bh^4. \tag{15}$$

The model has two free parameters, a and b. The first one corresponds to σ^2, and the second one corresponds to $|Bg''(t)|^2$. By parameterizing ASR(h), we are assured of a unique minimum. We determine a and b by minimizing the least-squares problem:

$$\{a, b\} = \text{argmin}_{\{a,b\}} \sum_{h_j} \{\text{ASR}(h_j) - [a\hat{V}(h_j) + bh_j^4]\}^2, \tag{16}$$

where the h_j are equispaced in h with a cutoff value of h_j chosen such that $C_R(h_{cutoff}) \sim 2C_R(h_{Rice})$. We then select the halfwidth to minimizes $aV(h) + bh^4$.

This fitting procedure gives more stable halfwidth estimates than the penalized goodness-of-fit methods do.

The solid red line in Fig. 7 is the actual EASE. The black line is the Rice functional minus σ^2. The dashed green line is our estimate of the EASE based on the two-parameter fit to the ASR. Near the minimum, random oscillations can shift the location of the minimum by an appreciable amount. This effect is even more pronounced when the correction term is estimated empirically. The minimum of the fitted ASR is closer to the actual optimal halfwidth due to the stabilizing effect of the h values with larger EASE. Note that in this plot only the locations of the minima are important and not the quality of the fit.

4. Plug-in-Derivative Estimates of the Local Halfwidth

In this method, we begin by estimating $g''(t)$ and then inserting this estimate into Eq. (11). This estimation and insertion procedure is called a plug-in-derivative estimate. We describe a particular plug-in method based on the work of Müller and Stadtmüller. The scheme is somewhat complicated because it is optimized to give optimal performance when the sampling rate is very high.

To estimate $g''(t)$, we use a kernel which satisfies

$$\frac{1}{2} \sum_{j=-L}^{L} (t_{i+j} - t_i)^m w_j = \begin{cases} 0 & \text{if} \quad m = 0, 1, 3; \\ 1 & \text{if} \quad m = 2. \end{cases}$$

Equation (A1) presents an "optimal" kernel for estimating derivatives. The bias in the estimating $g''(t)h^2$ scales as $g^{(iv)}h^4$, while the variance decreases as Δ/h. Optimizing the bias-versus-variance trade-off yields an optimal halfwidth for estimating second derivatives, h_2, which is proportional to $\Delta^{1/9}$. The halfwidth for estimating $g''(t)$ is larger than that of the halfwidth for estimating $g(t)$ because the bias error increases more slowly (h^4 versus h^2).

Now we may plug the estimates of $g''(t)$ and σ^2 into Eq. (11) to get the variable halfwidth $h_{as}(t)$ for estimating $g(t)$. The described procedure is not complete because the halfwidth for estimating $g''(t)$ is not specified. Since the final estimate is not sensitive to the initial halfwidth, it is possible to input the halfwidth by hand. A better approach is to begin the estimation procedure using a goodness-of-fit estimator. In our numerical implementation, we use the fitted goodness-of-fit method to determine the halfwidth for estimating $g''(t)$. When the goodness-of-fit initialization is included, the combined method has three stages:

1. Goodness-of-fit initialization to determine h_2.
2. Kernel-smoother estimation of $g''(t)$.
3. Final kernel estimation using the plug-in estimate of $h_{as}(t)$.

To illustrate this multiple-stage estimation, Fig. 8 plots the second-derivative estimate of the chirp signal. The fitted ASR criterion yielded an optimal halfwidth

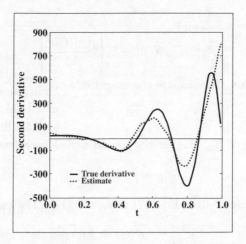

Fig. 8. Comparison of $g''(t)$ (red line) with its estimate (black line). Since the halfwidth is fixed at $h = 0.15$, we undersmooth the initial times and oversmooth the more rapidly varying segment near the end of the data.

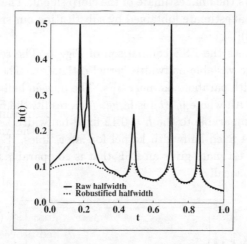

Fig. 9. Inferred value of the optimal kernel halfwidth using the plug-in estimate of Eq. (11). The solid black line is the simple plug-in, and the dotted red curve is the regularized plug-in using Eq. (17).

of 0.15 for the estimate of $g''(t)$. Since this halfwidth is time-independent, we undersmooth the initial times and oversmooth the more rapidly varying segment near the end of the data.

Figure 9 gives the plug-in estimate of the optimal halfwidth for smoothing $g(t)$. When $g''(t)$ is zero, the "optimal" halfwidth, $h_{as}(t)$, is infinite because Eqs. (10)–(12) neglect the higher-order bias, which in this case is proportional to $g^{(iv)}(t)h^4/(4!)$. We regularize Eq. (12) by replacing

$$|\widehat{g''(t)}|^2$$

with

$$|\widehat{g''}|^2 := |\widehat{g''}(t)|^2 + k_1 \left| \frac{\widehat{g^{(iv)}}(t)h^2}{12} \right|^2, \tag{17}$$

where k_1 is a constant. To estimate the magnitude of

$$|\widehat{g^{(iv)}}(t)|^2,$$

we invert the analog of Eq. (11) for the second-derivative estimate:

$$|\widehat{g^{(iv)}}|^2 \sim |\widehat{h_2}|^9.$$

Similarly, we eliminate h^2 from Eq. (17) using Eq. (11). The resulting expression is

$$|\widehat{g''}|^2 \leftarrow |\widehat{g''}(t)|^2 + k_2|\widehat{h_2}|^5.$$

We may use k_2 as a tuning parameter for the regularization of the kernel halfwidth. The red line in Fig. 9 shows the regularized halfwidth.

Figure 10 displays the final estimate of the chirp signal. The dotted lines are the error bars for the final estimate [obtained by substituting our smoothed estimate of $|g''(t)|^2$ into Eq. (12)].

Figure 11 continues the ESE comparison of Fig. 4. The red curve is the ESE of the multiple-stage variable-halfwidth kernel estimator. The other three curves are the fixed-halfwidth parabolic-kernel ESEs. The plug-in halfwidth automatically decreases the halfwidth where $|g''(t)|$ is large. As a result, the ESE of the multiple-stage estimate is comparable to the $h = 0.12$ fixed-halfwidth kernel for early times and to the $h = 0.04$ fixed-halfwidth kernel for later times. Essentially, the data-adaptive multistage estimate gives an ESE that is comparable to the best possible fixed-halfwidth ESE locally.

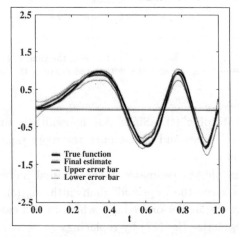

Fig. 10. Final estimate of the chirp signal using the multistep estimation procedure. The error bars are evaluated using Eq. (12).

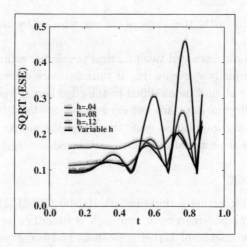

Fig. 11. Comparison of the ESE of the data-adaptive variable-halfwidth kernel smoother with the fixed-halfwidth kernel smoothers. The red curve is the ESE of the multiple-stage kernel estimator. The other three curves are the fixed-halfwidth parabolic-kernel ESEs. The plug-in halfwidth automatically decreases the halfwidth where $|g''(t)|$ is large. As a result, the multiple-stage-estimate's ESE is comparable to the smallest of the three ESEs of the fixed-halfwidth kernels at all time points.

5. Data-adaptive Smoothing

Local averaging is a common data-analysis technique. However, most of the practitioners make arbitrary and suboptimal choices about the amount of smoothing. In this tutorial, we determine the expected error by smoothing the data locally. Then we optimize the shape of the kernel smoother to minimize the error. Because the optimal estimator depends on the unknown function, our scheme automatically adjusts to the unknown function. By self-consistently adjusting the kernel smoother, the total estimator adapts to the data.

Goodness-of-fit estimators select a kernel halfwidth by minimizing a function of the halfwidth that is based on the average square residual fit error, ASR(h). A penalty term is included to adjust for using the same data to estimate the function and to evaluate the mean-square error. Goodness-of-fit estimators are relatively simple to implement, but the minimum (of the goodness-of-fit functional) tends to be sensitive to small perturbations. To remedy this sensitivity problem, we fit the mean-square error to a two-parameter model prior to determining the optimal halfwidth.

Plug-in-derivative estimators approximate the second derivative of the unknown function in an initial step, and then substitute this estimate into the asymptotic formula in Eq. (11). The multistage plug-in method is optimized to give the best possible performance when the sampling rate is very high. Figure 9 illustrates that the quality of fit is visibly enhanced using this plug-in estimate.

We caution that the analysis is based on a Taylor-series approximation of the unknown function. Essentially, we require that the sampling rate is sufficiently rapid in comparison with the time scale on which the unknown function is varying.

In the box, we have discussed two practical problems: estimation near the ends of the data and unequally spaced data. If your data are equispaced, the parabolic kernel in the interior should be modified to the edge kernel [see Eq. (A3)] near the ends of the data. If your data are not equispaced use local linear regression [see Eq. (A2)]. To reduce computation cost, the regression matrix can be modified by rank-one updates as the estimation point is advanced.

6. Further Reading

Two textbooks on kernel estimation are Hardle (1990) [1] and Müller (1980) [2]. The local-polynomial-regression approach is described in Hastie and Loader (1993) [3]. The equivalence of kernel smoothers and kernel regression is given in Sidorenko and Riedel (1993) [6]. This article also describes kernel estimation near the ends of the data. The goodness-of-fit approach to selecting kernel halfwidths is discussed in Hardle, Hall, and Marron (1988) [4]. The multistage data-adaptive scheme is described in Müller and Stadtmüller (1987) [5] and in Brockman, Gasser, and Hermann (1994) [7].

Special cases:

Kernel shapes, end points, and unequally spaced data

Kernel smoothers are local averages of the measurement. Thus a key question is "Which weighting is best?" We interpret "best" to mean minimizing the asymptotic ESE as given by Eq. (10). For equispaced data in the continuum limit, the answer is the parabolic weighting: $\kappa(t) = 3/4(1 - t^2)$. This parabolic weighting is optimal provided that the kernel halfwidth satisfies Eq. (11). More generally, to estimate the qth derivative to $O(h^2)$, the limiting shape of the optimal kernel is

$$\kappa(t) = \frac{(2q + 1)!}{2^{q+1}(q!)} [P_q(t) - P_{q+2}(t)], \tag{A1}$$

where $P_q(t)$ and $P_{q+2}(t)$ are the Legendre polynomials. To estimate the second derivative, Eq. (A1) reduces to $(105/16)(-1 + 6t^2 - 5t^4)$. In (A1), the estimation point is at $t = 0$ and the kernel support is $[-1, 1]$.

Does the kernel shape really matter?

For the parabolic kernel, the kernel parameters in Eq. (8b) are $B = 1/10$ and $C = 3/5$. In contrast, the rectangular kernel has $B = 1/6$ and $C = 1/2$. From Eqs. (11)–(12), the optimal halfwidth for the parabolic kernel is 27% longer than that of the rectangular kernel, but the ESE of the rectangular kernel is 6% larger than that of the parabolic kernel. The parabolic kernel outperforms the rectangular kernel because the bias error increases rapidly as the $t_{i+j} - t_i$ increases. The parabolic kernel compensates for this increasing bias by downweighting the points that are farthest away from the estimation point. More importantly, a rectangular window causes the estimate $\hat{g}(t)$ to be discontinuous in t while the parabolic kernel estimate is continuous in t.

What if the data points are not equispaced?

When the times between data points $t_{i+j} - t_i$ vary, the kernel shape needs to be modified to enforce the moment conditions. A simple way to do this is to reformulate the kernel smoother as a local linear (polynomial) regression. In other words, we fit the data $\{y_{i-L} \ldots y_{i+L}\}$ to a local polynomial, in this case $y_{i+j} \sim g(t) + g'(t)(t_{i+j} - t)$. We need to estimate both $g(t)$ and $g'(t)$ from the data. We do this by a weighted local regression:

$$\{\hat{g}(t), \hat{g'}(t)\} = \text{argmin}_{\{a,b\}} \sum_{j=-L}^{L} W_j |a + b(t_{i+j} - t) - y_{i+j}|^2, \qquad (A2)$$

where W_j are arbitrary weighting factors. When the W_j are given a parabolic weighting, the local-polynomial-regression estimate is identical to the kernel-smoother estimate with a parabolic weighting in the equispaced case.

The local-polynomial-regression formulation has three advantages. First, it automatically enforces the moment conditions even when the data is not equispaced. Second, it can be quickly updated as the estimation point is advanced in time. Third, polynomial regression works well in two- and three-dimensional settings because it automatically handles boundary effects.

How should the kernel weighting be modified near the boundary?

The parabolic kernel is optimal in the center of the domain. When the estimation point t_i is at or near the boundary, a nonsymmetric kernel should be used. We define the touch point t_p as the point where $h_{as}(t_p) = t_p$ [with h_{as} given in Eq. (11)]. When the estimation point is beyond the touch point, we fix the the kernel halfwidth to $h = h_{as}(t_p)$. We then adjust the kernel shape to maintain the moment conditions while keeping the ESE small.

We consider kernel estimates of the qth derivative near the left boundary of the data. We normalize the kernel weights near the boundary:

$$W_j(t) = \frac{\Delta}{h^{q+1}} G\left(\frac{t}{h} - 1, \frac{t_j}{h} - 1\right).$$

The function $G(z, y)$ is defined for $-1 \leq y \leq 1$ and $-1 \leq z \leq 0$. In the continuum limit, the boundary kernel that minimizes the ESE [under the assumption that $h(t) = h_0(t)$, i.e., that the kernel halfwidth is optimal] is

$$G(z, y) = \frac{(2q + 1)!}{2^{q+1}(q!)} \{P_q(y) + (2q + 3)z P_{q+1}(y) + [(2q + 3)z^2 - 1]P_{q+2}(y)\}. \qquad (A3)$$

When the data points are not equispaced, the simplest approach is to use local polynomial regression with the halfwidth fixed at $h = h_{as}(t_p)$. A simple weighting near the boundary is a parabola centered about the estimation point:

$$W_j(t) = 1 - \left(\frac{t - t_j}{t - 2t_p}\right)^2 .$$

What if the variance σ^2 is unknown?

We estimate it through three point residuals:

$$\hat{\sigma}^2 = \frac{2}{3(N-2)} \sum_{i=2}^{N-1} \left(y_i - \frac{y_{i+1} + y_{i-1}}{2}\right)^2 . \tag{A4}$$

Equation (A4) neglects the bias error from $g''(t)$. This is reasonable because only adjacent points are used.

Acknowledgments

The presentation has benefitted from a critical reading by Joe Keller. We thank Jim Matey for reformatting the figures. This work was funded by the U.S. Department of Energy. Grant DE-FG02-86ER53223.

References

[1] W. Hardle, *Applied nonparametric regression* (Cambridge University Press, Cambridge, MA, 1990).
[2] H. G. Müller, *Nonparametric regression analysis of longitudinal data* (Springer Verlag, Berlin, 1980).
[3] T. Hastie and C. Loader, "Local regression: automatic kernel carpentry", Statistical Science **8**, 120–143 (1993).
[4] W. Hardle, P. Hall and S. Marron, "How far are automatically chosen smoothing parameters from their optimum?" J. Amer. Stat. Assoc. **83**, 86–95 (1988).
[5] H. G. Müller and U. Stadtmüller, "Variable bandwidth kernel estimators of regression curves", Annals of Statistics **15**, 182–201 (1987).
[6] A. Sidorenko and K. S. Riedel, "Optimal boundary kernels and local polynomial regression", unpublished (1993).
[7] T. Brockman, Th. Gasser and E. Hermann, "Locally adaptive bandwidth choice for kernel regression estimators", J. Amer. Stat. Assoc. **88**, 1302–1309 (1994).

Authors' present address: S.A.C. Capital Management, LLC 540 Madison Ave, New York, NY 10022, USA.

E-ARCH MODEL FOR IMPLIED VOLATILITY TERM STRUCTURE OF FX OPTIONS

YINGZI ZHU

Citibank, N.A., 909 Third Avenue,
29th Floor, Zone 1, New York, NY 10043, USA

MARCO AVELLANEDA

Courant Institute of Mathematical Sciences,
251 Mercer Street, New York, NY 10012, USA

We construct a statistical model for term structure of implied volatility of currency options based on daily historical data for 13 currency pairs in a 19-month period. We examine the joint evolution of 1 month, 2 month, 3 month, 6 month and 1 year 50 Δ options in all the currency pairs. We show that from these five observable variables, there exist three uncorrelated state variables (principal components) which account for the parallel movement, slope oscillation, and curvature of the term structure and which explain, on average, the movements of the term structure of volatility to more than 95% in all cases. We test and construct an exponential ARCH, or E-ARCH, model for each state variable. One of the applications of this model is to produce confidence bands for the term structure of volatility.

1. Introduction

With the rapid innovation and growth of the derivative securities, the management of volatility risk in its many forms has become an important topic for both researchers and practitioners. This is not only because of the extreme sensitivity of derivatives to market volatility, but also that they provide risk managers a new tool to managing volatility risk which is otherwise not manageable. There have been several approaches taken to this end, each with its own advantages and pitfalls. One consists in using "implied tree" models (Dupire, 1994; Rubinstein, 1994 Derman and Kani, 1994); a more traditional approach consists managing the Vegas corresponding to different maturities. Other models use the notion of stochastic volatility. Hull and White (1987), for example, treat the spot volatility as an exogenous random source, while Engle and collaborators (Engle and Noh, 1994; Engle and Mezrich, 1995; Engle and Rosenberg, 1995) analyze the volatility of the underlying process using heteroskedastic auto-regressive models (the Autoregressive Conditionally Heteroskedastic, or ARCH–GARCH family). Other approaches involve the use of confidence bands for future volatility movements (Avellaneda and Parás, 1996).

271

In this article, we contribute to the theoretical understanding of the volatility of option prices by studying empirically the dynamics of the *term-structure of implied volatilities* of currency options. We use a Principal Component Analysis (PCA) (Judge, 1988) combined with ARCH techniques to derive a statistical model for the evolution of the term structure of volatility. Thus, the present statistical analysis is not on the volatility of the underlying asset, as in traditional work (see Engle, 1994, 1995), but rather on the implied volatilities, which provide a "dimensionless" representation of the currency options market.

Using historical data on the implied volatility of options on 13 currency pairs for the period January 1, 1995 to July 30, 1996, we develop a three-factor term-structure model which appears to be applicable to all studied currency pairs. A similar methodology was used by other authors in the study of term structure of interest-rates (Litterman and Scheinkman, 1991). There are, however, important structural differences between interest rates and implied volatilities. The main difference is that the term-structure of volatility is a stochastic process which is far from equilibrium. As an illustration of this, Fig. 1 shows an "equilibrium" AR model fitted using least squares and maximum likelihood techniques compared against real data. One clearly observes more structure in the real data than in the equilibrium model, and unlike the latter, the real implied volatility data exhibits a trend. A formal test for non-stationarity or "trend" present in time series is the unit root test (Enders, 1995). We apply this test to the term-structure of volatility, and the results fail to reject the null hypothesis of non-stationarity in all cases. On the other hand, we will show that in similar vein to interest rates, most of the variance on the term structure of volatility can be explained in terms of three factors: level movement ($\approx 90\%$), slope ($\approx 5\%$) and curvature ($\approx 1\%$).

The implied volatility processes exhibit strong heteroskedasticity, i.e., the volatility of volatility is not constant. Therefore, we propose a class of 3-factor exponential ARCH, or E-ARCH, models to describe its dynamics. Based on the analysis of Secs. 2 and 3, the example in Fig. 2 suggests that this model predicts the real movement of implied volatilities much better than naïve AR models.

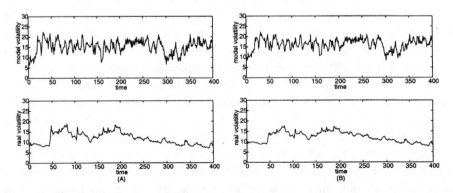

Fig. 1. AR Model vs. Real Term Volatility for USD/JPY. (A) 30 day implied Volatility, (B) 60 day implied Volatility. The equilibrium model doesn't reflect the structure of the real data.

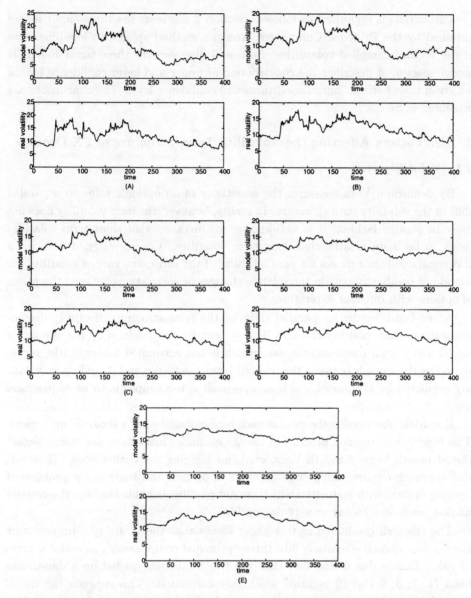

Fig. 2. E-ARCH Model vs. Real Term Volatility for USD/JPY. (A) 1 month, (B) 2 month, (C) 3 month, (D) 6 month, (E) 12 month. This model captures the structure of the real data.

As an application of this E-ARCH model, we present a method for calculating conditional confidence bands for the motion of the volatility curve. The method is illustrated with the aforementioned dataset. One possible application of such confidence bands could be for "statistical arbitrage", or alternatively, in the context of the Uncertain Volatility Model (Avellaneda and Parás, 1996), where option hedges are computed based on a proposed range for the future spot volatility.

This article is organized as follows. Section 2 discusses the three-factor model obtained by the Principal Component Analysis method applied to log-differences of the vector of implied volatilities. Section 3 discusses the three-factor E-ARCH model. Section 4 describes the construction of upper and lower confidence bands, for given time horizon, initial conditions and confidence level. The conclusions are presented in Sec. 5.

2. Risk Factors Affecting the Volatility Term-Structure of FX Options

2.1. *Volatility risk*

By definition, Vega measures the sensitivity of an option's value to a parallel shift in the volatility term structure. In reality, however, the term volatility does not move in parallel fashion: it is well-known, for instance, that short-term volatility tends to be more volatile than long-term volatility. Consequently, if we used a statistical one-factor model for parallel shifts of the term structure of volatility, we would not be able to explain the relative changes or correlations between the prices of options with different expirations.

A one-factor model for parallel shifts of the term-structure effectively decomposes the overall volatility risk into (i) a systematic risk modeled by the one-factor model and (ii) an unsystematic risk, which is not accounted for explicitly, represented by the spreads between the realized term volatilities and the reference volatility predicted by the model. Such an approach is too simplistic to be of practical use.

Empirical observation shows that each term volatility has a separate movement. This is why it is common practice in foreign-exchange markets to use "term Vegas" (i.e., 1-month Vega, 6-month Vega, etc.) for hedging the option book. However, this approach requires using large numbers of options and leads to the problem of hedging options with maturities which are not readily quoted in the over-the-counter market, such as a 75-day or a 47-day option.

The research conducted in this paper shows that the volatility term-structure can be decomposed essentially into three "principal components", or major sources of risk. This is due to the fact that the five maturities quoted on a day-to-day basis (1, 2, 3, 6 and 12 months) are highly correlated. This suggests the use of a multi-factor (three-factor) model to explain the fluctuations of the curve. This type of model offers the advantage of giving a better framework for hedging the book by hedging the exposure to each component rather than looking at individual expiration dates. It offers a solution to the aforementioned problems associated with "intensive" Vega-hedging, since fewer options will be involved if we hedge according to the principal components. We expect, in general, to hedge more than 95% of the risk in this way, in terms of a measure that will be made precise below.

The advantage of using a multi-factor model is that the underlying factors are not merely the quoted volatilities for standard maturities, i.e., we take into account existing statistical correlations between volatilities. In the point of view of hedging, we are only concerned with the sensitivity of the portfolio to each factor.

2.2. *The three-factor model*

The Principal Component Analysis (PCA) approach for analyzing a time-series consists in studying the covariance matrix of successive shocks. If we view the term-structure of volatility as a 5-dimensional vector, $(\sigma^{(1)}, \sigma^{(2)}, \ldots, \sigma^{(5)})$, where the $\sigma^{(i)}$'s represent the 1, 2, 3, 6 and 12 month volatility, respectively, then we should analyze a 5×5 symmetric matrix of squares and cross-products of volatility changes. The approach that we take in this paper is to analyze the covariance of the differences of the logarithm of the implied volatilities $A = a_{ij}$, which are defined as

$$a_{ij} = \sum_t (\log \sigma^{(i)}(t+1) - \log \sigma^{(i)}(t)) \times (\log \sigma^{(j)}(t+1) - \log \sigma^{(j)}(t)),$$

where $1 \le i, j \le 5$ and t ranges over the number of days observed.[a] Other possible candidates for analyzing the principal components of the term-structure could be successive differences of the volatilities (instead of the logarithms), the logarithm of the data or simply the data itself (in the latter two cases, the sample mean of the term structure enters the calculation). The reason for working with differenced data is the following: in conducting a unit root test on the implied volatility movements of FX rates for 13 currency pairs, we could not reject in any of the cases a unit root null-hypothesis. Hence, we concluded that the implied volatility curve does not behave like a stationary processes. Since, by convention, the PCA analysis leads to factors that have mean zero and constant variance, it is reasonable to select variables that are statistically stationary. The differenced data, by inspection or the same test as above, is stationary. Moreover, to preclude the possibility of having negative model volatilities, we chose to work with the differenced logarithms rather than the differences of the volatilities.

Let us denote by $Y_t^{(j)}$, the jth differenced logarithm of implied volatility at time t and by $\mathbf{v_1}, \mathbf{v_2}, \ldots, \mathbf{v_5}$ the five normalized eigenvectors of the sample covariance matrix of $\{Y_t^{(j)}\}_{j=1}^5$. We can define the coordinates of the vectors $Y_t^{(1)}, \ldots, Y_t^{(5)}$ in the orthonormal frame defined by the eigenvectors, *viz.*,

$$V_t^{(i)} = \sum_j v_{ij} Y_t^{(j)}$$

or

$$Y_t^{(j)} = \sum_i v_{ij} V_t^{(i)}.$$

In this formulation, the random variables $V^{(i)}$ are statistically uncorrelated linear combinations of the $Y^{(j)}$. This suggests the following *ansatz* for the term-structure

---

[a]The data used were daily closing prices posted electronically by brokers.

Table 1. Eigenvectors and normalized eigenvalues for USD/JPY.

	V_1	V_2	V_3	V_4	V_5
1M	0.0636	0.3257	-0.0004	-0.6406	0.6925
2M	-0.5004	-0.6825	-0.0394	0.1533	0.5087
3M	0.7505	-0.1794	0.2818	0.4107	0.3956
6M	-0.0457	0.3897	-0.7288	0.4900	0.2737
12M	-0.4246	0.4941	0.6229	0.3967	0.1740
Eig-value	0.0083	0.0104	0.0045	0.0310	0.9459

of volatility:

$$\sigma_t^{(j)} = \sigma_{t-1}^{(j)} \exp\left(\sum_i v_{ij} V_{t-1}^{(i)}\right),\tag{1}$$

where the statistics of the processes $V_t^{(i)}, i = 1, \ldots, 5$, will be determined in the next section.

Numerical values for the components of $\{\mathbf{v}_i\}_{i=1}^5$ and their corresponding normalized eigenvalues are shown in Table 1 for USD/JPY,[b] using the period from January 1, 1995 to July 30, 1996 with daily observations. The eigenvalue normalization is made such that their total sums are equal to 1. Each normalized eigenvalue represents the importance of the corresponding component for explaining the variance of the curve. An important consequence of the PCA analysis is that, in all 13 cases, *the variability of the term-structure of volatility is explained to more than 95% by just 3 components or eigenvectors.*

Figure 3 exhibit the *factor sensitivities* for USD/JPY, USD/DEM and CAD/USD. The plotted curves represent the percentage change in term-vol for a one standard deviation shock in the corresponding factor. Only the three factors with the largest eigenvalues are shown. We used cubic splines to generate a smooth curve interpolating between the five standard maturities.

Table 4 in Appendix shows that the first (largest eigenvalue) factor accounts for about 90% of the variance on average. We observe in Fig. 3 that the percentage changes caused by this first factor are positive and relatively flat across all maturities. Thus, this first factor corresponds approximately to a parallel movement of the term-structure of volatility. Note however that the sensitivity curve of the first factor is downward-sloping for most currencies, which is consistent with the fact that the longer term volatility is less volatile than shorter term volatility.

The second factor, which explains about 5% of the variance on average, corresponds to the variation of the slope of the term-structure. It "lowers" the short-term volatilities and "raises" the long-term volatilities.

---

[b]Table 4 in the Appendix shows results for all the 13 currency pairs.

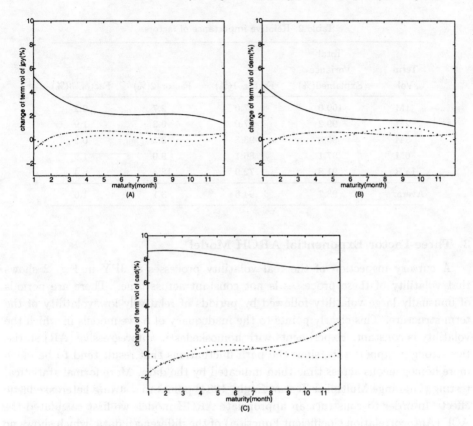

Fig. 3. Factor Sensitivities for (A) USD/JPY, (B) USD/DEM and (C) USD/CAD. Solid line is Factor 1, dot-dash line is Factor 2, and dotted line is Factor 3.

The third factor explains about 1% of the variability on average. We can view it as "twist component" of the term-structure curve: it tends to lower the term volatility for short and long maturities and raise it in the middle.

Table 2 exhibits the relative importance of the three factors of USD/JPY[c] in explaining the variation of each of the five volatilities corresponding to standard maturities.

We see that, on average, for all the currencies, the three factors account for more than 95% of the total variance. Although they explain the shorter term volatility better than the longer term volatility, the longer term volatility is much less volatile.

Based on this analysis and Eq. (1), we shall consider in the sequel the three-factor model for the volatility term-structure

$$\sigma_t^{(j)} = \sigma_{t-1}^{(j)} \exp\left(\sum_1^3 v_{ij} V_{t-1}^{(i)}\right), \qquad (2)$$

where $V_t^{(i)}, i = 1, 2, 3$, represent the factors with the largest eigenvalues.

[c]Table 5 in the Appendix shows the results for all 13 currency pairs.

Table 2. Relative importance of factors.

Term Vol	Total Variance Explained(%)	Factor 1(%)	Factor 2(%)	Factor 3(%)
1M	100.0	97.0	2.7	0.2
2M	99.2	97.0	0.3	1.9
3M	96.8	93.3	3.3	0.2
6M	97.1	86.1	9.0	1.9
1YR	91.8	72.9	12.4	6.4
Average	98.7	94.6	3.1	1.0

3. Three-Factor Exponential ARCH Model

A cursory inspection of the real volatility processes of JPY in Fig. 2 shows that volatility of these processes is not constant across time. There are periods of unusually large volatility followed by periods of relatively low volatility of the term-structure. This clearly points to the inadequacy of naïve models in which the volatility is constant. Experiments with homoskedastic autoregressive (AR) statistics strongly support this, since the path fluctuations that result tend to be much more homogeneous across time than indicated by the data. More formal statistical testing (Lagrange Multiplier Test of Engle, 1984) points to a strong heteroskedastic effect. In order to construct an appropriate ARCH model, we first calculated the ACF (Autocorrelation Coefficient Function) of the differenced data, which shows no autocorrelation effect to any order. In fact, the ACF's resemble those of simulated white noise. Once we have an appropriate AR model (in this case, no AR coefficients), the TR-square statistic of a regression of residuals for each $V^{(i)}$ is used, and it suggests that an ARCH(2) model is appropriate for most currency pairs. Therefore, we adopt the following model for each of the three factors:

$$V_t^{(i)} = a^{(i)} + \epsilon_t^{(i)}$$

$$\epsilon_t^{(i)} \sim N\left(0, h_t^{(i)}\right) \tag{3}$$

$$h_{t+1}^{(i)} \equiv E_t \epsilon_{t+1}^{(i)2} = \alpha_0^{(i)} + \alpha_1^{(i)} \left(\epsilon_{t-1}^{(i)}\right)^2 + \alpha_2^{(i)} \left(\epsilon_2^{(i)}\right)^2 ,$$

where $i = 1, 2, 3$. E_t is expectation conditional on time t. The parameters a, α_0, α_1 and α_2 are then determined by the Maximum Likelihood method for each of the three factors. These values are shown in the Appendix.

Figure 2 shows the model vs. the real data for USD/JPY. We see that the model selects the correct range of motions of the term-structure and captures some of the finer details of the series.

We believe that the present Exponential ARCH(2) model is appropriate for simulating term-volatility movement because it captures two important features of the real processes: the unit root effect, or non-stationarity, and the heteroskedasticity of the process. These hypotheses are strongly supported by statistical testing.

One should not think of the 3-factor Exponential ARCH(2) model as a "local volatility model" in the sense of Hull and White. Thus, it cannot be used directly in a derivatives pricing model as the dynamics for the spot volatility. Nevertheless, it gives us a rather realistic version of how the prices of options with different maturities are correlated in the currency markets, and thus provides useful information for hedging a book with a spectrum of options.

To feed the market information directly into a derivatives pricing model, we would need to model the local volatility variations. To this effect, we could use the 3-factor E-ARCH to construct dynamics for the "forward" volatility processes, i.e., the 1-month–2-month, 2-month–3-month volatilities, etc. Another approach, which we explore below, is to determine "confidence bands" for the term-structure of volatilities, which could be used as inputs in the Uncertain Volatility Model (Avellaneda and Parás, 1996). The next section describes a methodology for computing confidence bands.

4. Application: Confidence Bands

Consider the model of Eq. (3) for $V^{(1)}, V^{(2)}$ and $V^{(3)}$. Let us make a change of variables expressing volatility term-structure in its original representation. Using the notations of the last section, we have

$$\Delta \log \sigma_t^{(j)} = \sum_{i=1}^{3} v_{ij} a^{(i)} + \sum_{i=1}^{3} v_{ij} \epsilon_t^{(i)}$$

$$= A^{(j)} + \sum_{i=1}^{3} v_{ij} \epsilon_t^{(i)} \quad (j = 1, \dots, 5).$$

Our goal is, for given confidence level, to find the upper and lower bounds for $\log \sigma_s^{(i)}, i = 1, \dots, 5, s \in [0, t]$. To this end we

(i) approximate the processes by their continuous time version;
(ii) assume that the drift terms $A^{(i)}$'s are small compared with diffusion coefficient $\sum_{i=1}^{3} v_{ij}^2$.

Both approximations are reasonable. Typically $A^{(i)}$ is of order 10^{-4}, while

$$\langle \Delta \log \sigma_t \rangle = \sum_i v_{ij}^2$$

is of order 10^{-2} ($\langle \cdot \rangle$ is the quadratic variation).

In principle, one could use Monte-Carlo to find the joint distribution of $\log \sigma_{min}^{(i)}$ and $\sigma_{max}^{(i)}$. This involves finding a two-dimensional histogram for each standard maturity, but this is time-consuming. Instead, we used Monte-Carlo simulation to find the distribution of the quadratic variation process of $\log \sigma_t^{(i)}$, i.e.,

$$P(\langle \log \sigma_t \rangle \in dT),$$

which involves calculating only a one-dimensional histogram. Then, we used a time-change to transform, for each i, $\log \sigma_t^{(i)}$ to a Brownian motion with drift.[d] We then used the formula of Brownian passage time to compute the probability of the time-changed $\log \sigma_t$ exits a band. This is a well-known formula for the distribution of Brownian passage time with drift, often used to generate closed-form solutions for double-barrier options (Karatzas and Shreve, 1991).

First, we fix some notations. W_t is a Brownian motion starting from x, with drift μ. T_0 and T_b are the exiting time of W_t from 0 and b, respectively. If $P^{(\mu)}[T_0 \wedge T_b \leq t]$ is the probability of the process $\{\mu s + W_s\}_{s=0}^t$ exiting the band $[0, b]$, then

$$P^{(\mu)}[T_0 \wedge T_b \leq t] = \sum_{n=0}^{\infty} e^{\alpha_+}[1 - N(A_+) + e^{-2\alpha_+} N(A_-)]$$

$$- \sum_{n=0}^{\infty} e^{\alpha_-}[1 - N(B_+) + e^{-2\alpha_-} N(B_-)]$$

$$+ \sum_{n=0}^{\infty} e^{\alpha_-}\left[1 - N\left(A_+ + \frac{b}{\sqrt{t}}\right) + e^{-2\alpha_- + 2b\mu} N\left(A_- + \frac{b}{\sqrt{t}}\right)\right]$$

$$- \sum_{n=0}^{\infty} e^{\alpha_+ + b\mu}\left[1 - N\left(B_+ - \frac{b}{\sqrt{t}}\right) + e^{-2\alpha_+ - 2b\mu} N\left(B_- - \frac{b}{\sqrt{t}}\right)\right],$$

where

$$N(x) = \frac{1}{\sqrt{2\pi}} \int_{-\infty}^{x} e^{-\frac{z^2}{2}} dz$$

and

$$\alpha_\pm = -(2nb \pm x)\mu$$

$$A_\pm = -\mu\sqrt{t} \pm \frac{2nb + x}{\sqrt{t}}$$

$$B_\pm = -\mu\sqrt{t} \pm \frac{2nb - x}{\sqrt{t}}.$$

The terms in the series decay like $\exp(-\alpha n^2)$, with $\alpha = \frac{2b^2}{t}$. Evaluation of the sum of the first two or three terms is sufficient in practice.

---

[d]Although in this case the drift is not constant, due to assumption 2 that the drift is small compared to typical T, this conditional distribution won't change much if we spread the drift out evenly during time $0 \to T$. When T is small, i.e., the conditional motion is basically drift, the band is big enough that the probability won't be affected by the spreading out. Moreover, when T is small, constant drift is fairly a good approximation anyway.

Table 3. Implied vol 95% band. T is the time period for which different bands are valid. $\overline{\sigma}$ and $\underline{\sigma}$ are the upper bound and lower bound respectively.

	T		1M	2M	3M	6M	12M
	15	$\overline{\sigma}$	9.9781	9.4833	9.0959	8.9046	13.5821
		$\underline{\sigma}$	6.0989	6.7484	7.0336	7.3649	4.9966
	30	$\overline{\sigma}$	10.9664	10.1199	9.5547	9.2323	9.2005
		$\underline{\sigma}$	5.5504	6.3236	6.6938	7.1008	7.4829
USD−AUD	60	$\overline{\sigma}$	12.5266	11.0881	10.2405	9.7172	9.5792
		$\underline{\sigma}$	4.8612	5.7710	6.2416	6.7414	7.1829
	90	$\overline{\sigma}$	13.8944	11.9040	10.8097	10.1157	9.8876
		$\underline{\sigma}$	4.3845	5.3750	5.9094	6.4709	6.9547
	120	$\overline{\sigma}$	15.1229	12.6170	11.2958	10.4492	10.1433
		$\underline{\sigma}$	4.0299	5.0709	5.6516	6.2597	6.7754
	15	$\overline{\sigma}$	18.6214	16.6631	16.0034	14.4959	13.8153
		$\underline{\sigma}$	5.3403	6.4607	7.3952	8.6314	9.7241
	30	$\overline{\sigma}$	24.7961	20.7054	19.1007	16.3118	14.9680
		$\underline{\sigma}$	4.0007	5.1903	6.1854	7.6594	8.9649
USD−CHF	60	$\overline{\sigma}$	34.2818	26.4360	23.3269	18.6939	16.4289
		$\underline{\sigma}$	2.8726	4.0410	5.0389	6.6597	8.1488
	90	$\overline{\sigma}$	46.4759	33.2947	28.1467	21.1912	17.8945
		$\underline{\sigma}$	2.1064	3.1936	4.1588	5.8571	7.4653
	120	$\overline{\sigma}$	56.1850	38.4087	31.6422	22.9463	18.8931
		$\underline{\sigma}$	1.7328	2.7561	3.6851	5.3941	7.0571
	15	$\overline{\sigma}$	15.5809	15.4993	15.3015	14.4975	13.6618
		$\underline{\sigma}$	4.3031	4.5380	4.7065	5.0866	5.4010
	30	$\overline{\sigma}$	19.0959	18.8207	18.4282	17.1042	15.8199
		$\underline{\sigma}$	3.4960	3.7194	3.8893	4.2926	4.6481
USD−DEM	60	$\overline{\sigma}$	28.3647	27.4421	26.4692	23.5806	21.0344
		$\underline{\sigma}$	2.3459	2.5418	2.6979	3.1034	3.4865
	90	$\overline{\sigma}$	36.6813	35.0758	33.4948	29.0573	25.3024
		$\underline{\sigma}$	1.7990	1.9704	2.1122	2.4972	2.8791
	120	$\overline{\sigma}$	50.8531	47.9054	45.1775	37.9016	32.0228
		$\underline{\sigma}$	1.2953	1.4400	1.5630	1.9111	2.2713
	15	$\overline{\sigma}$	6.0702	5.6669	5.8365	5.0656	5.1074
		$\underline{\sigma}$	2.3736	2.6795	2.7385	3.1564	3.4521
	30	$\overline{\sigma}$	7.1700	6.4581	6.7021	5.5293	5.4984
		$\underline{\sigma}$	2.0066	2.3488	2.3830	2.8902	3.2053
DEM−CHF	60	$\overline{\sigma}$	9.1467	7.8282	8.1736	6.2656	6.1022
		$\underline{\sigma}$	1.5682	1.9334	1.9509	2.5479	2.8859
	90	$\overline{\sigma}$	11.0733	9.1044	9.5463	6.9102	6.6207
		$\underline{\sigma}$	1.2915	1.6588	1.6678	2.3079	2.6578
	120	$\overline{\sigma}$	12.9096	10.2714	10.8350	7.4873	7.0821
		$\underline{\sigma}$	1.1045	1.4672	1.4672	2.1278	2.4827

Table 3 (*continued*)

	T		1M	2M	3M	6M	12M
DEM−FRF	15	$\bar{\sigma}$	4.8144	4.6203	4.7438	5.0590	5.1818
		$\underline{\sigma}$	0.8281	1.0439	1.3136	1.7752	1.9728
	30	$\bar{\sigma}$	6.6910	6.0860	6.0197	6.1571	6.2193
		$\underline{\sigma}$	0.5944	0.7907	1.0331	1.4563	1.6415
	60	$\bar{\sigma}$	10.6467	8.9752	8.4334	8.1206	8.0486
		$\underline{\sigma}$	0.3719	0.5338	0.7345	1.1008	1.2652
	90	$\bar{\sigma}$	14.9690	11.9164	10.7845	9.9504	9.7342
		$\underline{\sigma}$	0.2633	0.4003	0.5721	0.8957	1.0435
	120	$\bar{\sigma}$	20.8403	15.7589	13.7252	12.1167	11.6729
		$\underline{\sigma}$	0.1883	0.3013	0.4477	0.7334	0.8679
DEM−ITL	15	$\bar{\sigma}$	14.5603	13.0722	12.0480	11.0815	10.4499
		$\underline{\sigma}$	4.3889	4.8904	5.3072	5.7698	6.1192
	30	$\bar{\sigma}$	18.1806	15.6822	14.0276	12.5067	11.5391
		$\underline{\sigma}$	3.5107	4.0730	4.5550	5.1082	5.5376
	60	$\bar{\sigma}$	24.9891	20.3243	17.4252	14.8850	13.3318
		$\underline{\sigma}$	2.5483	3.1376	3.6618	4.2855	4.7862
	90	$\bar{\sigma}$	32.0328	24.8852	20.6415	17.0511	14.9160
		$\underline{\sigma}$	1.9834	2.5584	3.0870	3.7354	4.2718
	120	$\bar{\sigma}$	39.1518	29.3086	23.6716	19.0319	16.3451
		$\underline{\sigma}$	1.6191	2.1688	2.6882	3.3417	3.8929
DEM−JPY	15	$\bar{\sigma}$	11.0145	11.1917	11.4694	11.7281	11.2358
		$\underline{\sigma}$	5.1023	6.1512	7.0585	7.8577	8.8987
	30	$\bar{\sigma}$	12.7203	12.5137	12.5610	12.6586	11.7450
		$\underline{\sigma}$	4.4149	5.4984	6.4425	7.2799	8.5118
	60	$\bar{\sigma}$	15.6032	14.6631	14.2883	14.1115	12.5038
		$\underline{\sigma}$	3.5942	4.6875	5.6591	6.5298	7.9931
	90	$\bar{\sigma}$	18.1937	16.5169	15.7446	15.3323	13.1124
		$\underline{\sigma}$	3.0781	4.1570	5.1315	6.0095	7.6201
	120	$\bar{\sigma}$	20.8027	18.3292	17.1362	16.4517	13.6622
		$\underline{\sigma}$	2.6883	3.7420	4.7109	5.6002	7.3114
USD−FRF	15	$\bar{\sigma}$	15.5483	14.4455	13.6527	12.8797	12.2370
		$\underline{\sigma}$	4.6258	5.7136	6.5904	7.7481	8.8269
	30	$\bar{\sigma}$	19.2527	17.0004	15.5144	14.0962	12.9726
		$\underline{\sigma}$	3.7254	4.8451	5.7888	7.0693	8.3187
	60	$\bar{\sigma}$	26.9794	21.9913	19.0005	16.2384	14.2082
		$\underline{\sigma}$	2.6430	3.7293	4.7084	6.1184	7.5806
	90	$\bar{\sigma}$	34.2630	26.4156	21.9281	17.9700	15.1665
		$\underline{\sigma}$	2.0693	3.0918	4.0642	5.5128	7.0883
	120	$\bar{\sigma}$	41.9166	30.8467	24.7406	19.5792	16.0226
		$\underline{\sigma}$	1.6825	2.6370	3.5894	5.0456	6.6975

Table 3 (*continued*)

	T		1M	2M	3M	6M	12M
USD–GBP	15	$\bar{\sigma}$	12.2662	11.7865	11.4443	11.2836	11.4482
		σ	4.6945	5.6929	6.5996	8.1575	9.0802
	30	$\bar{\sigma}$	14.7347	13.5512	12.7100	12.0034	11.9589
		σ	3.8993	4.9435	5.9327	7.6605	8.6863
	60	$\bar{\sigma}$	18.9455	16.4122	14.6803	13.0765	12.7119
		σ	3.0191	4.0686	5.1197	7.0178	8.1602
	90	$\bar{\sigma}$	22.9590	19.0154	16.3953	13.9706	13.3411
		σ	2.4801	3.5004	4.5693	6.5556	7.7644
	120	$\bar{\sigma}$	27.1407	21.5973	18.0421	14.7839	13.8886
		σ	2.0888	3.0721	4.1389	6.1827	7.4478
GBP–DEM	15	$\bar{\sigma}$	8.9646	8.5452	8.0783	7.5907	7.5108
		σ	3.4126	3.8027	4.3093	4.7422	5.1174
	30	$\bar{\sigma}$	10.6893	10.0164	9.1372	8.3295	8.1053
		σ	2.8302	3.2445	3.8100	4.3213	4.7417
	60	$\bar{\sigma}$	13.8544	12.4780	10.8378	9.4682	8.9995
		σ	2.1837	2.6051	3.2124	3.8009	4.2698
	90	$\bar{\sigma}$	17.0060	14.8488	12.4031	10.4760	9.7741
		σ	1.7792	2.1897	2.8072	3.4348	3.9306
	120	$\bar{\sigma}$	20.1387	17.1396	13.8636	11.3890	10.4642
		σ	1.5025	1.8975	2.5117	3.1589	3.6708
USD–JPY	15	$\bar{\sigma}$	14.5472	13.4890	13.0636	12.8907	12.7556
		σ	5.3224	6.4092	7.3488	8.5502	9.6576
	30	$\bar{\sigma}$	17.5806	15.5250	14.5580	13.9374	13.4512
		σ	4.4036	5.5670	6.5926	7.9064	9.1568
	60	$\bar{\sigma}$	22.7916	18.8156	16.8861	15.5093	14.4728
		σ	3.3960	4.5907	5.6805	7.1021	8.5081
	90	$\bar{\sigma}$	28.1021	21.9741	19.0323	16.9034	15.3508
		σ	2.7536	3.9286	5.0372	6.5136	8.0193
	120	$\bar{\sigma}$	33.5397	25.0451	21.0559	18.1734	16.1270
		σ	2.3066	3.4449	4.5506	6.0559	7.6312

This calculation gives the conditional probability

$$P(\log \sigma_s \in [b_1, b_2], s \in [0, t] \,|\, \langle \log \sigma_t \rangle = T),$$

where again $\langle \cdot \rangle$ denotes the quardratic variation. Therefore,

$$P(\log \sigma_s \in [b_1, b_2], s \in [0, t])$$

$$= \int_0^\infty P(\log \sigma_s \in [b_1, b_2] \,|\, \langle \log \sigma_t \rangle = T) \cdot P(\langle \log \sigma_t \rangle \in dT).$$

For a given confidence level, say 95%, set

$$P(\log \sigma_s \in [b_1, b_2], s \in [0, t]) = 0.95.$$

There are many pairs of $[b_1, b_2]$ which satisfy the above equation. We simply choose the initial position x in such a way that the probabilities for exiting the band on both sides are equal, namely

$$P_x[\tau_{b_1} \leq \tau_{b_2}] = P_x[\tau_{b_1} \geq \tau_{b_2}],$$

where τ_{b_i} is the exit time of the process $\log \sigma_s$ from boundary b_i. The subscript x denotes the initial condition. In practice, however, we approximate it by the following equation:

$$x = E\{y \,|\, (P_y[\tau_{b_1} \leq T_{b_2} \,|\, \langle \log \sigma_t \rangle] = P_y[\tau_{b_1} \geq T_{b_2} \,|\, \langle \log \sigma_t \rangle])\}$$

Table 3 shows the 95%-confidence bands for 11 currency pairs[e] over different time periods T.[f] We find that the bands for the term-structure of volatility form a "cone": short maturities have wider confidence intervals than long maturities for any given time-window t. The bands were calculating using a specific initial condition: we used the first three days in the dataset as initial condition for the ARCH(2) processes. Note that this is a *conditional* term volatility band, i.e., different initial conditions give rise to different bands, which depend on the positions of the five term volatilities over the past three days.

5. Conclusion

The analysis of the term structure of implied volatilities of currency options for 13 currency pairs shows that the movements of the term-structure are explained to more than 95% with a three-factor model. The factors are derived by a Principal Component Analysis of the sample covariance matrix of the changes in the log-differences of the implied volatility of options with the five standard maturities.

A heteroskedastic model for the evolution of the three factors driving the volatility curve was derived. We found that the ARCH(2) model was consistent with the data for each currency pair.

Finally, this model was used to calculate confidence bands for the term-structure over different periods of time. In all cases, these bands are "cone-like", in the sense that the confidence intervals become narrower as the option's expiration date increases.

---

[e]CAD and DEMESP are more suitable for jump-diffusion processes; here we exclude these two cases.

[f]Table 3 shows time periods over 15, 30, 60, 90 and 120 days.

Appendix A. Table for Eigenvectors and Normalized Eigenvalues

Table 4. Eigenvectors and normalized eigenvalues.

	V_1	V_2	V_3	V_4	V_5
1M	0.1142	0.4142	0.0460	−0.5382	0.7236
2M	−0.4633	−0.7075	−0.1437	0.0342	0.5128
3M	0.7490	−0.2424	0.1727	0.4704	0.3594
6M	−0.1624	0.4320	−0.6588	0.5485	0.2282
12M	−0.4300	0.2871	0.7165	0.4324	0.1796
USD−AUD	0.0188	0.0321	0.0083	0.0769	0.8640
1M	−0.0694	0.3003	−0.6623	−0.4984	0.4669
2M	0.4755	−0.7004	0.1263	−0.2163	0.4696
3M	−0.7708	−0.1139	0.4121	−0.0714	0.4670
6M	0.4172	0.6321	0.4795	0.0918	0.4337
12M	−0.0301	−0.0817	−0.3817	0.8314	0.3943
USD−CAD	0.0017	0.0037	0.0137	0.0299	0.9510
1M	0.0759	−0.0207	0.3928	−0.6209	0.6738
2M	−0.5324	0.1068	−0.6602	0.0715	0.5139
3M	0.7051	−0.3374	−0.2571	0.3867	0.4164
6M	0.0458	0.7409	0.3507	0.5007	0.2745
12M	−0.4600	−0.5705	0.4698	0.4574	0.1819
USD−CHF	0.0105	0.0061	0.0181	0.0389	0.9264
1M	−0.0475	−0.0420	0.3554	−0.6285	0.6889
2M	−0.2964	0.0972	−0.7846	0.1394	0.5174
3M	0.7962	−0.1924	0.0149	0.4055	0.4055
6M	−0.2304	0.6707	0.4284	0.4973	0.2576
12M	−0.4722	−0.7085	0.2728	0.4169	0.1639
USD−DEM	0.0092	0.0067	0.0164	0.0286	0.9392
1M	0.4129	−0.1051	0.0190	−0.6588	0.6198
2M	−0.8486	−0.1054	−0.0935	−0.0395	0.5084
3M	0.1731	0.7438	−0.0096	0.4367	0.4754
6M	0.1311	−0.4650	0.6982	0.4448	0.2853
12M	0.2495	−0.4565	−0.7095	0.4194	0.2239
DEM−CHF	0.0289	0.0511	0.0118	0.0718	0.8365
1M	−0.1898	0.0173	−0.4174	−0.6681	0.5858
2M	0.6912	0.0364	0.5089	−0.0687	0.5072
3M	−0.6426	−0.3050	0.4953	0.2457	0.4340
6M	−0.1057	0.7812	−0.1920	0.4722	0.3445
12M	0.2492	−0.5432	−0.5335	0.5153	0.3044
DEM−FRF	0.0147	0.0092	0.0231	0.0663	0.8867
1M	0.1694	−0.0293	−0.4132	−0.6475	0.6169
2M	−0.6962	0.1148	0.4819	0.0002	0.5196
3M	0.6906	0.1368	0.5042	0.2598	0.4273

Table 4 (*continued*)

	V_1	V_2	V_3	V_4	V_5
6M	−0.0535	−0.7700	−0.2453	0.4878	0.3258
12M	−0.0830	0.6118	−0.5317	0.5248	0.2465
DEM−ITL	0.0138	0.0091	0.0201	0.0651	0.8918
1M	0.2524	0.0513	−0.5393	−0.4509	0.6630
2M	−0.6978	0.3048	0.3807	−0.0423	0.5230
3M	0.2860	−0.6892	0.5105	0.0863	0.4185
6M	−0.1191	−0.0939	−0.4304	0.8454	0.2775
12M	0.5945	0.6486	0.3440	0.2696	0.1867
DEM−JPY	0.0139	0.0105	0.0255	0.0580	0.8921
1M	0.7092	−0.0628	−0.3035	0.0210	0.6328
2M	0.0301	−0.0089	0.4937	0.8515	0.1739
3M	0.0172	−0.0205	0.7702	−0.5221	0.3654
6M	−0.4656	0.7045	−0.1877	0.0305	0.5007
12M	−0.5282	−0.7065	−0.1888	0.0329	0.4302
USD−DEMESP	0.0102	0.0006	0.1170	0.3123	0.5600
1M	0.0276	−0.0619	0.4520	−0.5713	0.6817
2M	−0.4377	0.0285	−0.7365	0.0075	0.5149
3M	0.8053	0.1097	−0.1839	0.3708	0.4100
6M	−0.3771	0.4926	0.4367	0.5938	0.2680
12M	−0.1298	−0.8606	0.1696	0.4284	0.1737
USD−FRF	0.0089	0.0100	0.0168	0.0331	0.9312
1M	−0.6982	0.0842	−0.0409	0.0291	0.7092
2M	0.6039	0.5603	−0.1952	−0.1075	0.5212
3M	0.3022	−0.4638	0.7241	−0.1018	0.3986
6M	0.2249	−0.4441	−0.3961	0.7392	0.2210
12M	0.0766	−0.5163	−0.5282	−0.6563	0.1332
USD−GBP	0.0548	0.0495	0.0310	0.0182	0.8465
1M	0.1346	0.5494	0.0637	−0.5264	0.6316
2M	−0.5690	−0.6063	0.0036	−0.1297	0.5402
3M	0.7627	−0.4038	0.0520	0.2803	0.4171
6M	−0.1523	0.2882	−0.7277	0.5268	0.2943
12M	−0.2306	0.2906	0.6809	0.5916	0.2207
GBP−DEM	0.0166	0.0232	0.0095	0.0535	0.8973
1M	0.0636	0.3257	−0.0004	−0.6406	0.6925
2M	−0.5004	−0.6825	−0.0394	0.1533	0.5087
3M	0.7505	−0.1794	0.2818	0.4107	0.3956
6M	−0.0457	0.3897	−0.7288	0.4900	0.2737
12M	−0.4246	0.4941	0.6229	0.3967	0.1740
USD−JPY	0.0083	0.0104	0.0045	0.0310	0.9459

Appendix B. Table for Relative Importance of Factors

Table 5. Relative importance of factors.

	Term Volatility	Total Variance Explained(%)	Factor 1(%)	Factor 2(%)	Factor 3(%)
	1M	100.0	94.2	4.6	1.1
	2M	98.3	91.8	0.0	6.5
	3M	92.4	79.0	12.0	1.3
	6M	94.8	57.5	29.6	7.7
	1YR	85.3	53.0	27.3	5.0
AUD	Average	97.3	86.4	7.7	3.2
	1M	99.8	93.8	3.4	2.7
	2M	99.0	98.2	0.7	0.1
	3M	99.5	98.3	0.1	1.1
	6M	99.0	97.2	0.1	1.7
	1YR	100.0	86.7	12.1	1.2
CAD	Average	99.5	95.1	3.0	1.4
	1M	100.0	95.9	3.4	0.6
	2M	98.8	95.6	0.1	3.1
	3M	96.6	92.6	3.4	0.7
	6M	96.1	82.0	11.4	2.6
	1YR	91.1	65.3	17.3	8.5
CHF	Average	98.3	92.6	3.9	1.8
	1M	100.0	97.1	2.5	0.5
	2M	99.7	95.6	0.2	3.8
	3M	96.3	93.5	2.8	0.0
	6M	95.4	82.1	9.3	4.0
	1YR	85.4	68.5	13.5	3.3
DEM	Average	98.4	93.9	2.9	1.6
	1M	98.6	89.8	8.7	0.2
	2M	91.2	90.9	0.1	0.2
	3M	99.6	81.5	5.9	12.2
	6M	93.7	68.4	14.3	11.1
	1YR	89.4	57.5	17.3	14.6
DEM−CHF	Average	95.9	83.7	7.2	5.1
	1M	99.8	89.9	8.7	1.2
	2M	97.1	94.5	0.1	2.5
	3M	96.2	91.0	2.2	3.1
	6M	95.4	83.1	11.7	0.7
	1YR	96.7	74.7	16.0	6.0
DEM−FRF	Average	97.6	88.7	6.6	2.3

Table 5 (*continued*)

	Term Volatility	Total Variance Explained(%)	Factor 1(%)	Factor 2(%)	Factor 3(%)
	1M	99.9	91.6	7.4	0.9
	2M	97.3	95.5	0.0	1.9
	3M	96.2	90.9	2.5	2.9
	6M	95.4	81.0	13.3	1.0
	1YR	95.7	66.6	22.1	7.0
DEM−ITL	Average	97.7	89.2	6.5	2.0
	1M	99.8	95.1	2.9	1.8
	2M	97.0	95.5	0.0	1.4
	3M	96.4	92.2	0.2	3.9
	6M	99.8	59.6	36.0	4.1
	1YR	80.4	65.2	8.8	6.3
DEM−JPY	Average	97.6	89.2	5.8	2.5
	1M	97.9	93.3	0.1	4.5
	2M	100.0	6.2	83.3	10.5
	3M	100.0	32.6	37.1	30.3
	6M	98.3	95.3	0.2	2.8
	1YR	97.2	93.1	0.3	3.8
DEMESP	Average	98.9	56.0	31.2	11.7
	1M	100.0	96.8	2.4	0.8
	2M	99.3	95.8	0.0	3.5
	3M	96.5	93.4	2.7	0.3
	6M	95.7	78.3	13.7	3.8
	1YR	82.1	66.6	14.4	1.1
FRF	Average	98.1	93.1	3.3	1.7
	1M	100.0	94.0	0.1	5.9
	2M	99.5	86.2	5.8	7.5
	3M	90.1	80.7	6.4	3.0
	6M	78.5	60.2	14.2	4.0
	1YR	63.4	33.4	29.3	0.7
GBP	Average	95.1	84.7	5.0	5.5
	1M	99.9	94.2	3.9	1.8
	2M	98.1	94.7	0.3	3.1
	3M	94.4	89.8	2.4	2.2
	6M	94.6	77.8	14.9	1.9
	1YR	92.4	62.7	26.9	2.8
GBP−DEM	Average	97.4	89.7	5.3	2.3

Table 5 (*continued*)

Term Volatility	Total Variance Explained(%)	Factor 1(%)	Factor 2(%)	Factor 3(%)
1M	100.0	97.0	2.7	0.2
2M	99.2	97.0	0.3	1.9
3M	96.8	93.3	3.3	0.2
6M	97.1	86.1	9.0	1.9
1YR	91.8	72.9	12.4	6.4
JPY Average	98.7	94.6	3.1	1.0

Appendix C. Table for ARCH(2) Parameters

Table 6. Estimated ARCH(2) parameters and errors.

	α_0	α_1	α_2	a
	0.0000	0.2013	0.0018	−0.0001
	(0.0000)	(0.0028)	(0.0008)	(0.0000)
	0.0001	0.2587	0.0011	−0.0010
	(0.0000)	(0.0021)	(0.0002)	(0.0000)
	0.0011	0.1473	0.0006	−0.0002
AUD	(0.0000)	(0.0018)	(0.0003)	(0.0000)
	0.0000	0.3556	0.1798	−0.0002
	(0.0000)	(0.0044)	(0.0035)	(0.0000)
	0.0002	0.0609	0.0547	−0.0005
	(0.0000)	(0.0025)	(0.0027)	(0.0000)
	0.0020	0.7520	0.0727	−0.0055
CHF	(0.0000)	(0.0090)	(0.0026)	(0.0001)
	0.0000	0.5276	0.1496	−0.0004
	(0.0000)	(0.0054)	(0.0026)	(0.0000)
	0.0001	0.0315	0.0973	−0.0004
	(0.0000)	(0.0024)	(0.0027)	(0.0000)
	0.0023	0.8640	0.0515	−0.0074
DEM	(0.0000)	(0.0103)	(0.0020)	(0.0001)
	0.0002	0.4792	0.0007	0.0002
	(0.0000)	(0.0051)	(0.0011)	(0.0000)
	0.0003	0.2658	0.0005	0.0005
	(0.0000)	(0.0042)	(0.0025)	(0.0000)
	0.0037	0.3706	0.0007	−0.0035
DEM−CHF	(0.0000)	(0.0051)	(0.0015)	(0.0002)

Table 6 (*continued*)

	α_0	α_1	α_2	a
	0.0004	0.1937	0.0005	−0.0003
	(0.0000)	(0.0050)	(0.0000)	(0.0000)
	0.0013	0.0852	0.0005	−0.0003
	(0.0000)	(0.0028)	(0.0018)	(0.0001)
	0.0121	0.4344	0.0504	−0.0069
DEM−FRF	(0.0001)	(0.0064)	(0.0027)	(0.0003)
	0.0001	0.4247	0.0009	0.0005
	(0.0000)	(0.0045)	(0.0025)	(0.0000)
	0.0005	0.2409	0.0006	−0.0003
	(0.0000)	(0.0031)	(0.0033)	(0.0001)
	0.0067	0.2269	0.1043	−0.0031
DEM−ITL	(0.0000)	(0.0036)	(0.0034)	(0.0002)
	0.0001	0.1336	0.0000	−0.0001
	(0.0000)	(0.0030)	(0.0054)	(0.0000)
	0.0002	0.1548	0.0009	0.0004
	(0.0000)	(0.0031)	(0.0009)	(0.0000)
	0.0029	0.1724	0.0431	−0.0016
DEM−JPY	(0.0000)	(0.0035)	(0.0022)	(0.0001)
	0.0001	0.4138	0.1652	−0.0003
	(0.0000)	(0.0041)	(0.0034)	(0.0000)
	0.0002	0.0138	0.0508	−0.0006
	(0.0000)	(0.0027)	(0.0022)	(0.0000)
	0.0036	0.1601	0.4218	−0.0068
FRF	(0.0000)	(0.0040)	(0.0085)	(0.0002)
	0.0002	0.2189	0.0009	−0.0003
	(0.0000)	(0.0036)	(0.0018)	(0.0000)
	0.0002	0.2121	0.0009	0.0008
	(0.0000)	(0.0029)	(0.0020)	(0.0000)
	0.0035	0.2532	0.0535	−0.0057
GBP	(0.0000)	(0.0042)	(0.0021)	(0.0002)
	0.0001	0.1965	0.2524	−0.0002
	(0.0000)	(0.0032)	(0.0036)	(0.0000)
	0.0003	0.1020	0.0000	−0.0002
	(0.0000)	(0.0043)	(0.0016)	(0.0000)
	0.0061	0.0636	0.0007	−0.0001
GBP−DEM	(0.0000)	(0.0018)	(0.0017)	(0.0002)
	0.0001	0.2095	0.0006	0.0001
	(0.0000)	(0.0031)	(0.0026)	(0.0000)
	0.0001	0.0314	0.0973	−0.0004
	(0.0000)	(0.0033)	(0.0036)	(0.0000)
	0.0036	0.2808	0.1029	−0.0007
JPY	(0.0000)	(0.0042)	(0.0023)	(0.0002)

References

[1] M. Avellaneda and A. Parás, "Managing the volatility risk of portfolios of derivative securities: The Lagrangian Uncertain Volatility model", *Applied Mathematical Finance* **3** (1996) 21–52.

[2] E. Derman and I. Kani, "Riding on the Smile", *Risk* **7** (February 1994) 32–39.

[3] B. Dupire, "Pricing with a Smile", *Risk* **7** (January 1994) 18–20.

[4] W. Enders, *Applied Econometric Time Series*, John Wiley & Sons, Inc., 1995.

[5] F. Robert Engle, "Wald, Likelihood Ratio, and Lagrange Multiplier Tests in Econometrics" in *Handbook of Econometrics, Vol. II* (eds. Z. Griliches and M. D. Intriligator), North-Holland, Amsterdam, 1984.

[6] J. Noh, R. F. Engle and Alex Kane, "Forecasting Volatility and Option Prices of the S&P 500 Index", *The Journal of Derivatives* (Fall 1994).

[7] R. Engle and J. Mezrich, "Grappling With GARCH", *Risk* **8** (September 1995) 112–117.

[8] R. Engle and J. V. Rosenberg, "Garch Gamma", *The Journal of Derivatives* **2** (1995) 47–59.

[9] J. Hull and A. White, "The Pricing of Options on Asset with Stochastic Volatilities", *Journal of Finance* (June 1987) 281–300.

[10] G. G. Judge, W. E. Griffiths, R. C. Hill, H. Lutkepohl and T. C. Lee, *Introduction to the Theory and Practice of Econometrics*, 2nd ed. New York: John Wiley, 1988.

[11] I. Karatzas and S. Shreve, Brownian Motion and Stochastic Calculus, Springer-Verlag, 1991.

[12] R. Litterman and J. Scheinkman, "Common Factors Affecting Bond Returns", *The Journal of Fixed Income* (June 1991) 55–61.

[13] M. Rubinstein, "Implied Binomial Trees", *Journal of Finance* **49** (1994) 771–818.

A TEST OF EFFICIENCY FOR THE CURRENCY OPTION MARKET USING STOCHASTIC VOLATILITY FORECASTS

DAJIANG GUO

Centre Solutions, Zurich Group, One Chase Manhattan Plaza,
New York, NY 10005, USA
E-mail: dajiang.guo@centresolutions.com
and
Institute for Policy Analysis, University of Toronto,
Toronto, Canada, M5S 3G6
E-mail: guod@chass.utoronto.ca

The conditional volatility of foreign exchange rates can be predicted with GARCH models, and with implied volatility extracted from currency options. This paper investigates whether the difference in these predictions is economically meaningful. In an efficient market, after accounting for transaction costs and risk, no trading strategy should earn abnormal risk-adjusted returns. In the absence of transaction costs, both the delta-neutral and the straddle trading strategies lead to significant positive economic profits against the option market, regardless of which volatility prediction method is used. The agent using the Implied Stochastic Volatility Regression method (ISVR) earns larger profits than the agent using the GARCH method. However, after accounting for transaction costs assumed to equal one percent of market prices, observed profits are not significantly different from zero in most trading strategies; the exception is for an agent using the ISVR method with a 5% price filter.

1. Introduction

Several studies on stock index options conclude that volatility changes are statistically predictable. However, it is inconclusive as to whether this predictability is economically meaningful. Harvey and Whaley (1992) observe that while the implied volatility delivers precise forecasts for the S&P 100 index options, abnormal returns are not possible in a trading strategy that takes transaction costs into account. This suggests that predictable time-varying volatility is consistent with market efficiency. In contrast, Engel, Kane and Noh (1993) calculate the economic value of volatility forecasts from implied volatility and GARCH models on S&P 500 index options. They find that the GARCH model earns greater profit than does the implied volatility regression model even after taking transaction costs into account.

These studies have focused on stock index options, whereas this paper focuses on foreign currency options. An advantage in using currency options is that the simultaneous recording of exchange rates and option prices alleviates the "nonsynchronous trading problem" encountered in the studies of stock index options. In this paper, prices of the U.S. dollar/German mark (dollar/mark) options traded on the Philadelphia Stock Exchange (PHLX) are analyzed. The dollar/mark currency options are chosen for two reasons: they are highly volatile; and they are the most frequently traded.

In an efficient market, after accounting for transaction costs and risk, no trading strategy should earn abnormal returns. If predictions of volatility changes can be used to generate abnormal risk-adjusted profits, the market efficiency hypothesis will be challenged. If no economic profits are generated, the market efficiency hypothesis is supported. The efficiency of the currency option market is tested based on the performance of the GARCH and the Implied Stochastic Volatility Regression (ISVR) volatility prediction methods. Out-of-sample daily volatilities for the dollar/mark rate are predicted with the ISVR and GARCH methods. The GARCH volatility prediction method is based on the GARCH models of Engle (1982) and Bollerslev (1986). The implied volatilities are extracted from the stochastic volatility model of Hull and White (1987). The hypothesis that market volatility changes are unpredictable is first tested and rejected. The resulting regression formula is modified to predict daily out-of-sample volatility.

A combination of implied volatility and GARCH volatility may only improve volatility forecasts marginally. Recent study by Guo (1996) suggests that the implied variance extracted using the Hull and White (1987) stochastic volatility model from PHLX currency options is an efficient, but biased forecast of the future variance of the exchange rate, and that the MA(60) and GARCH variance forecasts do not contain significant incremental information. These results are consistent with the findings of Fleming (1993) and Jorion (1995) that implied volatility is a dominant, but biased estimator in terms of ex ante forecasting power.

There are at least two reasons to use the Hull and White (1987) model. First, the time-varying nature of exchange rate volatility suggests that the estimated implied volatility from the constant volatility Black–Scholes (1973) model is subject to specification error. Descriptive Statistics of the dollar/mark spot rate returns has a negative skewness and excess kurtosis. This means that the dollar/mark return does not follow lognormal distribution as assumed by the Black–Scholes model. Second, implied volatility estimated from the Hull and White (1987) model has good statistical performance in predicting future realised variance (Guo (1996)).

To get an operational option pricing model, Hull and White (1987) assumes that the volatility risk is independently distributed with the aggregate consumption growth, thus the risk premium of volatility is zero. This assumption may not be true in light of the facts that at-the-money implied volatility is usually observed to be higher than realised volatility. Though the risk premium of volatility may not be zero, it will not affect the trading results of this paper which based on implied volatility estimates. Similar as the practice of option traders using Black–Scholes implied volatility matrix as inputs in their pricing models, this paper uses the Hull and White (1987) model to extract information about market expectation of future volatility from traded option prices, then use the implied volatility forecasts to improve trading performance. In doing so, we take market expectation of future volatility as reflecting an aggregate view of the market participates towards the future movements of exchange rate; the risk premium of volatility and the transaction costs are already embedded in the implied volatility estimates. This approach is different from Melino and Turnbull (1990), where they first estimate the parameters of the stochastic volatility process from historical exchange rate data, then adjust for risk premium parameters in option pricing.

With the prediction of daily volatilities, theoretical option prices are computed and compared with the observed market prices. Corresponding positions are taken to cash in the profit. Two types of dynamic trading strategies are considered: delta-neutral trading and straddles trading. The investment amounts are all standardized to $100, and agents do not reinvest the profit the next day. Filters are also introduced to improve the profitability of each strategy. Each strategy is evaluated under two situations: with no transaction cost, and with transaction cost.

In the absence of transaction cost, agents using either the GARCH or the ISVR volatility prediction method earn significant positive economic profits against the option market, for both trading strategies. An agent using the ISVR method earns larger profits than an agent using the GARCH method in all three filters. This suggests that traders who can execute the delta-neutral and the straddle trading strategies at the observed market transaction prices can lock in significant profits during the period examined in this paper. However, for a nonmember of an exchange, especially an individual investor who faces higher transaction costs, the magnitude of the correctly predicted volatility changes is generally not large enough to allow for abnormal risk-adjusted profits. After accounting for transaction costs, the economic profits are not significantly different from zero for most trading strategies. The exception is for an agent using the ISVR method with a 5% price filter.

Section 2 describes data source, an option pricing model, and the procedure to estimate the implied volatility. Section 3 introduces the two volatility prediction models. Section 4 compares the actual and predicted option prices. Section 5 develops dynamic trading strategies to capture the abnormal risk-adjusted profits. Section 6 concludes.

2. Implied Volatility Estimation

2.1. *Data description*

Trading records for foreign currency options are taken from the "Foreign Currency Options Pricing History" database of the Philadelphia Stock Exchange (PHLX). The record for each option trading contains: the date of trade, currency and option identification, maturity symbol, strike price, time of trade, and the option premium per contract. The trade-by-trade option prices cover the period from January 1983 to March 1993. Most of the options are American with maturity of less than one year.

Trading prices for American style options on the dollar/mark exchange rates are considered. The dollar/mark currency options are chosen for two reasons: they are highly volatile, and they are the most frequently traded. To provide a manageable dataset, only trading prices for American call and put options on dollar/mark currencies are considered, from January 2, 1991 to March 25, 1993.

The interest rates used are the daily closing quotes of Euro Deposits in terms of the U.S. dollar, and the German mark. The maturities are of 30, 60, and 90 days. The proxies used for intermediate rates are the rates whose maturities are the closest to the option expiration date.

2.2. *Hull and white (1987) stochastic volatility model*

The stochastic nature of exchange rate volatility suggests that the estimated implied volatility from the constant volatility Black–Scholes (1973) model is subject to specification error. In the literature various stochastic volatility models were developed by Scott (1987), Hull and White (1987,1988), Wiggins (1987), Melino and Turnbull (1990), and Heston (1993). These models make it theoretically attractive to recover the market expectation of future variance from a logically consistent stochastic volatility model.

To meet this challenge, this paper follows Guo (1996) by adopting the model of Hull and White (1987) in the extraction of the implied variance, and seeks to explore the economic implication of the estimated stochastic implied variance. It is assumed that:

(a) The market is frictionless: trading takes place continuously in time, there are no transaction costs, taxes, or short sale restrictions.

(b) The instantaneous risk-free interest rate r_F and domestic/foreign interest differential $b = r_D - r_F$ are known and nonstochastic.

(c) The state variables are the underlying asset $S(t)$ and the instantaneous variance $V(t)$, and the Hull and White (1987) model is based on the following stochastic processes:

$$dS(t) = \phi S(t)dt + \sigma S(t)dw, \tag{1}$$

$$dV(t) = \mu V(t)dt + \xi V(t)dz. \tag{2}$$

The variable ϕ is a parameter that may depend on $S(t)$, $\sigma(t)$, and t. The variables μ and ξ are assumed to be constant. dz and dw are Wiener processes with instantaneous correlation ρ. Notice that the instantaneous variance $V(t) = \sigma^2(t)$ follows a geometric Brownian motion, so ξ is the volatility of variance parameter.

If the instantaneous variance is uncorrelated with aggregate consumption (i.e., the random variance has no systematic risk) and the instantaneous correlation ρ between the underlying asset and its variance process is zero, the Hull and White option price C^{HW} (their Eq. (8)) is the integral of the Black–Scholes price $C^{BS}(\bar{V})$ over the distribution of the mean variance $\bar{V}$,

$$C^{HW}(S_t, \sigma_t^2) = \int C^{BS}(\bar{V}) h(\bar{V}|\sigma_t^2) d\bar{V}, \qquad (3)$$

where h is the conditional distribution of $\bar{V}$, and $\bar{V}$ is the mean variance over the time interval $[t, T]$,

$$\bar{V} = \frac{1}{T} \int_t^T \sigma^2(t) dt.$$

Hull and White (1987) suggested that if the instantaneous variance is uncorrelated with aggregate consumption and the instantaneous correlation between the underlying asset and its variance process is zero, the Hull and White option price $C^{HW}(S_t, \sigma_t^2)$ is the integral of the Black–Scholes price $C^{BS}(\bar{V})$ over the distribution of the mean variance. Furthermore, Hull and White (1987) proposes a power series approximation technique based on expanding Black–Scholes price $C^{BS}(\bar{V})$ in a Taylor series about its expected average variance $E(\bar{V})$,

$$C^{HW}(S_t, \sigma_t^2) \approx C^{BS}(E(\bar{V})) + \frac{1}{2} \frac{\partial^2 C^{BS}(E(\bar{V}))}{\partial \bar{V}^2} E(\bar{V}^2)$$

$$+ \frac{1}{6} \frac{\partial^3 C^{BS}(E(\bar{V}))}{\partial \bar{V}^3} E(\bar{V}^3), \qquad (4)$$

where $E(\bar{V}^2)$ and $E(\bar{V}^3)$ are the second and third central moments of $\bar{V}$. The second and third terms in the equation are essential to capture the so called "smile" effect for PHLX foreign currency options (Bodurtha and Courtadon (1987)). For each day t, the daily implied variance, $\bar{V}_t$, is estimated by nonlinear least squares. It is then considered to be the unbiased estimator of the market expectation of future variance $E(\bar{V}) = \bar{V}_t$ in the Hull and White framework (Fleming (1993) and Guo (1996)).

Because the main objective is to design trading strategy that based on the predictive power of implied variance over future variance, it is necessary to estimate the implied variance from options with maturities that can match option maturity. In addition, increasing the number of option prices used to extract the implied variance is also useful in further reducing the measurement error. Therefore, option contracts with maturities ranging from 30 to 90 calendar days (or 20 to 60 trading days) are pooled in the daily implied variance estimation. The number of trades in a day varies significantly across the sample. The average number of trades is around 100 for the dollar/mark options.

The quality of the approximation given by Eq. (4) depends on the values of the parameters μ and ξ. Hull and White (1987) suggest that for $\mu = 0$ and sufficiently small values of $\xi^2(T - t)$ (ξ with approximate range of 1 to 4), this approximation is very accurate.[a] Values of ξ are estimated from the GARCH(1,1) parameters

---

[a]Hull and White (1987) argued that the choice of $\mu = 0$ is empirically justified on the grounds that, for any nonzero μ, options of different maturities would exhibit markedly different implied volatilities, but this was never observed empirically.

Table 1. In-sample estimation of GARCH models.

This table reports the descriptive statistics of spot mark/dollar exchange rate. It also reports the in-sample estimation of parameters for GARCH models in various conditional variance specifications. The GARCH models are from Engle (1982) and Bollerslev (1986). The series studied is the change in the logarithm of the daily dollar/mark exchange rate. The sample period is from January 4, 1991 to March 25, 1993. The t-statistics in parentheses are computed using the robust inference procedure of Bollerslev and Wooldridge (1988). The various specifications of conditional variances are as follows:

$$R_t = \lambda_0 + \epsilon_t \quad \epsilon_t | \epsilon_{t-1}, \epsilon_{t-2}, \ldots \sim N(0, h_t^2)$$

$$\text{ARCH}(1) h_t^2 = \alpha_0 + \alpha_1 \epsilon_{t-1}^2$$

$$\text{ARCH}(2) h_t^2 = \alpha_0 + \alpha_1 \epsilon_{t-1}^2 + \alpha_2 \epsilon_{t-2}^2$$

$$\text{GARCH}(1,1) h_t^2 = \alpha_0 + \alpha_1 \epsilon_{t-1}^2 + \beta_1 h_{t-1}^2$$

Panel A: Descriptive statistics of spot exchange rate.

Year	Series	Mean	Std.	Skewness	E. Kurtosis	Min.	Max.
1991	$S(t)$	0.6045	3.8237×10^{-2}	0.6077	−0.7268	0.5448	0.6916
	$\log(S(t))$	−0.5054	6.2251×10^{-2}	0.5224	−0.8058	−0.6073	−0.3688
	$S(t)/S(t-1)$	15.9055	0.1324	−0.2188	1.2271	15.4248	16.2837
	$\log(S(t)/S(t-1))$	-0.0199×10^{-3}	0.1326	−0.2588	1.2793	−0.4886	0.3733
1992	$S(t)$	0.6424	3.2259×10^{-2}	0.5933	−0.6718	0.5944	0.7192
	$\log(S(t))$	−0.4437	4.9574×10^{-2}	0.5215	−0.7697	−0.5203	−0.3297
	$S(t)/S(t-1)$	15.9027	0.1383	−0.4541	1.3465	15.4301	16.4118
	$\log(S(t)/S(t-1))$	-3.8247×10^{-3}	0.1386	−0.4949	1.3587	−0.4831	0.4979

Panel B: GARCH parameter estimation.

Variance	λ_0	α_0	α_1	α_2	β_1	Log. Likelihood
ARCH(1)	0.0017	0.0091	0.1294			342.942
	(0.324)	(3.68)	(2.12)			
ARCH(2)	0.0043	0.0114	0.264	0.027		344.942
	(0.96)	(9.08)	(3.89)	(0.687)		
GARCH(1,1)	0.0013	0.00149	0.0947		0.81913	345.499
	(0.216)	(2.016)	(2.87)		(12.84)	

estimates reported in Table 1. Nelson (1991) shows that as the time interval goes to zero, the GARCH parameter α_1 approaches $\xi \sqrt{dt/2}$. Using daily exchange rates data from January 1991 to March 1993, this suggests an estimate of ξ of 2.126 for the dollar/mark rate.

Because the currency options used in this paper are American-style options, it is important to adjust for the early exercise premium. This is done by using the Barone-Adesi and Whaley (1987) quadratic approximation method.

2.3. *Implied volatility estimation procedure*

Suppose that the option market is informationally efficient and that option traders incorporate their expectation of future volatility into the actual trading, then the market expectation of future volatility should be contained in the market prices. The *benchmark* implied volatilities are usually estimated by matching the model prices to the *at-the-money* options prices, and then inverting the Black–Scholes model. As discussed earlier, an advantage of using the stochastic volatility model is to capture the "smile" effect. The approximation (4) of the Hull and White (1987) model allows extraction of the implied volatility from currency options with a wider range of moneyness. The single day averaged implied volatility is estimated by pooling together options with strike price/spot price ratio (moneyness) between 0.8 to 1.2.

The method to extract implied volatility follows the approach of Whaley (1982) and Lamoureux and Lastrapes (1993). For each day t in the sample, the daily average variance, $\hat{V}_t = E(\bar{V}|\Omega_t)$, is estimated by nonlinear least squares. That is, $\hat{V}_t$ is chosen to minimize the distance between observed option prices on day t and the predicted option prices from the Hull and White (1987) model

$$\min_{\hat{V}_t} SSE(\hat{V}_t) = \sum_i [CP_{t,i} - CP_{t,i}^{HW}]^2, \qquad (5)$$

where i is an index over observations in a given day; $CP_{t,i}$ is the observed market price; $CP_{t,i}^{HW}$ is the theoretical option price given the interest rate differential $r_D - r_F$, strike price/spot price ratio $(X/S)_{t,i}$, and the remaining life of an option $\tau_{t,i}$. The number of trading quotes in a day varies significantly across the sample, but the average number of quotes per day is about 100 for the dollar/mark options.

3. Volatility Prediction Methods

This section introduces two volatility prediction methods: the GARCH and the Implied Stochastic Volatility Regression methods. The difference between these two methods is that while the GARCH volatilities are estimated from the realized returns of the foreign exchange rates, the implied volatilities are estimated from the realized prices of the foreign exchange options.

3.1. *GARCH(1, 1) variance prediction*

The GARCH(1, 1) model is selected to calculate the GARCH variance forecasts from a class of GARCH models. Parameters estimates of ARCH(1), ARCH(2), and GARCH(1, 1) models are reported in Table 1. The GARCH(1,1) model of Bollerslev (1986) is specified as

$$R_t = \lambda_0 + \epsilon_t, \quad \epsilon_t|\epsilon_{t-1}, \epsilon_{t-2}, \ldots \sim N(0, h_t^2), \qquad (6)$$

$$h_t^2 = \alpha_0 + \alpha_1 \epsilon_{t-1}^2 + \beta_1 h_{t-1}^2, \qquad (7)$$

$$h_{t+k,t}^2 = \alpha_0 + \alpha_1 E[\epsilon_{t+k-1}^2 | \omega_t] + \beta_1 h_{t+k-1,t}^2$$

$$= \alpha_0 + (\alpha_1 + \beta_1) h_{t+k-1,t}^2, \quad k = 1, \ldots, N, \tag{8}$$

where $R_t = \sqrt{252} \ln(S_t/S_{t-1})$ is the change in the logarithm of the exchange rate ratios from time $t-1$ to t, ϵ_t is assumed to be conditionally normal with mean zero and conditional variance h_t^2, ω_t is the information set. The exchange rate is measured as U.S. dollars per Deutsche mark. Daily observations from January 1991 to March 1993 are used. The parameter estimates in the variance equation are constrained to be positive, $\alpha_0 > 0$, $\alpha_1 > 0$, $\beta_1 > 0$, and $\alpha_1 + \beta_1 < 1$. The GARCH models are estimated by Maximum Likelihood; the t-statistics are computed using the robust inference procedure proposed by Bollerslev and Wooldridge (1992).

For the GARCH(1, 1) model, the estimated α_1 and β_1 are 0.095 and 0.819 respectively, with t-statistics of 2.87 and 12.84. The results are consistent with previous findings that the conditional heteroscedasticity in daily spot rates can be represented by a GARCH(1,1) specification (Hsieh (1988, 1989), Baillie and Bollerslev (1989)). Exchange rate changes appear to have zero unconditional means, and almost zero serial correlations. The excess kurtosis is consistent with the documented fat-tail features of the exchange rates.

To ensure that the forecast horizon of the GARCH(1, 1) variance is identical to that of the option maturity, the T-day GARCH variance is constructed by recursive substitution of the one period ahead GARCH(1,1) variance forecast $h_{t+1,t}^2$ to obtain a k-period ahead prediction of the conditional variance $h_{t+k,t}^2$, then the predictions are averaged over the forecast period. Parameters estimated from the past two years are used as starting values for the GARCH specification to reestimate the sample variance at each time t.

By mathematical induction, we have

$$h_{t+k,t}^2 = \alpha_0 \left[\frac{1 - (\alpha_1 + \beta_1)^{k-1}}{1 - (\alpha_1 + \beta_1)} \right] + (\alpha_1 + \beta_1)^{k-1} h_{t+1,t}^2. \tag{9}$$

Therefore, the T period ahead average GARCH variance forecast is calculated as

$$\sigma_{G,t}^2 = \frac{1}{T} \left(\sum_{k=1}^{T} h_{t+k,t}^2 \right). \tag{10}$$

3.2. *Implied stochastic volatility regression method*

This section first tests the hypothesis that the changes of implied volatility are unpredictable. After rejecting this hypothesis, the dynamic properties of the implied volatility is investigated and the resulting regression formula is modified to predict one period ahead out-of-sample volatility.

Summary statistics of the implied volatilities for call and put options are reported in Table 2. The average implied volatility is 0.126 for calls, and 0.125 for puts. The positive and significant serial correlation of the implied volatility

Table 2. Summary statistics of volatility measures.

This table reports the mean, the standard deviation (S.D.), the coefficient of variation (C.V.), and the autocorrelations for various volatility measures. $\sigma_{H,t}$ is the historical volatility, $\sigma_{G,t}$ is the 60-day average GARCH(1,1) volatility forecast, $\sigma_{IHW,P}$ and $\sigma_{IHW,C}$ are implied put and call volatilities from the model of Hull and White (1987). The sample period is from January 4, 1991 through March 25, 1993.

	$\sigma_{H,t}$	$\sigma_{G,t}$	$\sigma_{IHW,P}$	$\Delta\sigma_{IHW,P}$	$\sigma_{IHW,C}$	$\Delta\sigma_{IHW,C}$
Mean	0.114	0.132	0.125	-0.12×10^{-4}	0.126	-0.118×10^{-5}
S.D.	0.008	0.0075	0.022	0.011	0.018	0.0065
C.V.	0.07	0.057	0.171		0.145	
ρ_1	0.992	0.92	0.878	-0.247	0.938	-0.161
ρ_2	0.981	0.827	0.816	-0.008	0.897	-0.101
ρ_3	0.968	0.743	0.756	-0.036	0.868	-0.010
ρ_4	0.952	0.655	0.704	-0.062	0.84	-0.01
ρ_5	0.934	0.581	0.668	-0.024	0.813	-0.033
ρ_8	0.877	0.465	0.587	-0.053	0.714	-0.051
ρ_{12}	0.786	0.343	0.514	-0.221	0.667	0.046
ρ_{16}	0.677	0.289	0.513	0.054	0.599	0.049
ρ_{20}	0.557	0.184	0.427	-0.029	0.485	-0.004
ρ_{24}	0.430	0.143	0.356	-0.027	0.386	-0.036
Unit Root Test Stat.	-0.299	-3.26	-3.44	-5.72	-3.57	-5.96

indicates substantial persistence in the level of volatility. The autocorrelation declines gradually at long lags. However, the modified Dickey–Fuller tests reject the unit root hypothesis of the implied volatilities at the 5% significant level. The first order correlation of the first difference of the daily implied volatility estimates is -0.247 for puts, -0.161 for calls. The autocorrelation and the Dickey–Fuller tests demonstrate that the estimated implied volatilities are stationary and mean-reverting.

The changes of the implied call and put volatilities are regressed over several variables in the information set, including lagged implied call and put volatilities, lagged changes of interest rate differentials, and lagged exchange rates. The regression is specified as follows:

$$\Delta V_t = \alpha_0 + \alpha_1 D_{t,1} + \alpha_2 D_{t,5} + \sum_{i=1}^{3} \beta_i \Delta V_{P,t-i} + \sum_{j=1}^{3} \gamma_j \Delta V_{C,t-j}$$

$$+ \sum_{k=0}^{2} \delta_k R_{t-k} + \delta_{ABS}|R_t| + \theta_1 DF_{t-1} + \epsilon_t \,, \tag{11}$$

where ΔV_t represents either $\Delta V_{C,t}$ or $\Delta V_{P,t}$; $R_t = \sqrt{252} \cdot \log(S_t/S_{t-1})$ is the annualized change in the logarithm of the exchange rate; $DF_t = RD_t - RF_t$ is the domestic-foreign interest rate differential based on the daily quotes of 90-day T-bills; $D_{t,1}$ and $D_{t,5}$ are dummy variables for Monday and Friday.

Table 3. Dynamic properties of the implied call and put volatilities.

This table reports regressions of the changes in implied call and put volatilities on information variables, such as the include lagged implied volatilities for call and put options, lagged values of interest rate differentials, contemporaneous and lagged changes of exchange rates. The implied call and put volatilities are extracted from the model of Hull and White (1987). The sample period is from January 4, 1991 to March 25, 1993. The interest rate differential is the day-to-day difference of the 90-day domestic and foreign T-bill rates. All t-statistics are based on White (1980) heteroscedasticity-consistent standard errors. Out-of-sample correct prediction is the percentage of times that the sign of volatility changes being correctly predicted.

Variables	Chg. in put VAR		Chg. in call VAR	
	Coefficient	t-ratio	Coefficient	t-ratio
Intercept	−0.0017	−2.45	−0.002	−3.96
Monday	0.0018	1.58	0.00072	1.04
Friday	−0.0026	−2.32	−0.0008	−1.21
Chg. implied put volatility $(t-1)$	−0.318	−7.123	0.142	5.33
Chg. implied put volatility $(t-2)$	−0.133	−2.76	0.073	2.54
Chg. implied put volatility $(t-3)$	−0.09	−2.04	0.025	0.94
Chg. implied call volatility $(t-1)$	0.22	3.02	−0.27	−6.18
Chg. implied call volatility $(t-2)$	0.089	1.19	−0.22	−4.98
Chg. implied call volatility $(t-3)$	0.08	1.12	−0.093	−2.18
Log. Spot Rate Chg. (t)	−0.003	−0.87	−0.003	−1.61
Log. Spot Rate Chg. $(t-1)$	0.0006	0.294	0.0006	0.29
Log. Spot Rate Chg. $(t-2)$	0.003	0.89	−0.004	−2.01
Abs. Log. Spot Chg. (t)	0.019	1.82	0.017	1.57
Chg. Interest Rate Diff. (t)	0.0034	0.705	0.002	0.72
Chg. Interest Rate Diff. $(t-1)$	0.003	0.6116	−0.0024	−0.83
Sample size (days)	510		510	
$\bar{R}^2$	0.255		0.185	
Durbin–Watson	2.05		2.01	
Likelihood value	1739.58		2020.4	
AIC–LOG	−9.18		−10.224	
SCHWARZ–LOG	−9.07		−10.105	
Out-of-sample correct direction	67.5%		63.3%	

Table 3 reports the results of these regressions. All t-statistics are based on the heteroscedasticity consistent standard errors of White (1980). The Monday dummy has no significant effect in either the call or put regressions; while the Friday dummy only has a significant positive coefficient for the put volatility.

The largest explanatory power in both regressions comes from the corresponding lagged values of implied call or put volatilities. These regressions also confirm that the implied volatilities are mean-reverting processes. Notice the significant negative estimates of coefficients $(\gamma_1, \gamma_2, \gamma_3)$ in the call regression $(\Delta V_{C,t})$, and $(\beta_1, \beta_2, \beta_3)$ in the put regression $(\Delta V_{P,t})$.

The inclusion of the contemporaneous and the lagged values of the log of exchange rate changes enables us to study the intertemporal relationship between the exchange rates and the volatility changes. In addition, to capture the possible asymmetry of the contemporaneous relationship, the absolute value of the contemporaneous exchange rate change $|R_t|$ is added to the regression. The estimated coefficients of the contemporaneous, the lagged one and lagged two period exchange rates $(\delta_0, \delta_1, \delta_2)$, are all statistically insignificant. The estimated coefficient for $|R_t|$ is also statistically insignificant, which suggests that there is no asymmetric relationship between the size of exchange rates changes and the contemporaneous changes in its volatility. The estimates of θ_0 and θ_1 suggest that the contemporaneous and lagged values of interest rates differentials have no significant explanatory power in predicting out-of-sample volatility changes.

Generally speaking, these regressions show that the changes in both the implied call and put volatilities are predictable. The $\bar{R}^2$ is 0.185 for the call regression and 0.255 for the put regression. The implied volatility exhibits no strong relationship with the contemporaneous exchange rate. There is no evidence of asymmetric relationship between the size of exchange rates changes and the contemporaneous volatility. Also the contemporaneous and intertemporal interest rate differentials have no significant explanatory power in predicting the changes of the implied call or put volatilities.

Based on the above evidence, the regression equation is modified to predict the one-period ahead volatility,

$$\Delta V_{C,t} = \hat{\alpha}_0 + \hat{\alpha}_1 D_{t,1} + \hat{\alpha}_2 D_{t,5} + \sum_{i=1}^{3} \hat{\beta}_i \Delta V_{P,t-i} + \sum_{j=1}^{3} \hat{\gamma}_j \Delta V_{C,t-j}, \qquad (12)$$

$$\Delta V_{P,t} = \hat{\alpha}_0 + \hat{\alpha}_1 D_{t,1} + \hat{\alpha}_2 D_{t,5} + \sum_{i=1}^{3} \hat{\beta}_i \Delta V_{P,t-i} + \sum_{j=1}^{3} \hat{\gamma}_j \Delta V_{C,t-j}, \qquad (13)$$

where $\Delta V_{C,t}$ and $\Delta V_{P,t}$ are changes of call and put option implied volatilities, respectively.

4. Pricing Options Using Predicted Volatility

In order to price the foreign currency options, it is necessary to obtain the predictions of daily volatilities. As discussed in Sec. 3, two approaches to predict volatilities are the ISVR and the GARCH methods. For the predictive tests of this paper, the daily volatility is predicted by both the GARCH and the ISVR methods. Along with the predicted volatilities, the current exchange rates and the interest rates, the option prices are computed using the Hull and White (1987) model.

Descriptive statistics for the actual and the predicted option prices are reported in Table 4. The first part of the table shows that for both the GARCH and the ISVR methods, the predicted prices of calls and puts are on average higher than the

Table 4. Out-of-sample comparison of the actual and predicted option prices.

This table reports the out-of-sample tests for the actual and predicted option prices. The time t volatility is predicted by the Implied Stochastic Volatility Regression (ISVR) model and the GARCH(1,1) model. The implied stochastic volatilities are estimated from the model of Hull and White (1987). The average GARCH variance forecast is based on the GARCH models of Engle (1982) and Bollerslev (1986). The Wald statistic is for the joint test of $H_0 : \alpha_0 = 0, \alpha_1 = 1$, and the t-statistics are reported in each parentheses. The sample period is from January 4, 1991 to March 25, 1993.

Summary statistics of the actual and predicted option prices.

	Type	Observation	Mean	Std.
Market	call	227,589	0.675×10^{-2}	0.416×10^{-2}
ISVR	call	227,589	0.685×10^{-2}	0.417×10^{-2}
GARCH	call	227,589	0.704×10^{-2}	0.402×10^{-2}
Market	put	409,892	0.908×10^{-2}	0.653×10^{-2}
ISVR	put	409,892	0.935×10^{-2}	0.692×10^{-2}
GARCH	put	409,892	0.914×10^{-2}	0.616×10^{-2}

$$\text{Market Price}_t = \alpha_0 + \alpha_1 \text{ Model Price}_t + \epsilon_t$$

Call options (227,589 observations)

Volatility	α_0	α_1	$\bar{R}^2$	Wald statistic
ISVR	0.204×10^{-4}	0.9824	0.9677	427.1
	(6.77)	(2612.7)		
GARCH	-0.122×10^{-3}	0.9759	0.8863	554.3
	(−20.55)	(1332.4)		

Put options (409,892 observations)

Volatility	α_0	α_1	$\bar{R}^2$	Wald statistic
ISVR	0.441×10^{-3}	0.9239	0.9600	956.3
	(126.96)	(3100.3)		
GARCH	-0.205×10^{-3}	1.0152	0.9163	376.7
	(−38.39)	(2092.9)		

observed market prices. To examine in detail the differences between the observed market prices and the model prices, the following regressions are estimated separately for calls and puts

$$C_t = \alpha_0 + \alpha_1 \hat{C}_t + \epsilon_t, \tag{14}$$

where C_t is the market price, $\hat{C}_t$ is the model price, and ϵ_t is the error term.

The model provides an unbiased estimate of the actual prices if $\alpha_0 = 0$ and $\alpha_1 = 1$. This hypothesis is tested using the Wald statistic that is asymptotically distributed as $\chi^2(2)$. The large number of observations (227,589 calls, 409,892 puts) implies that even small mispricings are evident in the regressions. Table 4 shows that for all cases of call and put options, the null hypothesis of unbiasedness can be overwhelmingly rejected.

In prices predicted by the ISVR method, the $\hat{\alpha}_1$ coefficient is always less than one and $\hat{\alpha}_0$ is always positive. This suggests that the model prices based on the ISVR method tend to underpredict low priced (less than $\frac{\hat{\alpha}_0}{1-\hat{\alpha}_1}$) calls and puts, and overpredict high-priced calls and puts.

The model prices predicted by the GARCH method suggest a slightly different story. For calls, the estimated $\hat{\alpha}_1$ is less than one and the estimated $\hat{\alpha}_0$ is negative. This indicates that the predicted prices from the GARCH method tend to overpredict calls. For puts, the $\hat{\alpha}_1$ is larger than one and $\hat{\alpha}_0$ is negative. This suggests that the predicted prices obtained from the GARCH method overvalue puts that are less than $\frac{\hat{\alpha}_0}{1-\hat{\alpha}_1}$, and undervalue puts that are higher than $\frac{\hat{\alpha}_0}{1-\hat{\alpha}_1}$).

5. Economic Profits for Predictable Volatility Changes

Sections 3 and 4 indicate that the conditional volatility of the dollar/mark exchange rate is statistically predictable using either the ISVR or the GARCH model. This section is devoted to the investigation of whether this predictability is significant enough to generate abnormal profit opportunities. In an efficient market, after accounting for transaction costs and risk, no trading strategy should earn abnormal returns. The market efficiency hypothesis is tested based on the so-called *ex post* trading strategy, which assumes that the agent can trade at the market prices that indicate deviations from the model prices (see Galai (1977), Shastri and Tandon (1986)). Furthermore, the trading strategies require the option positions to be hedged against the exchange rate movements. Two hedging methods are developed: delta-neutral hedging and straddle hedging. In calculating the hedged returns, I assume that the model of Hull and White (1987) is valid.

5.1. *Delta-neutral trading strategy*

The delta-neutral portfolio consists of selling or buying options and taking positions on holding or selling foreign currencies. Generally speaking, if the hedging position can be adjusted continuously, the *delta-neutral* trading strategy works well for the Black–Scholes (1973) model. In practice, since the hedged position can only be adjusted at discrete intervals, practitioners also use *gamma-hedging*. Furthermore, *vega-hedging* becomes necessary against the stochastic volatility of the underlying asset. However, in most cases, the benefits of the *gamma-hedging* and the *vega-hedging* are marginal, and only *delta-hedging* is considered here.[b]

---

[b]Hull and White (1987b) compares th relative performance of various hedging schemes. They find that the delta-gamma hedging performs well when the traded option has a constant implied volatitily and a short time to maturity, but it can perform far worse than delta-hedging other situations. They also find that delta-vega hedging outperforms other hedging schemes when the traded option has a non-constant implied volatitily and a long time to maturity.

The *delta-hedging* trading strategy proceeds as follows. On day t, each agent applies a volatility prediction method (ISVR or GARCH) to forecast volatility and compute the theoretical option price. The agent is allowed to change the position daily by buying the option if it is undervalued, or selling it if it is overvalued. The restriction imposed is that $100 worth of options and foreign currency are always bought and sold, and that the agent does not reinvest the profit the next day. Next, the agent compares the theoretical price C_t^M with the actual trading price C_t^A. If $C_t^M > C_t^A$, so the option appears underpriced, the agent decides to buy the option, and delta-hedges the position by buying or selling foreign currency. The hedged position is liquidated at day $t + 1$, and the agent obtains the return. If $C_t^M < C_t^A$, so the option appears overpriced, the agent sells options, delta-hedges the position by buying or selling foreign currency, and invests the $100 plus the proceeds from selling the options on a risk-free asset. On day $t + 1$, the hedged position is liquidated, and the agent obtains the return. The return formula for various strategies are as follows:

(1) if the agent buys call option at price C_t^A and sells foreign currency at price S_t, the absolute return is

$$AR_{H,t+1} = n_c \cdot [(C_{t+1}^A - C_t^A) - \delta_c(S_{t+1} - S_t)] + 200 \cdot r_D, \qquad (15)$$

where δ_c is the *delta* of a American call on one unit of foreign currency, and $n_c = |100/(C_t^A - \delta_c S_t)|$,

(2) if the agent sells call option and buys foreign currency, the absolute return is

$$AR_{H,t+1} = n_c[-(C_{t+1}^A - C_t^A) + \delta_c(S_{t+1} - S_t)], \qquad (16)$$

where $n_c = 100/(-C_t^A + \delta_c S_t)$,

(3) if the agent buys put option and buys foreign currency, the absolute return is

$$AR_{H,t+1} = n_p[(P_{t+1}^A - P_t^A) - \delta_p(S_{t+1} - S_t)], \qquad (17)$$

where δ_p is the *delta* of a American put on one unit of foreign currency, $n_p = 100/(P_t^A - \delta_p S_t)$,

(4) if the agent sells put option and sells foreign currency, the absolute return is

$$AR_{H,t+1} = n_p[-(P_{t+1}^A - P_t^A) + \delta_p(S_{t+1} - S_t)] + 200 \cdot r_D, \qquad (18)$$

where $n_p = |100/(-P_t^A + \delta_p S_t)|$.

With the initial investments of $100, the relative return for $1 of investment is

$$RR_{H,t+1} = AR_{H,t+1}/100. \qquad (19)$$

Table 5 reports the summary statistics of annualized return of the delta-neutral trading strategy for the full sample period. Before accounting for transaction

Table 5. Daily profits for delta-neutral trading strategies.

This table reports summary statistics of the daily relative returns of delta-neutral trading strategies. The daily volatilities are predicted from the Implied Stochastic Volatility Regression and the GARCH methods. With the prediction of daily volatilities, theoretical option prices are computed and subsequently compared with the market prices. The investment amounts are all standardized to $100 worth of the portfolio, and the agent does not reinvest the profit the next day. Each strategy is evaluated under the situations of no transaction cost, and with transaction cost. Sharpe ratios are based on the strategies' sample standard deviations of returns. For the same sample period, the annualized return for the Standard and Poor 500 Stock Index is 13.7%, with a Sharpe ratio of 0.75.

Filter	No. of days	Option Type	Volatility Forecast	No Transaction Cost		With Transaction Cost	
				Annum Return (%)	Sharpe ratio	Annum Return (%)	Sharpe ratio
0%	510	call	ISVR	13.92	2.60	−1.52	−2.34
	510	put	ISVR	14.93	3.28	−1.27	−18.61
	510	call	GARCH	10.12	0.94	−5.31	−2.54
	510	put	GARCH	5.06	−0.48	−11.13	−5.04
2%	508	call	ISVR	18.72	3.94	3.54	−0.74
	510	put	ISVR	17.46	3.92	1.77	−1.39
	510	call	GARCH	12.14	1.4	−3.04	−1.99
	510	put	GARCH	6.33	0.09	−9.61	−4.51
5%	496	call	ISVR	27.58	5.92	12.40	1.72
	506	put	ISVR	22.26	4.85	7.08	0.32
	509	call	GARCH	15.69	2.13	0.759	−1.32
	509	put	GARCH	8.10	0.56	−7.34	−3.62

costs, the agent using the ISVR method makes more profit than the agent using the GARCH method. For the ISVR method, the annualized returns are 13.92% for calls and 14.93% for puts, with Sharpe ratio of 2.6 and 3.28, respectively. For the GARCH method, the annualized returns are 10.12% for calls and 5.06% for puts, with Sharpe ratio of 0.94 and −0.48, respectively. For the same sample period, the annualized return for the Standard and Poor 500 Stock Index is 13.7%, with a Sharpe ratio of 0.75.

To test whether the price deviations are large enough to cover the transaction costs, filters with values of 2% and 5% of market prices are applied. Under these filters, options are only traded when the predicted price deviation is larger than the filter value. When the value of the filter increases, the number of trades decreases, and the agent is allowed to invest in the risk-free asset on no trading days. The results of using different filters are also reported in Table 5. Increasing the value of the filter increases the profitability. For the whole sample period, the daily average returns for agents using the ISVR and the GARCH methods are significantly greater than zero, but the returns for agents using the ISVR method are larger than returns for agents using the GARCH method in all three filters. In particular, with a 5% filter and no transaction costs, the agent using the ISVR method earns an annualized return of 27.58% for calls, and 22.56% for puts, while the agent using the GARCH ethod earns merely 15.67% for calls and 8.09% for puts.

To demonstrate that the delta-neutral trading strategy may lead to abnormal profits, the effect of transaction costs must be included. Studies by Black–Scholes (1972), Galai (1976), Shastri and Tandon (1986), and Harvey and Whaley (1991) on option trading strategies find that after accounting for transaction costs, the documented positive excess profits are generally not significantly different from zero. I assume that the transaction costs consist of a round-trip cost of one tick plus commissions. The per contract transaction charge for currency options at the PHLX is $0.28 for Customer Execution, $0.23 for Firms, $0.07 for Registered Option Trader and Specialist, and $0.05 for Floor Brokerage Transaction. Each dollar/mark option contract contains 62,500 marks. The minimum tick size on the PHLX is $.0001 for currency options, or $6.25 per contract. It is likely that an option identified as overpriced (underpriced) will have traded at the ask (bid), so actual selling (buying) this option in an arbitrage trade would most likely occur at a price one tick lower (higher). For most of the options selected (near-the-money, less than six months maturity), the option premium for each contract is between $400 to $800. For illustration purpose, the cost of changing investment position is assumed to be 1% of option prices.[c] For example, the net return for buying calls and selling foreign currency with transaction costs becomes

$$NAR_{H,t+1} = n_c \cdot [(C_{t+1}^A - C_t^A) - \delta_c(S_{t+1} - S_t) - 0.01 \cdot (C_{t+1}^A + C_t^A)] + 200 \cdot r_D .$$
(20)

It is evident that abnormal profits are dramatically reduced after accounting for transaction costs. For a one day holding period, transaction costs equal to 1% of the option price eliminate the hedged return for the GARCH method, while the ISVR method can only lock in a significant profit with a 5% filter.

5.2. Straddle trading strategy

A straddle is a pair of call and put options with the same maturity and striking price. An advantage of straddle trading is that it provides a natural hedge to the straddle holder since the call and put positions offset each other. Straddle data are collected from the daily trading record of dollar/mark options. Only options with at least fifteen days to expiration are considered. To avoid using deep-in or out-of-money options, options with moneyness (S/X) within the range of $[0.85, 1.15])$ are selected.

---

[c]This assumption is more justified to institutional investors for the following reasons: (1) the bid/ask spread can be expressed in this way; (2) commission is approximately proportional to trading volume; (3) the effect of putting in an order on the market is positively related to the size of the order. Although absolute cost can be viewed as fixed cost (e.g., computer, salary), it can normally be deducted on an annual basis; and its number is relatively independent of trading volume.

The straddle trading strategy proceeds as follows. Assume that investments of $100 worth of straddles are always bought and sold, and agents do not reinvest the profit the next day. On day t, the agent applies either the ISVR or the GARCH method to forecast future volatility, and calculates the straddle price. The model price $C_t^M + P_t^M$ is compared with the actual closing price $C_t^A + P_t^A$. If $C_t^M + P_t^M > C_t^A + P_t^A$, so the straddle appears underpriced, the agent buys $100 worth of straddles. If $C_t^M + P_t^M < C_t^A + P_t^A$, so the straddle appears overpriced, the agent sells the straddle, and invests in a risk-free asset with the initial $100 plus the proceeds (assumed to be $100) obtained from selling the straddles. On day $t + 1$, the agent liquidates the position and claims the return.

If the agent buys the straddles, the absolute return is

$$ AR_{B,t+1} = n_s \cdot [(C_{t+1}^A - C_t^A) + (P_{t+1}^A - P_t^A)], \tag{21} $$

where $n_s = 100/(C_t^A + P_t^A)$.

If the agent sells the straddles, the absolute return is

$$ AR_{S,t+1} = -n_s \cdot [(C_{t+1}^A - C_t^A) + (P_{t+1}^A - P_t^A)] + 200 \cdot r_D, \tag{22} $$

where $n_s = 100/(C_t^A + P_t^A)$.

In calculating returns from day t to day $t + 1$, it is necessary to find straddle pairs that have the same date of maturity and striking prices. Since many straddle pairs satisfy this criterion, the profit or loss from each straddle pair is calculated and the average is taken as a one day return. Again, a practical trading strategy will be to trade options when the price deviation is greater than the transaction costs. Filters with values of 0, 2%, and 5% of market prices are adopted to pick up the large price deviations. Under these filters, the number of days of trading straddles is reduced. To compare the performance of the ISVR and the GARCH methods, the agents can invest in a risk-free asset when no trade happens.

Table 6 reports the averaged return from trading straddles with and without transaction costs. Before transaction costs, the annualized return is 257.81% for the ISVR method, and 230.48% for the ISVR method, with Sharpe ratio of 3.34 and 2.88, respectively. This shows that the average trading profits are significantly greater than zero under both the ISVR and the GARCH methods. The returns from the ISVR method are greater than from the GARCH method for all three filters. For example, with a 5% filter, the ISVR method earns an average return of 316.76%, whereas the GARCH method makes 266.16%.

After accounting for transaction costs equal to 1% of the straddle prices, profits become negative regardless of the filter rule, for both the ISVR or the GARCH methods. The net returns from straddle trading after transaction cost of 1% per straddle is calculated as

$$ NAR_{B,t+1} = n_s \cdot \{[(C_{t+1}^A - C_t^A) + (P_{t+1}^A - P_t^A)] - 0.01 \cdot [(C_{t+1}^A + C_t^A) $$

$$ + (P_{t+1}^A + P_t^A)]\}. \tag{23} $$

Table 6. Daily profits for straddle trading strategies

This table reports summary statistics of the daily relative returns of straddle trading strategies. The daily volatilities are predicted from the Implied Stochastic Volatility method and the GARCH method. With the prediction of daily volatilities, theoretical option prices are computed and subsequently compared with the market prices. The investment amounts are all standardized to $100, and agents do not reinvest the profit the next day. Each strategy is evaluated under the situation of no transaction cost and with transaction cost. Sharpe ratio are based on the strategies' sample standard deviations of returns. For the same sample period, the annualized return for the Standard and Poor 500 Stock Index is 13.7%, with a Sharpe ratio of 0.75.

Filter	No. of days	Volatility Forecast	No Transaction Cost		With Transaction Cost	
			Annum Return (%)	Sharpe ratio	Annum Return (%)	Sharpe ratio
0%	502	ISVR	257.81	3.34	−247.69	−3.37
	502	GARCH	230.48	2.88	−273.24	−3.61
2%	486	ISVR	287.41	3.75	−217.83	−2.97
	490	GARCH	239.59	2.94	−265.39	−3.43
5%	485	ISVR	316.76	4.01	−190.01	−2.52
	488	GARCH	266.16	3.25	−239.08	−3.04

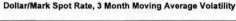

Dollar/Mark Spot Rate, 3 Month Moving Average Volatility

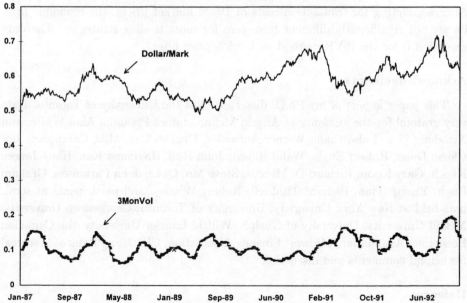

6. Conclusion

The conditional volatility of the dollar/mark exchange rate can be predicted both with the implied volatility extracted from foreign currency options, and with the GARCH models. The major difference between these two methods is that while the GARCH method uses *ex post* volatility of past returns, the ISVR method uses the implied volatility extracted from the foreign currency options. This paper investigates whether this volatility predictability is large enough to be economically meaningful, in the sense of identifying the existence of abnormal profit opportunities. In an efficient market, after accounting for transaction costs and risk, no trading strategy should earn abnormal returns.

This paper finds that in the absence of commission costs and the bid-ask spreads, both the delta-neutral and the straddle trading strategies yield significant positive economic profits regardless of which volatility prediction method is used. Increasing the value of the filter serves to raise the profitability of the trading strategies. Therefore, traders who can execute the delta-neutral and the straddle trading strategies at the observed transaction prices can lock in significant profits during the period considered here. Furthermore, the agent using the ISVR method earns greater profits than the agent using the GARCH method. This is consistent with Guo (1996), who finds that implied variance dominate the GARCH model in forecasting future realized variance of the foreign exchange rate.

However, for a nonmember of an exchange, especially individual investors who face higher transaction costs, the magnitude of the correctly predicted volatility changes is generally not large enough to generate abnormal risk-adjusted profits. After accounting for transaction costs of 1% of market prices, the economic profits are not significantly different from zero for most trading strategies. The only exception is for the ISVR method with a 5% price filter.

Acknowledgements

This paper is part of my Ph.D. dissertation at the University of Toronto. I am very grateful for the guidance of Angelo Melino, James Pesando, Alan White, and Xiaodong Zhu. I also thank Werner Antweiler, Charles Cao, Mike Campolieti Jin-Chuan Duan, Robert Engle, Walid Hejazi, John Hull, Raymond Kan, Hans-Jurgen Knoch, Gary Koop, Richard D. Marcus, Steve Mo, Qi Li, John Parkinson, Graham Pugh, Yisong Tian, Richard Timbrell, Robert Whaley, and participants at seminars held at New York University, University of Toronto, Georgetown University, McGill University, University of Guelph, Wilfrid Laurier University, the Canadian Economics Association, Midwest Finance Association, CIBC, and Bank of Montreal for helpful comments and discussion.

Refences

R. Baillie and T. Bollerslev, "The Message in Daily Exchange Rates: A Conditional-Variance Tale", *Journal of Business and Economic Statistics* **7** (1989) 297–306.

G. Barone-Adesi and R. Whaley, "Efficient Analytic Approximation of American Option Values", *Journal of Finance* **42** (1987) 301–320.

F. Black and M. Scholes, "The pricing of Options and Corporate Liabilities", *Journal of Political Economy* **81** (1973) 637-659.

J. Bodurtha and G. Courtadon, "Tests of the American Option Pricing Model in the Foreign Currency Option Market", *Journal of Financial and Quantitative Analysis* **22** (1987) 153–167.

T. Bollerslev, "Generalized Autoregressive Conditional Heteroscedasticity", *Journal of Econometrics* **51** (1986) 307–327.

T. Bollerslev and J. Wooldridge, "Quasi Maximum Likelihood Estimation and Inference in Dynamic Models with Time Varying Covariances", *Econometric Reviews* **11** (1992) 143–172.

D. Dickey and W. Fuller, "Distribution of the Estimators for Autoregressive Time Series with a Unit Root", *Journal of the American Statistical Association* **74** (1979) 427–431.

J.-C. Duan, "A Unified Theory of Option Pricing with Stochastic Volatility — from GARCH to Diffusion", manuscript (1994) McGill University.

R. F. Engle, "Autoregressive Conditional Heteroskedasticity With Estimates of the Variance of U.K. Inflation", *Econometrica* **50** (1982) 987–1008.

R. Engle, A. Kane, and J. Noh, "A Test of Efficiency for the S&P Index Option Market Using Variance Forecasts" NBER working paper (1993) 4520.

D. Galai, "Tests of Market Efficiency and the Chicago Board Options Exchange", *Journal of Business* **50** (1977) 167–197.

D. Guo, "The Predictive Power of Implied Stochastic Variance from Currency Options", *Journal of Futures Markets* **16** (1996) 915–942.

C. R. Harvey and R. Whaley, "Market Volatility Prediction and the Efficiency of the S&P 100 index Option Market", *Journal of Financial Economics* **31** (1992) 43–73.

S. Heston, "A Close Form Solution for Options with Stochastic Volatility", *Review of Financial Studies* **6** (1993) 327–343.

J. Hull and A.White, 1987a,"The Pricing of Options on Assets with Stochastic Volatility", *Journal of Finance* **42** (1987a) 281–300.

J. Hull and A. White, "Hedging the Risks from Writing Foreign Currency Options", *Journal of International Money and Finance* **6** (1987b) 131–152.

D. A. Hsieh, "The Statistical Properties of Daily Foreign Exchange Rates: 1974–1983", *Journal of International Economics* **24** (1988) 129–145.

D. Hsieh, "Modelling Heteroscedasticity in Daily Foreign Exchange Rates", *Journal of Business and Economic Statistics* **7** (1989) 307–317.

C. G. Lamoureux and W. D. Lastrapes, "Forecasting Stock Return Variance: Toward an Understanding of Stochastic Implied Volatilities", *The Review of Finacial Studies* **6** (1993).

A. Melino and S. Turnbull, "Pricing Foreign Currency Options with Stochastic Volatility", *Journal of Econometrics* **45** (1990) 7–39.

S. M. Phillips and C. W. Smith, "Trading Costs for Listed Options: The Implication for Market Efficiency", *Journal of Financial Economics* **8** (1980) 179–201.

K. Shastri and K. Tandon, "Valuation of Foreign Currency Options: Some Empirical Tests", *Journal of Financial and Quantitative Analysis* **21** (1986) 144–160.

R, Whaley, "Valuation of American Call Options on Dividend-Paying Stocks: Empirical Tests", *Journal of Financial Economics* **10** (1982) 29–58.

J. B. Wiggins, "Option Values Under Stochastic Volatility: Theory and Empirical Estimates", *Journal of Financial Economics* **19** (1987) 351–372.

PORTFOLIO-BASED RISK PRICING: PRICING LONG-TERM PUT OPTIONS WITH GJR-GARCH(1,1)/JUMP DIFFUSION PROCESS

SERGEI ESIPOV and DAJIANG GUO*

*Centre Solutions, a member of Zurich Financial Services Group,
One Chase Manhattan Plaza, New York, NY 10005, USA*
E-mail: sergei.esipov@centresolutions.com
E-mail: dajiang.guo@centresolutions.com

1. Introduction

Fleming (1993) and Guo (1996) show that the implied variances extracted from the stochastic volatility model of Hull and White (1987) is a dominant, but still a biased estimator in terms of *ex ante* forecasting power in the stock index and foreign currency option markets. There are at least three potential explanations for the rejection of the *unbiasedness* restriction: the nonzero risk premium for distributed risk, transaction costs in dynamic hedging, and potential inefficiency of the option market.

The existence of the risk premium of volatility in the option prices is empirically demonstrated by Guo (1998), who found that the market price of volatility risk is non-zero and time varying. The size of the estimated risk premium implies that the compensation for volatility risk is a significant component of the risk premium in the currency market. A nonzero risk premium on the nontraded volatility process may account for the puzzling empirical evidence that the implied volatility is a biased forecast of future volatility (Fleming (1993), Guo (1996), Jorion (1995)).

Option prices observed from liquid option markets contain rich information about the market's expectation of future distribution of the underlying asset and profit and loss distributions of the option underwriter. An option under-writer who sold a put option will dynamically hedge his/her position by shorting a certain amount of stock according to the Black–Scholes's greeks. In a standard

*We thank Alexander Adamchuk, Marco Avellaneda, Peter Carr, Dan D. Dibartolomeo, Xiaodong Zhu and participants at seminars held at New York University, the University of Chicago, Southern Finance Association for helpful comments and discussions. We are especially thankful for the encouraging business environment provided by Richard Timbrell, our colleagues, Tony Huang, Todd Fonner, and others at Centre Solutions. Opinions expressed in this article are the authors' and do not necessarily correspond to the view of Centre Solutions, and Zurich Financial Services Group. Remaining errors are ours.

Black–Scholes world, if the underlying asset is lognormally distributed, and continuous frictionless hedging is possible, the full hedging (replication) cost would be a well-defined number, thus it is the value of the option contract. However, in reality, none of the Black–Scholes assumption is true, the underlying volatility could be time varying, the hedging invokes transaction costs, and can only be done discretely. Thus, the full hedging cost has a distribution. The option price charged should reflect the average cost of hedging, and a risk premium that compensates the marginal risk the contract brings for the existing portfolios, or the profit and loss distribution of the financial institution.

While the complete list of risk factors is extensive, here we focus on three major concerns an option underwriter has to consider: (1) non-log-normal distribution of the underlying asset, such as stochastic volatility and/or jumps, (2) transaction costs in dynamic hedging, (3) risk premium for unhedgeable remaining basis risk. With these complications, the final profit and loss (P&L) to the option underwriter is no longer certain. The underwriter has to develop a formal procedure which would allow him/her to relate the P&L distribution at maturity to a (dollar) value of the derivative security. It is clear that charging the limit loss is overly expensive. This article proposes a portfolio-based pricing method to evaluate risk and systematically consider risk premium. The risk premium is charged to satisfy risk management and return on risk capital requirements. The P&L distributions are priced based on Value-at-Risk and return on capital approach. The pricing-generating process is tested on the example of the Standard and Poor 500 index (SPX) put options.

We use a GJR-GARCH(1,1)/Jump-Diffusion model to simulate the underlying SPX process. This is a minimal model which is capable of reproducing most of the empirically observed regularities of the volatility process (mean-reverting, persistence, and leverage effect) as close as possible. Time series analyses reflected in extensive literature on ARCH and GARCH models have already determined the statistical properties of the reproducible distribution of the SPX stock index. The index exhibits time varying volatility, nonzero skewness and leptokurtosis (Baillie and Bollerlev (1989), Engle and Lee (1997)). In addition, stock market downside jumps are added from a compound Poisson–Beta distribution with maximum loss of up to 25% in a day, thus, the 1987 stock market crash has a positive probability in the model. Smaller jumps are more frequent.

We can exactly fit the implied volatility "smile" or "smirk" curve of the one-year SPX options (liquid Exchange market) in term of slope and volatility values. It is then used to price five-year out-of-the-money put (non-liquid over-the-counter market). In total, the implied volatility is about 10%–40% higher than the historical one. This numbers are consistent with the behavior of S&P 500 market.

This method can be applied to price long term stocks, currency, and commodity options. The result of the present study can be subsequently used by academia to improve positive option pricing models, and by practitioners to structure over-the-counter derivative products and perform firm-wide risk management.

The paper proceeds as follows. Section 2 is devoted to a portfolio-based pricing method for risk evaluation; it systematically considers risk premium. Section 3 estimates the model parameters needed to simulate S&P 500 index. Section 4 describes pricing of short and long term put options, and Sec. 5 concludes this paper.

2. Portfolio-Based Risk Pricing

The pricing technique presented in this section is based on historical probabilities, and correlations with the existing portfolio. This technique combines some actuarial methods for computing the risk premiums and Value-at-Risk approach. Calibration procedure used in the conventional risk-neutral approach is replaced here by a set of constraints on the values of the risk premiums expressed through a return on allocated capital.

An important characteristic of a portfolio is its Value-at-Risk, VaR. Notwithstanding the growing popularity of this term, its meaning is far from being unique. We adopt the definition where VaR is the 99%-ile (or 1% left limit) of P&L distribution of the net cash flows associated with a given time horizon. The conventional approach to determine such VaR is a Monte-Carlo simulation of all the contracts under management. The situation simplifies when one focuses on a single contract which cash flows are smaller than the scale of the whole portfolio. This is usually the case for the so-called "Market portfolio", i.e. the collection of contracts traded on a market. The same assumption holds true for large financial institutions. Under the assumption the change of VaR caused by adding a single contract is much smaller than VaR itself. It follows from here that the detailed shape of the P&L distribution added to the portfolio is irrelevant, the small parameter being the ratio of scales.

The P&L distribution of a Market portfolio is assumed to be normal, $N(u, \Xi)$. Adding a contract with probability distribution $p(x)$ to this portfolio shifts the VaR (99%-ile) by

$$\Delta\text{VaR} \approx \langle x \rangle + \zeta(\sigma_x^2/2\Xi + \rho\sigma_x) \approx \langle x \rangle - \zeta\rho\sigma_x \,,$$

where $\langle x \rangle$ is the expected value of P&L, σ_x is the corresponding standard deviation, ρ is the correlation between the P&L of the Market portfolio and that of the contract. Ξ is the standard deviation of the Market portfolio. The shift of VaR is usually a negative number, indicating a loss to the portfolio, $\zeta = N^{-1}(0.01) \approx 2.326$ is a constant.

In most markets, the market premiums charged for contracts are smaller than the change of VaR, and additional risk capital has to be allocated and invested. If the actual premium is Pr, one has to allocate $-\Delta\text{VaR} - \text{Pr}$ to keep the VaR limit unchanged. This leads to the expected continuous return

$$\mu = r + \frac{1}{T} \ln \left(1 + \frac{\text{Pr} + \langle x \rangle}{-\langle x \rangle + \zeta\rho\sigma_x - \text{Pr}} \right) \,,$$

where r is the risk-free rate. Solving this equation for Pr gives

$$\text{Pr} = -\langle x \rangle + \zeta \rho \sigma_x [1 - e^{-(\bar{\mu}-r)T}],$$

where $\bar{\mu}$ is the market-*required* return on risk capital.

In summary, the minimum premium earned must ensure that: (1) the new portfolio (the contract plus the original portfolio) will meet the risk management requirement, that is, the VaR keeps at the same threshold as the original portfolio; (2) the return on capital requirement is met for the risk capital allocated to this new contract. The premium formula given above is far from being universal, it simply reflects some wide-spread capital allocation guidelines. It is worth noting here that a single universal formula can hardly exist for capital allocation, and attempts to formalize this issue usually sparkle a debate.

3. Parameters Estimation

3.1. *Parameters for the GJR-GARCH(1,1)/jump-diffusion model*

Volatility language is spoken to describe the dynamics of option markets. Volatility has well documented time series properties as mean-reversion, persistence, and leverage effect. Mean-reversion means that after experiencing positive or negative shocks, the volatility itself will converge back to its long term average value. Persistence means that a large movement of stock volatility tends to be followed by a large movement, while small volatility changes tend to be followed by subsequent small volatility changes. Leverage effect means that negative moves of the underlying have a greater impact on volatility than positive moves of equal amount. During the historical period of 1988–1998, the key descriptive statistical properties of SPX index are:

Ann. SPX mean return	13.86%
Ann. SPX volatility	11.95%
Daily Prob. Down Jump	0.90%
Daily Down Jump Size	6.30%
Historical Implied Volatility	13.53%
Current Implied Volatility	23%

The implied volatility changes over time as the underlying index moves and derivative securities are traded. The number 23% is an indicative single-number estimate for the range of implied volatilities observed during the day when calibration of the Simulator was performed (see below).

This article adopts a GJR-GARCH(1,1) model of Glosten, Jagannathan, and Runkle (GJR), modified by using an additional jump component to capture conservatively the large downside movement of stock index, especially large market corrections and crashes. While the adopted model generates rather realistic time series, and belongs to minimal models in its class, it is not necessarily unique. It would be of interest to test the expected stability our results for a range of different models for the underlying index.

First, the diffusion process for the stock index process is assumed to be a standard discrete Geometric Brownian Motion,

$$S_t - S_{t-1} = S_t(m + s_t\epsilon_t + q_t),$$

where m is the log-drift per time step, s_t is the volatility, and q_t is the jump component, respectively. The GJR-GARCH(1,1) volatility process is

$$s_t^2 = \omega(1 - \alpha - \beta - \gamma/2) + (\alpha + \gamma D_{t-1})\epsilon_{t-1}^2 + \beta s_{t-1}^2.$$

Here ε_t is a set of normally distributed random variables with zero mean and unit variance. $D_t = 1$ if $\epsilon_t < 0$ and $D_t = 0$ if $\epsilon_t > 0$. A positive γ reflects the "leverage effect". Description and discussion of other parameters can be found in the GJR article.

The parameterization of GJR-GARCH(1,1) model for the S&P 500 index return (Engle and Lee (1996)) is the following (the time step is one business day, and all units are adjusted correspondingly).

ω	$9.580 \cdot 10^{-7}$
α	0.0332
β	0.9122
γ	0.0925
m	0.00835

Unfortunately, the GARCH class models (designed to mimic auto-correlations) are not satisfactory in generating large deviations. In view of this it is important to add the jump component. We didn't modify the GJR parameters while adding q_t term. As a result the following quantities were observed in simulations.

Ann. SPX expected return (observed)	13.9%
Ann. SPX return volatility (assumed)	17.0%
Daily Probability Down Jump (observed)	0.9%
Daily Max Down Jump Size (assumed and observed)	25.0%
Daily Mean Down Jump Size (observed)	6.82%

4. Pricing S&P 500 Index Put Options

4.1. *Option parameters for one-year European puts*

On the day when the test described below was performed the market implied volatility was approximately 23% per year for a one-year European puts. The table below describes the quantities used for simulating the put option.

Forward SPX	1.0618
SPX spot	1.0000
SPX strike	1.0618
SPX Strike/Forward SPX	1
Option Maturity (yrs)	1
Risk-free rate per annum	6.0%
Selected Volatility for Delta-Hedging	23.0%
Proportional Transaction Costs	0.05%

4.2. *From distributions to prices*

In the ideal Black–Scholes world, where the underlying index value follows a Geometric Brownian Motion, and hedging can be performed continuously, the full accumulated hedging cost has a fixed value, which should be equal to the theoretical Black–Scholes option price. However, when the underlying spot process does not follow log-normal distribution, and hedging is discrete, the hedging cost is a random variable depends on the realization of each pass. Usage of greeks, however advanced, at best generates an over-hedge.

The portfolio-based risk pricing can be applied to determine the price of the put contract along the lines described in Sec. 2 above. Now σ_x denotes the standard deviation of P&L associated with the position of writing option and hedging, $\langle x \rangle$ refers to the Expected Cost, and ρ is the correlation of the "option + hedge" position with the rest of the Market portfolio ($\rho = 50\%$ is used everywhere below). We have tested *four* different versions of the premium formula (cf. Sec. 2 above):

(1) Annual return on allocated capital, $\bar{\mu} = 30\%$.

$$\mathrm{Pr}_1 = \text{Expected Cost} + 2.326 \cdot 0.5 \cdot \sigma_x \cdot \left[1 - e^{-(\bar{\mu}-r)T}\right].$$

(2) μ is infinite or no capital is allocated.

$$\mathrm{Pr}_2 = \text{Expected Cost} + 2.326 \cdot 0.5 \cdot \sigma_x.$$

(3) Maximum Observed Cost is used, which is the 100th percentile of the hedging cost distribution.

$$\mathrm{Pr}_3 = \text{Maximum Observed Cost} = \text{MOC}.$$

(4) 1% of the Maximum Observed Cost is charged on top. Allocated capital is not invested.

$$\mathrm{Pr}_4 = \mathrm{Pr}_1 + 0.01 \cdot \text{MOC}.$$

Put price

The benchmark Black–Scholes (BS) price for a one-year at-the-money ($X/F = 1$) European put is 9.16% per annum with implied vol of 23%. If we use Pr_4, the

X/F	(1)	(2)	(3)	(4)
0.75	0.0145	0.0288	0.184	0.0164
1	0.085	0.104	0.168	0.0867
1.25	0.269	0.285	0.453	0.274

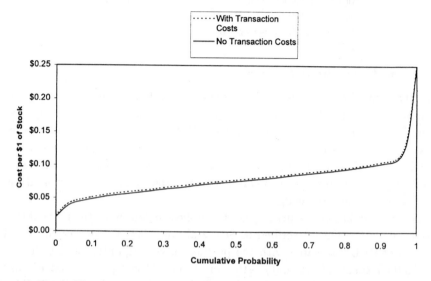

Fig. 1. The cost of writing a one-year put option and hedging it $X/F = 1$.

suggested price for a put underwriter is 8.67% per annum. BS $-$ Pr$_4$ = 0.49%, which is 5.35% lower than the BS price. The following prices are generated by different pricing methods at different strike levels. Numbers (1)–(4) refer to the four pricing formulae given above.

Figure 1 shows the cumulative probability distribution function of the costs of writing a one-year put option and performing the Black–Scholes delta-hedge with implied volatility 23% on a daily basis. This strategy is insufficient for loss immunization. The probability distribution of the hedging P&L has been studied analytically by Esipov and Vaisburd (1998). Evolution of this distribution obeys a certain integro-differential equation. In the ideal Black–Scholes world this distribution becomes a delta-function centered at the Black–Scholes Price.

4.3. Volatility smile

For out-of-the-money SPX put options, the implied volatilities exhibit a "smile" or "smirk" effects for different Strike/Forward ratios. Conventionally, these volatilities are fitted from the observed market prices by using the BS formula. Similarly, implied volatilities can be inferred from the option prices generated by simulators. If market implied volatility smile curve is reproduced by the Simulator, one may conclude that this Simulator captures the price-generating mechanism

X/F	(1)	(2)	(3)	(4)
0.75	0.2569	0.3249	0.8682	0.2667
1	0.2136	0.2615	0.4248	0.2178
1.25	0.2204	0.2728	0.7107	0.2363

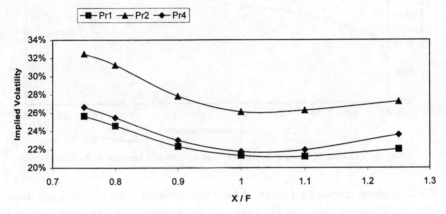

Fig. 2. Implied volatility "smile" from Simulator, one-year put.

of the put options in the market place. For the four pricing methods described above, the fourth one provides the best fit of the observed implied volatility "facial expression" of SPX put options.

Figure 2 shows the implied volatility "smile" curve for different pricing formulae.

4.4. *Comparison with historical market price of puts in 1988–1997*

This section focuses on the underlying assumptions and performance of the Simulator by comparing its outputs with historical mid-market prices observed in 1988–1997. The empirical study of SPX returns during 1988–1997 suggests a three-month moving average volatility of 11.95% with a max of 22.4% and a min of 5.8%, and volatility of volatility is around 28%. During the period of 1988–1997, averaged annual implied volatility of the SPX put is 16.73%, or 40% higher than the actual historical volatility.

The benchmark BS price for a one-year European put is 3.97% per annum (with implied volatility of 16.73%) per $1 worth of underlying stock value. Without transaction costs, the average hedging cost for a put underwriter is 3.21% per annum, that is, an average absolute value difference of 0.35%. In relative terms, the average hedging cost is 8.86% lower than the BS price. Finally, adding risk premium to the average hedging cost, the model Pr_2 suggests to charge a zero capital allocation premium of 3.74% per annum.

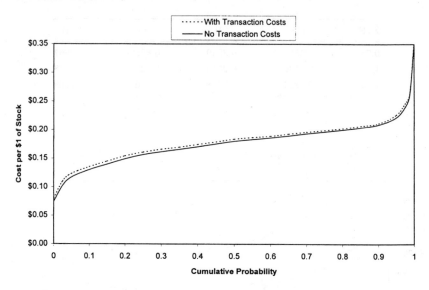

Fig. 3. The cost of writing a five-year put option and hedging it $X/F = 1$.

With assumed transaction costs, the average hedging costs is 3.84% per annum, which is 0.13% lower than the BS price. After charging risk premium, model Pr_2 suggests a price of 3.96% per annum.

We conclude that the historical market prices of one-year put options are reproducible by the Simulator equipped with an appropriate risk–premium formula based on Value-at-Risk approach.

4.5. *Five-year European puts*

The Simulator which can reproduce the price-generating process of the put options in the market place is itself an asset. It is used here to price five-year puts, which are considered to be non-liquid, over-the-counter contracts.

Premium

Because the one-year market implied volatility smile curve can be reproduced by the simulator, that is, it is successfully captured the price-generating process of the put option in the market place. These parameters are used to price five-year long term put (non-liquid, over-the-counter market).

If the objective average annual volatility of the SPX index in the five-year period is assumed to be 17%, implied (hedging) volatility is assumed to be 23%, risk-free interest rate is 6%, the benchmark Black–Scholes price (BS) for a five-year at-the-money ($X/F = 1$) European put (no dividend) is 20.29%. If we use Pr_1, the suggested price for a put underwriter is 19.44%. BS-$Pr_1 = 0.85\%$, that is 4.19% lower than the BS price. The following prices are generated by different pricing methods at different Strike/Forward (X/F) levels.

X/F	(1)	(2)	(3)	(4)
0.75	0.081	0.0904	0.5637	0.0866
1	0.1944	0.2052	0.6422	0.2008
1.25	0.3562	0.3667	0.7087	0.3633

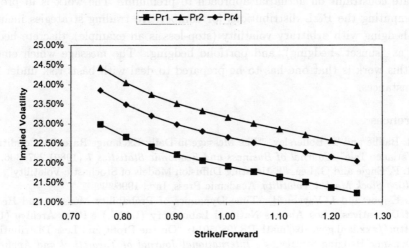

Fig. 4. Implied volatility "smile", five-year put.

Volatility Smile

Figure 4 shows the implied volatility "smirk" curve of different pricing model. It is important to note that the "smirk" is less friendly as compared to Fig. 2. The Simulator reproduces a well-known property of the volatility term-structure: the "smile" weakens at large maturities. This is a direct consequence of the portfolio based or VaR-based risk premium loading. The following table summarizes the "implied volatilities" generated from different pricing methods at different strike levels.

X/F	(1)	(2)	(3)	(4)
0.75	0.2299	0.2444	1.093	0.2386
1	0.2192	0.2316	0.8204	0.2266
1.25	0.2123	0.2244	0.6252	0.2205

5. Conclusion

Large financial institutions, which are permanently involved in selling and trading derivative securities, should pay more attention to estimating the profit and distributions associated with these contracts. While such general statement may be regarded as self-obvious in actuarial communities, its perspectives for financial quantitative analysis are far-reaching. We have shown that the profit and loss distributions may lead to consistent pricing which can potentially reproduce complex market effects such as volatility "smile" and its term structure. Merge of financial and actuarial worlds is taking place today, and its complete quantitative description is subject of future work. On the other side, the market price provides an ultimate constraint on actuarial approach to premium. The work is in progress on computing the P&L distributions for competitive trading strategies including delta-hedging with arbitrary volatility (stop-loss is an example), discrete hedging (such as "sunset"-hedging), and portfolio hedging. The message which emerges from this work is that one has to be prepared to deal with basis risk under most circumstances.

References

[1] R. Baillie and T. Bollerslev, "The Message in Daily Exchange Rates: A Conditional-Variance Tale", *Journal of Business and Economic Statistics* **7** (1989) 297–306.

[2] R. F. Engle and G. Lee, "Estimating Diffusion Models of Stochastic Volatility", *Modelling Stock Market Volatility*, Academic Press, Inc., 1996.

[3] S. Esipov and I. Vaysburd, "Time Dynamics of Probability Measure and Hedging of Derivatives", Los Alamos National Laboratory (LANL) e-Print Archive (1998), http://xxx.lanl.gov/abs/math.PR/9805014; "On the Profit and Loss Distribution of Dynamic Hedging Strategies", *International Journal of Theoretical and Applied Finance*, forthcoming (1999).

[4] J. Fleming, "The Quality of Market Volatility Forecasts Implied by S&P 100 Index Option Prices", unpublished Ph.D. dissertation, Fuqua School of Business, Duke University, 1993.

[5] L. R. Glosten, R. Jagannathan and D. E. Runkle, "On the Relation Between the Expected Value and the Volatility of the Nominal Excess Return on Stocks", *Journal of Finance* **48** (1993) 1779–1801.

[6] D. Guo, "The Predictive Power of Implied Stochastic Variance from Currency Options", *Journal of Futures Markets* **16** (1996) 915–942.

[7] D. Guo, "The Risk Premium of Volatility Implicit in Currency Options", *Journal of Business & Economic Statistics*, Vol. 16, No. 4, 1998.

[8] J. Hull, *Options, Futures, and Other Derivatives*, Prentice Hall, 1997.

[9] J. Hull and A. White, "The Pricing of Options on Assets with Stochastic Volatility", *Journal of Finance* **42** (1987) 281–300.

[10] P. Jorion, "Predicting Volatility in the Foreign Exchange Market", *Journal of Finance* **50** (1995) 507–528.

THE EXISTENCE OF EQUILIBRIUM IN A FINANCIAL MARKET WITH TRANSACTION COSTS

XING JIN*

Institute of Systems Science, Academia Sinica,
Beijing 100080, PR China

FRANK MILNE†

Department of Economics, Queen's University, Kingston, Ontario, Canada

This paper proves the existence of a general equilibrium in a financial model with transaction costs. A general equilibrium is shown to exist in a model with convex trading technology, in which the agents include consumers, production firms, brokers or dealers. When the trading technology is non-convex, an individual approximate equilibrium, introduced by Heller and Starr (1976), is proved in the above model. And, moreover, under a further assumption of finite p-convexity on the commodity excess demand correspondence, a general equilibrium for a non-convex exchange economy is obtained for an economy with consumers, brokers or dealers.

Keywords: Arbitrage, general equilibrium, transaction cost, individual approximate equilibrium, finite p-convexity.

1. Introduction

A number of authors have considered financial markets with transaction costs, particularly the impact of transaction costs on optimal portfolio selection (e.g., Magill and Constantinides (1976), Kandell and Ross (1983), Taksar, Klass and Assaf (1988), Duffie and Sun (1990), Fleming *et al.* (1989), Davis and Norman (1990)); and the pricing and hedging of derivative securities using the underlying stock and bond (e.g., Leland (1985), Boyle and Vorst (1992), Bensaid, *et al.* (1992), Edirisinghe, Naik and Uppal (1993), Constantinedes and Zariphopoulou (1995)).

More recently, some authors (e.g., Jouini and Kallal (1995), Ortu (1995), Milne and Neave (1996)) have investigated economies with transaction costs and the implications for asset prices and allocations. Jouini and Kallal (1995) use arbitrage methods as introduced by Harrison and Kreps (1979) to obtain a set of equivalent Martingale measures that are deduced from an economy with transaction costs and an absence of arbitrage. Ortu (1995) uses duality and linear programming methods in finite dimensions extending Jouini and Kallal's results. Milne and Neave

*Supported by National Natural Science Foundation of China.
†Supported by SSHRC of Canada.

(1996) formulate a competitive economy with multiple dates, uncertainty, a single physical commodity and a set of consumers trading assets through a set of broker/intermediaries, who have explicit transaction technologies. Their formulation draws upon an older literature in General Equilibrium theory that characterises transaction costs by discriminating between bought and sold commodities or assets (see Ostroy and Starr (1990) for a survey of this literature). Assuming the existence of an equilibrium, Milne and Neave characterise pricing and asset allocations for a number of special cases of their model, emphasising the model's flexibility in encompassing many cases discussed separately in the literature (e.g., broking, personal transaction costs, fixed and variable transaction costs, inventory-type models, incomplete markets). The Milne–Neave model is consistent with the formulations of Jouini–Kallal and Ortu, in providing a general primal formulation. The latter papers exploit no arbitrage/duality methods to obtain similar, or complementary results.

This paper constructs a more general version of Milne and Neave (1996) including many physical commodities, producers/firms and general assumptions on feasible consumption, production sets, and transaction technologies. We provide conditions that guarantee the existence of an equilibrium in this economy when the transaction technology is convex. To incorporate non-convexities (or fixed costs) in transactions, the model considers a modification introduced in an earlier general equilibrium literature (see Heller and Starr (1976)) that allows us to prove the existence of an approximate equilibrium. In addition we introduce a different method for proving existence in a version of our economy with non-convex transaction technologies. It is well-known that transactions often involve a fixed cost for each transaction, and some element of marginal cost: our non-convex technology deals with that case. We suppose a condition of finite p-convexity on the commodity excess demand correspondence. This assumption allows a limited degree of non-convexity in the asset trade technology, and yet a well-defined element of convexity in commodity demand to generalise the existence of an exact equilibrium.

The rest of this paper is organized as follows. In Sec. 2, we outline the basic model and introduce the concept of no arbitrage and its equivalent condition. In Sec. 3, we will prove some preliminary results and, finally, show the existence of equilibrium for the model with convex trading technology. Section 4 is devoted to individual approximate equilibrium in the model with non-convex trading technology. In Sec. 5, a concept of finite p-convexity is introduced and the general equilibrium is proved in an exchange economy. In Sec. 6, we conclude with a discussion of special cases and possible extensions of our model. The appendix includes a proof of one preliminary result.

2. The Model and No Arbitrage

Consider an economy with uncertainty characterized by a event tree such as that depicted in Fig. 1 of Duffie (1987). This tree consists of a finite set of nodes **E** and directed arcs $\mathbf{A} \subset \mathbf{E} \times \mathbf{E}$, such that $(\mathbf{E}, \mathbf{A})$ forms a tree with a distinguished root e_0. For any node $e \in \mathbf{E}$ other than the root node e_0, e^- designates the unique predecessor of e. The number of immediate successor nodes of any $e \in \mathbf{E}$ is denoted $\#e$. A node $e \in \mathbf{E}$ is terminal if it has no successor node. The immediate successor nodes of any non-terminal node $e \in \mathbf{E}$ are labeled $e^{+1}, \ldots, e^{+K}$, where $K = \#e$. The sub-tree with root e is denoted $\mathbf{E}(e)$. Particularly, $\mathbf{E} = \mathbf{E}(e_0)$. Suppose there are N securities and M commodities at any node $e \in \mathbf{E}$. Assume that all assets pay dividends at each node.

There are J production firms (indexed by j) with objective function $V_j(\cdot)$, each of whom chooses a production plan and a trade plan. A production plan of firm j is an array of numbers $y_{j,m}(e)$, one for each $m \in \mathbf{M} = \{1, \ldots, M\}$, and $e \in \mathbf{E}$ with the usual sign convention for inputs (non-positive) and outputs (non-negative). Thus a contingent production plan $y_j(e) = (y_{j,1}(e), \ldots, y_{j,M}(e))$ of firm j is a point of vector space $\mathbf{R}^M$. The set of all contingent production plans that are technologically feasible for firm j will be denoted by $Y_j \subseteq \mathbf{R}^{|E| \times M}$, where $|\mathbf{E}|$ is the total number of nodes of $\mathbf{E}$.

A trade plan of firm j is a vector $\theta_j = (\theta_j(e))_{e \in \mathbf{E}} = (\theta_j^B(e), \theta_j^S(e))_{e \in \mathbf{E}} = (\theta_{j,1}^B(e), \ldots, \theta_{j,N}^B(e), \theta_{j,1}^S(e), \ldots, \theta_{j,N}^S(e))_{e \in \mathbf{E}}$ in $\mathbf{R}_+^{2(|E| \times N)}$, where $\theta_{j,n}^B(e)(\theta_{j,n}^S(e))$ represents the accumulated purchase (sale) of asset n by firm j after trading at node e. Let $\theta_{j,0,n}^B$ and $\theta_{j,0,n}^S$ denote the initial shares of asset n bought and sold respectively by firm j just before trading begins at node e_0. And let $\theta_{j,0} = (\theta_{j,0,1}^B, \ldots, \theta_{j,0,N}^B, \theta_{j,0,1}^S, \ldots, \theta_{j,0,N}^S)$.

There are H brokers or intermediaries (indexed by h) with objective function $W_h(\cdot)$. They are intermediaries specializing in the transaction technology that transforms bought and sold assets. Let $\phi_{h,0} = (\phi_{h,0}^S, \phi_{h,0}^B) = (\phi_{h,0,1}^S, \ldots, \phi_{h,0,N}^S, \phi_{h,0,1}^B, \ldots, \phi_{h,0,N}^B)$ be the initial trading by broker h. $\phi_{h,n}^B(e)(\phi_{h,n}^S(e))$ be the accumulated number of bought (sold) asset n supplied by intermediary h after trading at e and denote $\phi_h = (\phi_h^S, \phi_h^B) = (\phi_h^S(e), \phi_h^B(e))_{e \in \mathbf{E}}$; and $z_h(e) = (z_{h,1}(e), \ldots, z_{h,M}(e))$ be the vector of contingent commodities used up in the activity of intermediation at date t and denote $(z_h(e))_{e \in \mathbf{E}}$ by z_h. For intermediary h, let $T_h(e) \subseteq \mathbf{R}_+^N \times \mathbf{R}_+^N \times \mathbf{R}_+^M$ denote its technology at node e.

There are I consumers, indexed by i, with endowment $\omega_{i,t}$, initial trading $\psi_{i,0}$, and utility function $U_i(\cdot)$ and consumption set $X_i = \mathbf{R}_+^{|E| \times M}$. The consumer i chooses a consumption plan $x_i = (x_{i,1}(e), \ldots, x_{i,M}(e))_{e \in \mathbf{E}} \in X_i$ and a portfolio plan $\psi_i = (\psi_i^B, \psi_i^S) \in \mathbf{R}_+^{2(|E| \times N)}$ which can be explained analogously to θ_j.

Now we turn to assets. At each node $e \in \mathbf{E}$, asset $n(n = 1, \ldots, N)$ has a buying price $B^n(e)$ and selling price $S^n(e)$ and dividend $d_n(e)$ denominated in the first commodity (numeraire). Suppose that at each node e, the asset n pays its dividend d_n (denominated in the numeraire commodity) and is then available for trade at prices $B_n(e)$ and $S_n(e)$.

Let $d = \{d(e) = (d_1(e), \ldots, d_N(e)) : e \in \mathbf{E}\}$, $B = \{B(e) = (B^1(e), \ldots, B^N(e)) : e \in \mathbf{E}\}$ and $S = \{S(e) = (S^1(e), \ldots, S^N(e)) : e \in \mathbf{E}\}$. A dividend process d^θ

generated by a generic strategy $\theta = (\theta^B, \theta^S)$ is defined by

$$d^\theta(e) = (\theta^B(e^-) - \theta^S(e^-))d(e) + \Delta\theta^S(e)S(e) - \Delta\theta^B(e)B(e), e \in \mathbf{E},$$

where $\theta(e_0^-) = (\theta^B(e_0^-), \theta^S(e_0^-))$ is the initial shares bought and sold by an agent just before the trading takes place at node e_0. Here $\Delta\theta^B(e) = \theta^B(e) - \theta^B(e^-)$; and $\Delta\theta^S(e) = \theta^S(e) - \theta^S(e^-)$. And define $\Delta\theta(e) = \theta(e) - \theta(e^-)$.

Let $p = (p)_{e \in \mathbf{E}} = (p_1(e), \ldots, p_M(e))_{e \in \mathbf{E}} (\in \triangle_0)$ denote the spot price of commodities, where $\triangle_0$ is the unit simplex of $\mathbf{R}^{M \times |\mathbf{E}|}$.

Now the problem of firm j can be specified as

$$\sup_{(\theta_j, y_j) \in \Gamma_j^1(p)} V_j(d^{\theta_j} + py_j), \qquad (*)$$

where $\Gamma_j^1(p)$ denote the set of feasible production-trade plans (θ_j, y_j) of firm j given price p, which satisfies

(2.1) y_j is in Y_j;
(2.2) $d^{\theta_j}(e) + p(e)y_j(e) \geq 0, \forall e \in \mathbf{E}.$

The maximization problem of broker h can be stated as

$$\sup_{(\phi_h, z_h) \in \Gamma_h^2(\gamma_h)} W_h(-d^{\phi_h} - pz_h), \qquad (**)$$

where $\Gamma_h^2(\gamma_h)(\gamma_h = ((\psi_i), (\theta_j), (\phi_{h' \neq h}), p)$ is the space of feasible trade-production plans $(\phi_h, z_h) = (\phi_h^S, \phi_h^B, z_h)$ given γ_h, which satisfies, at each node,

(2.3) $(\Delta\phi_h^B(e), \Delta\phi_h^S(e), z_h(e)) \in T_h(e)$ and $z_h(e) \geq 0, \forall e \in \mathbf{E}$;
(2.4) $-d^{\phi_h}(e) - p(e)z_h(e) \geq 0, \forall e \in \mathbf{E}$;
(2.5) $\sum_h \Delta\phi_h(e) \geq \sum_i \Delta\psi_i(e) + \sum_j \Delta\theta_j(e), \forall e \in \mathbf{E}.$

Comments: 1. The condition (2.5) requires that all consumers and all production firms buy and sell securities through brokers.

2. The productive firms and intermediary firms are treated similarly to consumers. Because of transaction costs, it is well-known that the Fisher separation theorem fails. Therefore we assume that each firm has an objective function (utility function) derived in some fashion from the preference of the owners. For example, we can either assume one-owner firms or draw on DeMarzo (1988) and argue that the objective is derived from a more complicated composition of owner preferences.

3. The intermediary formulation allows the agent to trade on its own account; or by interpreting the transaction technology more narrowly, it can be restricted to a pure broker with direct pass through of assets bought and sold (see Milne and Neave (1996) for further discussion).

The problem of consumer i is as follows:

$$\sup_{(\psi_i, x_i) \in \Gamma_i^3(\tau)} U_i(x_i), \qquad (***)$$

where $\tau = ((\phi_h, z_h), (\theta_j, y_j), p)$ and $\Gamma_i^3(\tau)$ is the set of feasible portfolio-consumption plans (ψ_i, x_i) given τ, which satisfies

(2.6) x_i is in X_i;

(2.7) $p(e)x_i(e) \leq p(e)\omega_i(e) + d^{\psi_i}(e) + \sum_j \alpha_{i,j} \bar{y}_j(e) + \sum_h \beta_{i,h} \bar{z}_h(e)$, $\forall e \in \mathbf{E}$.

Here $\bar{y}_j(e) = d^{\theta_j}(e) + p(e)y_j(e)$, $\bar{z}_h(e) = -d^{\phi_h}(e) - p(e)z_h(e)$; $\alpha_{i,j} \geq 0 (\sum_i \alpha_{i,j} = 1)$ is consumer i's initial share of the net cash flow of firm j; and $\beta_{i,h} \geq 0 (\sum_i \beta_{i,h} = 1)$ is consumer i's initial share of net cash flow of broker h.

Now we can define the abstract economy: $\mathcal{E} = (X_1 \times \mathbf{R}^{2(N \times |\mathbf{E}|)}, \ldots, X_I \times \mathbf{R}^{2(N \times |\mathbf{E}|)}, Y_1 \times \mathbf{R}^{2(N \times |\mathbf{E}|)}, \ldots, Y_J \times \mathbf{R}^{2(N \times |\mathbf{E}|)}, T_1', \ldots, T_H', \triangle_0, U_1(x_1), \ldots, U_I(x_I), V_1(d^{\theta_1} + py_1), \ldots, V_J(d^{\theta_J} + py_J), W_1(-d^{\phi_1} - pz_1), \ldots, W_H(-d^{\phi_H} - pz_H), \sum_{e \in \mathbf{E}} \pi(e)p(e)w(e), \Gamma_1^3(\tau), \ldots, \Gamma_I^3(\tau), \Gamma_1^1(p), \ldots, \Gamma_J^1(p), \Gamma_1^2(\gamma_1), \ldots, \Gamma_H^2(\gamma_H), \triangle_0)$, where $\pi(e)(e \in \mathbf{E})$ will be defined in Lemma 2.1 below, $w(e) = \sum_i x_i(e) + \sum_h z_h(e) - \sum_j y_j(e) - \sum_i \omega_i(e)$ and

$$T_h' = \{(\phi_h^S, \phi_h^B, z_h) : (\triangle \phi_h^B(e), \triangle \phi_h^S(e), z_h(e)) \in T_h(e), \forall e \in \mathbf{E}\}.$$

A point $e^* = ((\psi_i^*, x_i^*), (\theta_j^*, y_j^*), (\phi_h^*, z_h^*), p^*)$ is called an equilibrium solution of economy $\mathcal{E}$ given the market system (B, S, d) if e^* solves problems $(*)$, $(**)$ and $(***)$ and

$$\sum_i x_i^* + \sum_h z_h^* = \sum_j y_j^* + \sum_i \omega_i,$$

$$\sum_h \triangle \phi_h^* = \sum_i \triangle \psi_i^* + \sum_j \triangle \theta_j^*.$$

The following assumptions are made in the remainder of this paper.
For consumer i:

(A.1) $U_i(\cdot)$ is a continuous, concave and strictly increasing function.

For firm j:

(A.2) Y_j is a closed convex subset of $\mathbf{R}^{M \times |\mathbf{E}|}$ containing $-R_+^{M \times |\mathbf{E}|}$;

(A.3) $Y_j \cap \mathbf{R}_+^{M \times |\mathbf{E}|} = \{0\}$;

(A.4) $(\sum_j Y_j) \cap (-\sum_j Y_j) = \{0\}$;

(A.5) $V_j(\cdot)$ is a continuous, concave and strictly increasing function.

For broker h:

(A.6) For each e, $T_h(e)$ is a closed and convex set with $0 \in T_h(e)$.

(A.7) For any e and given $x = (x_1, \ldots, x_{2N}, z_1, \ldots, z_M) \in T_h(e)$, If $y = \sum_{n=1}^{2N} x_n \to \infty$, then $|(z_1, \ldots, z_M)| = \sum_{m=1}^M z_m \to \infty$.

(A.8) For each e, if $(\psi, z) \in T_h(e)$ and $z' \geq z$, then $(\psi, z') \in T_h(e)$ (free disposal).

(A.9) $W_h(\cdot)$ is a continuous, concave and strictly increasing function:

(A.10) The initial holdings $\psi_{i,0}, \theta_{j,0}, \phi_{h,0}$ of securities of all consumers, all firms and all brokers are taken as given and satisfy

$$\sum_i \psi_{i,0} + \sum_j \theta_{j,0} = \sum_h \phi_{h,0}.$$

The assumptions (A.1)–(A.6) are standard. (A.7) says that transactions must consume resources. (A.8) and (A.9) are standard. And (A.10) says that the initial net trading is zero.

Now we conclude this section by introducing the concept of no-arbitrage.

Let an agent have initial shares $\theta_0 = (\theta_0^B, \theta_0^S)$ just before the trading takes place at node e_0. Then his or her initial capital will be $d_0 = d(e_0)(\theta_0^B - \theta_0^S) + S(e_0)\theta_0^B - B(e_0)\theta_0^S$ if he or she sells the holding and buys back the short-sale. Given a price-dividend triple (B, S, d) for N securities, a trading strategy θ is an arbitrage if $-d_0 \geq 0$ and $d^\theta(e) \geq 0$, $\forall e \in \mathbf{E}$, and moreover, $-d_0 > 0$ or there exists at least one node $e \in \mathbf{E}$ such that $d^\theta(e) > 0$.

(A.11) The price-dividend triple (B, S, d) admits no-arbitrage.

The following equivalent condition of no arbitrage is similar to Proposition 2C of Duffie (1996) and will play an important role in the proof of market clearing later.

Lemma 2.1. *There is no arbitrage if and only if there is a strictly increasing linear function $F : \mathbf{R}^{|\mathbf{E}|+1} \to \mathbf{R}^1$ such that $F((-d_0, d^\theta)) \leq 0$ for any trading strategy $\theta \in \Theta$, where Θ denotes the space of trading strategies and is a closed and convex set.*

Proof. There is no arbitrage if and only if the cones $\mathbf{R}_+^{|\mathbf{E}|+1}$ and $M^0 = \{(-d_0, d^\theta) : \theta \in \Theta\}$ intersect precisely at zero. If there is no arbitrage, the theorem "Linear Separation of Cones" in Appendix B of Duffie (1996) implies the existence of a nonzero linear functional F such that $F(x) < F(y)$ for each x in M^0 and each nonzero y in $\mathbf{R}_+^{|\mathbf{E}|+1}$. Since M^0 is a cone, this implies $F(x) \leq 0$ for each x in M^0. And, moreover, $0 \in M^0$; thus $F(y) > F(0) = 0$ for each nonzero $y \in \mathbf{R}_+^{|\mathbf{E}|+1}$. That is, F is strictly increasing. The converse is immediate. □

In comparison with Proposition 2C in Duffie (1996), the next result shows the difference between the model without transaction costs and the model with transaction costs caused by the bid-ask spread at some node.

Lemma 2.2. *Suppose there is no arbitrage and $B^n(\tilde{e}) > S^n(\tilde{e})$ for security n at a certain node $\tilde{e}$. Then $F((-d_0, d^\theta)) \not\equiv 0$ over Θ.*

Proof. Suppose not. Then $F((-d_0, d^\theta)) = 0$ for each $\theta \in \Theta$. Since $F(\cdot)$ is a strictly increasing linear functional on $\mathbf{R}_+^{|\mathbf{E}|+1}$, this implies that there exists a vector $\pi = (\pi^0, (\pi^e)_{e \in \mathbf{E}}) \in \text{int}(\mathbf{R}_+^{|\mathbf{E}|+1})$ such that $F(x) = \pi^0 x_0 + \sum_{e \in \mathbf{E}} \pi^e x_e$ for each $x = (x_0, (x_e)_{e \in \mathbf{E}}) \in \mathbf{R}^{|\mathbf{E}|+1}$. Without loss of generality, we assume $N = n = 1$. For a fixed positive integer M, set

$$\theta^B(e) = 0, \forall e \in \mathbf{E} - \{\tilde{e}\}, \theta^B(\tilde{e}) = M \, ;$$

$$\theta^S(e) = 0, \forall e \in \mathbf{E} \, .$$

Then $d^\theta(e) = d(e)(\theta_0^B - \theta_0^S), \forall e \notin \mathbf{E}(\tilde{e}), d^\theta(\tilde{e}) = d(\tilde{e})(\theta_0^B - \theta_0^S) - MB^1(\tilde{e}), d^\theta(e) = d(e)(\theta_0^B - \theta_0^S + M), \forall e \in \mathbf{E}(\tilde{e}) - \{\tilde{e}\}$.

Hence

$$M\pi(\tilde{e})B^1(\tilde{e}) = -\pi^0 d_0 + \sum_{e \in \mathbf{E}} \pi(e)d(e)(\theta_0^B - \theta_0^S) + M \sum_{e \in \mathbf{E}(\tilde{e}) - \{\tilde{e}\}} \pi(e)d(e) \, .$$

Likewise, by setting

$$\theta^S(e) = 0, \forall e \in \mathbf{E} - \{\tilde{e}\}, \theta^S(\tilde{e}) = M \, ;$$

$$\theta^B(e) = 0, \forall e \in \mathbf{E} \, .$$

$$M\pi(\tilde{e})S^1(\tilde{e}) = \pi^0 d_0 - \sum_{e \in \mathbf{E}} \pi(e)d(e)(\theta_0^B - \theta_0^S) + M \sum_{e \in \mathbf{E}(\tilde{e}) - \{\tilde{e}\}} \pi(e)d(e) \, ,$$

and therefore,

$$M\pi(\tilde{e})(B^1(\tilde{e}) - S^1(\tilde{e})) = 2 \left[\sum_{e \in \mathbf{E}} \pi(e)d(e)(\theta_0^B - \theta_0^S) - \pi^0 d_0 \right] \, .$$

Thus $B^1(\tilde{e}) = S^1(\tilde{e})$ since M is arbitrary, which provides a contradiction and proves the conclusion of the lemma. $\qquad\square$

Therefore, from Proposition 2C of Duffie (1996), in a security market without friction and with no arbitrage, the initial value of any trading strategy is uniquely determined by the inner product of its future cash flows with a martingale measure. But, by the above two lemmas, this conclusion does not hold in the security market with transaction costs: the martingale measure is not unique (see Jouini and Kallal (1995) for similar observations).

3. Proof of Existence of Equilibrium

We will adopt the technique of proof used in Arrow–Debreu (1954). First of all, we will show that the set of attainable plans for economy $\mathcal{E}$ is bounded, and replace the original economy $\mathcal{E}$ by a bounded one. Secondly, we will show the continuity of the constrained correspondences.

For broker h, define

$Z_h = \{z_h$: there exists $(\phi_h^S, \phi_h^B) \geq 0$ such that $(\phi_h^S, \phi_h^B, z_h) \in T_h'\}$;

$\hat{Z}_h = \{z_h \in Z_h$: there exist $z_{h'} \in Z_{h'}$ for each $h' \neq h, x_i \in X_i$ for each i and $y_j \in Y_j$ for each j such that $w = \sum_i x_i + \sum_h z_h - \sum_j y_j - \sum_i \omega_i \leq 0\}$;

$\Phi_h = \{\phi_h = (\phi_h^S, \phi_h^B): \text{ there exists } z_h \in \hat{Z}_h \text{ such that } (\phi_h, z_h) \in T_h'\};$

$\hat{\Phi}_h = \{\phi_h = (\phi_h^S, \phi_h^B) \in \Phi_h: \text{ there exist } \phi_{h'} \in \Phi_{h'} \text{ for each } h' \neq h, \Delta\theta_j \geq 0 \text{ for each } j \text{ and } \Delta\psi_i \geq 0 \text{ for each } i \text{ such that } \sum_i \Delta\psi_i + \sum_j \Delta\theta_j \leq \sum_h \Delta\phi_h\}.$

For consumer i, define

$\hat{X}_i = \{x_i \in X_i: \text{ there exist } x_{i'} \in X_{i'} \text{ for each } i' \neq i, z_h \in Z_h \text{ for each } h \text{ and } y_j \in Y_j \text{ for each } j \text{ such that } w = \sum_i x_i + \sum_h z_h - \sum_j y_j - \sum_i \omega_i \leq 0\};$

$\hat{\Psi}_i = \{\psi_i = (\psi_i^B, \psi_i^S): \text{ there exist } \Delta\psi_{i'} \geq 0 \text{ for each } i' \neq i, \phi_h \in \Phi_h \text{ for each } h, \Delta\theta_j \geq 0 \text{ for each } j \text{ such that } \sum_i \Delta\psi_i + \sum_j \Delta\theta_j \leq \sum_h \Delta\phi_h\}.$

Likewise, we can define $\hat{Y}_j$ and $\hat{\Theta}_j$ for firm j.

By use of the technique of 3.3.1 of Arrow–Debreu (1954) and by Lemma A1, we can show the following result.

Lemma 3.1. *The sets $\hat{X}_i, \hat{Y}_j$ and $\hat{Z}_h$ are all compact and convex.*

In exactly the same method as Lemma 3.1 (and A1 in Appendix), we can show the following boundedness result of trade plan for broker h.

Lemma 3.2. *The set Φ_h is a compact and convex subset of $\mathbf{R}^{2(|\mathbf{E}|\times N)}$. And so is $\hat{\Phi}_h$, $\hat{\Psi}_i$ and $\hat{\Theta}_j$.*

Thus there exist cubes $C^1 (\subseteq \mathbf{R}^{|\mathbf{E}|\times M})$ and $C^2 (\subseteq \mathbf{R}_+^{2(|\mathbf{E}|\times N)})$ so that C^1 contains in its interior all $\hat{X}_i$, all $\hat{Y}_j$ and all $\hat{Z}_h$; C^2 contains all $\hat{\Psi}_i$, all $\hat{\Theta}_j$ and all $\hat{\Psi}_h$. Define $\tilde{X}_i = C^1 \cap X_i, \tilde{\Psi}_i = C^2, \tilde{Y}_j = C^1 \cap Y_j, \tilde{\Theta}_j = C^2, \tilde{Z}_h = C^1 \cap Z_h$ and $\tilde{\Phi}_h = C^2$. And let $\tilde{\Gamma}_j^1(p), \tilde{\Gamma}_h^2(\gamma_h)$ and $\tilde{\Gamma}_i^3(\tau)$ be the resultant modification of $\Gamma_j^1(p), \Gamma_h^2(\gamma_h)$ and $\Gamma_i^3(\tau)$ respectively.

We now turn to the proof of continuity of $\tilde{\Gamma}_j^1(p), \tilde{\Gamma}_h^2(\gamma_h)$ and $\tilde{\Gamma}_i^3(\tau)$. We only investigate the continuity of $\tilde{\Gamma}_h^2$, the continuity of the others can be shown similarly.

Lemma 3.3. *Given p, all ψ_i, all θ_j and all $\phi_{h'}(h' \neq h)$, and there exists $(\phi_h, z_h) \in T_h'$ such that $\sum_i \Delta\psi_i + \sum_j \Delta\theta_j \ll \sum_h \Delta\phi_h$ and $0 \ll -d^{\phi^h} - pz_h$. Then $\tilde{\Gamma}_h^2(\gamma_h)$ is continuous.*

Proof. Without loss of generality, we show the continuity of $\tilde{\Gamma}_1^2(\gamma_1)$. Let $\gamma_1^k = (\psi_1^k, \ldots, \psi_I^k, \theta_1^k, \ldots, \theta_J^k, \phi_2^k, \ldots, \phi_H^k, p^k) \to \gamma_1 = (\psi_1, \ldots, \psi_I, \theta_1, \ldots, \theta_J, \phi_2, \ldots, \phi_H, p)$. Consider a point $(\phi_1, z_1) \in \tilde{\Gamma}_1^2(\gamma_1)$, then

$$0 \leq -d^{\phi_1} - pz_1, \quad \sum_i \Delta\psi_i + \sum_j \Delta\theta_j \leq \sum_h \Delta\phi_h.$$

If $0 \ll -d^{\phi_1} - pz_1, \sum_i \Delta\psi_i + \sum_j \Delta\theta_j \ll \sum_h \Delta\phi_h$, then for k sufficiently large, $0 \ll -d^{\phi_1} - p^k z_1$, and $\sum_i \Delta\psi_i^k + \sum_j \Delta\theta_j^k \ll \sum_{h \neq 1} \Delta\phi_h^k + \Delta\phi_1$. By taking $(\phi_1^k, z_1^k) = (\phi_1, z_1)$, we prove the conclusion of the lemma.

If the above case does not hold, then there exist $\mathbf{E}_0 \subseteq \mathbf{E}$ and $\mathbf{E}_0', \widetilde{\mathbf{E}}_0' \subseteq \widetilde{\mathbf{E}} = \mathbf{E} \times \{1, \ldots, N\}$ (there is at least one nonempty set among $\mathbf{E}_0$ and $\mathbf{E}_0'$ and $\widetilde{\mathbf{E}}_0'$) such that

$$-d^{\phi_1}(e) - p(e)z_1(e) = 0, \ e \in \mathbf{E}_0;$$

$$-d^{\phi_1}(e) - p(e)z_1(e) > 0, \ e \in \mathbf{E} - \mathbf{E}_0;$$

$$\sum_i \triangle \psi_{i,q}^B(e) + \sum_j \triangle \theta_{j,q}^B(e) = \sum_h \triangle \phi_{h,q}^S(e), \ (e,q) \in \mathbf{E}_0',$$

$$\sum_i \triangle \psi_{i,q}^B(e) + \sum_j \triangle \theta_{j,q}^B(e) < \sum_h \triangle \phi_{h,q}^S(e), \ (e,q) \in \widetilde{\mathbf{E}} - \mathbf{E}_0',$$

and

$$\sum_i \triangle \psi_{i,q}^S(e) + \sum_j \triangle \theta_{j,q}^S(e) = \sum_h \triangle \phi_{h,q}^B(e), \ (e,q) \in \widetilde{\mathbf{E}}_0',$$

$$\sum_i \triangle \psi_{i,q}^S(e) + \sum_j \triangle \theta_{j,q}^S(e) < \sum_h \triangle \phi_{h,q}^B(e), \ (e,q) \in \widetilde{\mathbf{E}} - \widetilde{\mathbf{E}}_0'.$$

By assumption, we can choose $(\phi_1', z_1') \in T_1'$ such that

$$0 \ll -d^{\phi_1'} - pz_1',$$

and

$$\sum_i \triangle \psi_i + \sum_j \triangle \theta_j \ll \sum_{h \neq 1} \triangle \phi_h + \triangle \phi_1'.$$

Clearly

$$\triangle \phi_{1,q}'^S(e) > \triangle \phi_{1,q}^S(e), \ (e,q) \in \mathbf{E}_0',$$

and,

$$\triangle \phi_{1,q}'^B(e) > \triangle \phi_{1,q}^B(e), \ (e,q) \in \widetilde{\mathbf{E}}_0'.$$

Let

$$\lambda_k^1 = \min \left\{ 1, \frac{\triangle \phi_{1,q}'^S(e) - (\sum_i \triangle \psi_{i,q}^{k,B}(e) + \sum_j \triangle \theta_{j,q}^{k,B}(e) - \sum_{h \neq 1} \triangle \phi_{h,q}^{k,S}(e))}{\triangle \phi_{1,q}'^S(e) - \triangle \phi_{1,q}^S(e)} \right.$$

$$\left. : (e,q) \in \mathbf{E}_0' \right\};$$

$$\bar{\lambda}_k^1 = \min \left\{ 1, \frac{\triangle \phi_{1,q}'^B(e) - (\sum_i \triangle \psi_{i,q}^{k,S}(e) + \sum_j \triangle \theta_{j,q}^{k,S}(e) - \sum_{h \neq 1} \triangle \phi_{h,q}^{k,B}(e))}{\triangle \phi_{1,q}'^B(e) - \triangle \phi_{1,q}^B(e)} \right.$$

$$\left. : (e,q) \in \widetilde{\mathbf{E}}_0' \right\};$$

and

$$\lambda_k^2 = \min\left\{1, \frac{-d^{\phi'_1}(e) - p^k(e)z'_1(e)}{-d^{\phi'_1}(e) - p^k(e)z'_1(e) - (-d^{\phi_1}(e) - p^k(e)z_1(e))} : e \in \mathbf{E}_0\right\}.$$

Let $\lambda_k = \min(\lambda_k^1, \bar{\lambda}_k^1, \lambda_k^2)$ and $(\phi_1^k, z_1^k) = \lambda_k(\phi_1, z_1) + (1 - \lambda_k)(\phi'_1, z'_1)$. It is easy to verify that $\lambda_k^i \to 1 (i = 1, 2)$ and

$$-d^{\phi_1^k}(e) - p^k(e)z_1^k(e) \geq 0, \ e \in \mathbf{E}_0 ; \tag{3.1}$$

$$\sum_i \triangle\psi_{i,q}^{k,B}(e) + \sum_j \triangle\theta_{j,q}^{k,B}(e) \leq \sum_h \triangle\phi_{h,q}^{k,S}(e), \ (e, q) \in \mathbf{E}'_0 ;$$

and

$$\sum_i \triangle\psi_{i,q}^{k,S}(e) + \sum_j \triangle\theta_{j,q}^{k,S}(e) \leq \sum_h \triangle\phi_{h,q}^{k,B}(e), \ (e, q) \in \tilde{\mathbf{E}}'_0 ;$$

for k sufficiently large.

On the other hand, since $\alpha = \min_{e \in \mathbf{E} - \mathbf{E}_0}\{-d^{\phi_1}(e) - p(e)z_1(e)\} > 0$ and

$$\lim_{k \to \infty} (-d^{\phi_1^k}(e) - p^k(e)z_1^k(e)) = -d^{\phi_1}(e) - p(e)z_1(e).$$

Hence, for k sufficiently large,

$$-d^{\phi_1^k}(e) - p^k(e)z_1^k(e) \geq 0 : e \in \mathbf{E} - \mathbf{E}_0,$$

which, combining with (3.1), implies that for k sufficiently large,

$$-d^{\phi_1^k} - p^k z_1^k \geq 0.$$

Likewise, we can show that for k sufficiently large,

$$\sum_i \triangle\psi_i^k + \sum_j \triangle\theta_j^k \leq \sum_h \triangle\phi_h^k.$$

Consequently, $(\phi_1^k, z_1^k) \in \tilde{\Gamma}_1^2(\gamma_1^k)$ and converges to (ϕ_1, z_1), proving the continuity of $\tilde{\Gamma}_1^2(\cdot)$.   □

For firms and consumers, we have the following similar results.

Lemma 3.4. *For firm j, given any price $p \in \triangle_0$, assume there exist $\theta'_j \geq 0$ and $y'_j \in Y_j$ such that $0 \ll d^{\theta'_j} + py'_j$. Then $\tilde{\Gamma}_j^1(\cdot)$ is continuous.*

Lemma 3.5. *For consumer i, given any price $p \in \triangle_0$, assume there exist $\psi'_i \geq 0$ and $x'_i \in X_i$ such that*

$$px'_i \ll p\omega_i + d^{\psi'_i} + \max\left\{0, \sum_j \alpha_{i,j}\bar{y}_j + \sum \beta_{i,h}\bar{z}_h\right\}.$$

Then $\tilde{\Gamma}_i^3(\cdot)$ is continuous.

Remark 3.1. The conditions in Lemmas 3.4 and 3.5 will be satisfied if there exists a portfolio θ such that $d^\theta(e) > 0, \forall e \in \mathbf{E}$, which is the Assumption 2.1 of Ortu (1995) called the "Internality Condition". This condition can be guaranteed if the agent has a positive initial long position and has no initial short position and adopt a buy-hold strategy, that is, $\theta^B(e) = \theta_0^B > 0$ and $\theta^S(e) = \theta_0^S = 0$, $\forall e \in \mathbf{E}$. Observe that these conditions are extensions of well-known standard assumptions that ensure continuity of budget correspondences. Weaker conditions could be found, but these will suffice for our purpose.

Let

$$\mu_i = \mu_i(\tau) = \left\{(\psi_i, x_i) : U_i(x_i) = \sup_{(\bar{\psi}_i, \bar{x}_i) \in \tilde{\Gamma}_i^3(\tau)} U_i(\bar{x}_i)\right\};$$

$$\upsilon_j = \upsilon_j(p) = \left\{(\theta_j, y_j) : V_j(d^{\theta_j} + py_j) = \sup_{(\bar{\theta}_j, \bar{y}_j) \in \tilde{\Gamma}_j^1(p)} V_j(d^{\bar{\theta}_j} + p\bar{y}_j)\right\};$$

$$\tau_h = \tau_h(\gamma_h) = \left\{(\phi_h, z_h) : W_h(-d^{\phi_h} - pz_h) = \sup_{(\bar{\phi}_h, \bar{z}_h) \in \tilde{\Gamma}_h^2(\gamma_h)} W_h(-d^{\bar{\phi}_h} - p\bar{z}_h)\right\};$$

$$\bar{p} = \bar{p}(w) = \left\{p : \sum_{e \in \mathbf{E}} \pi(e)p(e)w(e) = \sup_{p' \in \triangle_0} \sum_{e \in \mathbf{E}} \pi(e)p'(e)w(e)\right\},$$

and

$$\Psi = \prod_{i=1}^I \mu_i \times \prod_{j=1}^J \upsilon_j \times \prod_{h=1}^H \tau_h \times \bar{p}.$$

By Berge's Maximum Theorem and standard methods, we can prove that the correspondences $\mu_i, \upsilon_j, \tau_h$ and $\bar{p}$ are upper hemi-continuous and convex valued. This implies Ψ is also upper hemi-continuous and convex valued.

The correspondence Ψ has been shown to satisfy the hypotheses of the Kakutani fixed point theorem, and therefore to have a fixed point, say $e^* = ((\psi_i^*, x_i^*), (\theta_j^*, y_j^*), (\phi_h^*, z_h^*), p^*)$. Especially, this fixed point satisfies

$$\sum_i \triangle\psi_i^*(e) + \sum_j \triangle\theta_j^*(e) \leq \sum_h \triangle\phi_h^*(e), \quad \forall e \in \mathbf{E}, \tag{3.2}$$

$$\sum_{e \in \mathbf{E}} \pi(e)p^*(e)w^*(e) \geq \sum_{e \in \mathbf{E}} \pi(e)p(e)w^*(e), \quad \forall p \in \triangle_0. \tag{3.3}$$

By assumption (A.10) and Lemma 2.1,

$$\sum_{e \in \mathbf{E}} \pi(e) p^*(e) w^*(e) \le \sum_{e \in \mathbf{E}} \pi(e) d^{\sum_i \psi_i^* + \sum_j \theta_j^* - \sum_h \phi_h^*}(e)$$

$$= -\pi_0 \left[d(e_0) \left(\sum_i \psi_{i,0} + \sum_j \theta_{j,0} - \sum_h \phi_{h,0} \right) \right.$$

$$+ S(e_0) \left(\sum_i \psi_{i,0}^B + \sum_j \theta_{j,0}^B - \sum_h \phi_{h,0}^S \right)$$

$$\left. - B(e_0) \left(\sum_i \psi_{i,0}^S + \sum_j \theta_{j,0}^S - \sum_h \phi_{h,0}^B \right) \right]$$

$$+ \sum_{e \in \mathbf{E}} \pi(e) d^{\sum_i \psi_i^* + \sum_j \theta_j^* - \sum_h \phi_h^*}(e)$$

$$= F \left(-d_0, d^{\sum_i \psi_i^* + \sum_j \theta_j^* - \sum_h \phi_h^*} \right) \le 0 \, ,$$

where

$$d_0 = d(e_0) \left(\sum_i \psi_{i,0} + \sum_j \theta_{j,0} - \sum_h \phi_{h,0} \right)$$

$$+ S(e_0) \left(\sum_i \psi_{i,0}^B + \sum_j \theta_{j,0}^B - \sum_h \phi_{h,0}^S \right) - B(e_0) \left(\sum_i \psi_{i,0}^S + \sum_j \theta_{j,0}^S - \sum_h \phi_{h,0}^B \right) .$$

Hence, from (3.3), $w^*(e) \le 0, \forall e \in \mathbf{E}$.

Let $\triangle y_J^*(e) = -w^*(e) \ge 0, \triangle \theta_J'^*(e) = \sum_h \triangle \phi_h^*(e) - \sum_i \triangle \psi_i^*(e) - \sum_j \triangle \theta_j^*(e) \ge 0$, $e \in \mathbf{E}$. And set

$$\bar{y}_J^* = y_J^* - \triangle y_J^*, \ \triangle \bar{\theta}_J^* = \triangle \theta_J^* + \triangle \theta_J'^* \, .$$

Clearly,

$$\bar{y}_J^* \in Y_J, \ \sum_i x_i^* = \sum_{j \ne J} y_j^* + \bar{y}_J^* + \sum_h z_h^* \, ,$$

and

$$\sum_i \triangle \psi_i^* + \sum_{j \ne J} \triangle \theta_j^* + \triangle \bar{\theta}_J^* = \sum_h \triangle \phi_h^* \, .$$

Moreover

$$p^*(\triangle y_J^*) = -p^* w^* = -d^{\sum_i \psi_i^* + \sum_j \theta_j^* - \sum_h \phi_h^*}$$

$$d^{\theta'^*} = -d^{\sum_i \psi_i^* + \sum_j \theta_j^* - \sum_h \phi_h^*} \, ,$$

this implies that

$$d^{\bar{\theta}_j} + p * \bar{y}_j^* = d^{\theta_j^*} + p^* y_j^*.$$

Finally, in exactly the same method as Arrow and Debreu (1954), it is not difficult to show $\bar{e}^* = ((x_i^*, \psi_i^*), (y_{j\neq J}^*, \theta_{j\neq J}^*), (\bar{y}_J^*, \bar{\theta}_J^*), (\phi_h^*, z_h^*), p^*)$ is an equilibrium point of the original Economy $\mathcal{E}$.

4. Non-Convex Production Economy

This section is devoted to a economy in which the trading technology of each broker is not necessarily convex so that we allow for fixed costs in trading assets. By using the technique of Heller and Starr (1976), we will show the existence of an individual approximate equilibrium defined by Heller and Starr (1976). An approximate equilibrium is generally defined as a price p^* and two allocations, a^* and $a^{*\prime}$. One, a^*, is the allocation desired by households, firms and brokers at this price, which may not clear the market. The other, $a^{*\prime}$, is an allocation obeying the market clearance condition although it need not represent agents' optimizing behaviour. The equilibrium is approximate of a modulus C if some suitably chosen norm of the difference between these two allocations is no larger than C. The desired allocation represents an approximate equilibrium in the sense that the failure to clear the market at this price is bounded by C. And, furthermore, the bound of the approximation improves as the number of the agents in the economy increases.

We will continue to make all the assumptions in Sec. 2 except the convexity of the broker's technology. We further assume the following:

(A.12) $B_\omega = Y \cap (X + Z - \omega)$ is bounded, where $X = \sum_{i \in I} X_i$, $\omega = \sum_{i \in I} \omega_i$, $Y = \sum_{j \in J} Y_j$ and $Z = \sum_{h \in H} Z_h$.

Since the assumptions of Theorem 1 of Hurwicz and Reiter (1973) can be easily verified through Assumptions A.4 and A.12, we can show the boundedness of $\hat{Z}_h, \hat{X}_i, \hat{Y}_j$; and, hence, $\hat{\Phi}_h, \hat{\Psi}_i$ and $\hat{\Theta}_j$ are all bounded.

In order to show that the equilibrium of the bounded economy is the equilibrium of the original economy, an additional assumption is required.

(A.13) There is a positive number L_0 such that $|z_h| \leq L_0, \forall z_h \in Z_h$.

That is, the quantity of commodities used in transaction of assets is limited. This is reasonable since a quantity larger than the total supply of the world is not feasible. So the feasible plan of the broker should satisfy the additional assumption (A.13). The cubes C^1 and C^2 used in defining the bounded economy can be chosen to be large enough to contain the feasible plan of any broker.

As in Heller and Starr (1976), in order to prove the continuity of $\tilde{\Gamma}_h^2(\gamma_h)$, we give the definition of local interior.

Definition 4.1. $\tilde{\Gamma}_h^2$ is said to be locally interior if for each $(\phi, z) \neq 0, (\phi, z) \in \tilde{\Gamma}_h^2(\gamma_h)$ there is (ϕ^*, z^*) so that

(i) $(\phi^*, z^*) \in \tilde{\Gamma}_h^2(\gamma_h)$.

(ii) $0 \ll -d^{\phi^*} - pz^*$.

(iii) There exists a continuous function $f : [0,1] \to \tilde{\Gamma}_h^2(\gamma_h)$ so that $f(0) = (\phi^*, z^*), f(1) = (\phi, z)$ and for all $\sigma(\in [0,1))$, $f(\sigma)$ satisfies the strict inequality in (ii).

(A.14) $\tilde{\Gamma}_h^2(\gamma_h)$ is locally interior.

Now we are in a position to show the existence of an individual approximate equilibrium of the economy with non-convexities. But we omit its proof since it can be obtained in the same method as Heller and Starr (1976). In the proof, we use the correspondence Ψ (defined in Sec. 3) instead of $\gamma(p)$ defined in Heller and Starr (1976). The boundedness of $R(\Psi)$ defined in Heller and Starr (1976) can be clearly guaranteed by the assumption (A.13).

Theorem 4.1. *Under the assumptions (A.1)–(A.14) omitting the convexity of the brokers' technologies, there exists an individual approximate equilibrium of modulus C which only depends on M, N, L_0 and $R(p^*)$, where p^* is an approximate equilibrium price. That is, there exist two vectors $a^* = (\phi_1^*, z_1^*, \ldots, \phi_H^*, z_H^*, \psi_1^*, x_1^*, \ldots, \psi_I^*, x_I^*, \theta_1^*, y_1^*, \ldots, \theta_J^*, y_J^*)$ and $a^{*\prime} = (\phi_1^{*\prime}, z_1^{*\prime}, \ldots, \phi_H^{*\prime}, z_H^{*\prime}, \psi_1^{*\prime}, x_1^{*\prime}, \ldots, \psi_I^{*\prime}, x_I^{*\prime}, \theta_1^{*\prime}, y_1^{*\prime}, \ldots, \theta_J^{*\prime}, y_J^{*\prime})$ such that*

(i) $a^{*\prime}$ *satisfies market clearance with respect to* p^*.

(ii) a^* *solves problems (*), (**) and (***) with respect to* p^*.

(iii) $(\phi_i^*, x_i^*) = (\phi_i^{*\prime}, x_i^{*\prime}), (\theta_j^*, y_j^*) = (\theta_j^{*\prime}, y_j^{*\prime}), i \in I, j \in J$.

(iv) $(\sum_h |(\psi_h^*, z_h^*) - (\psi_h^{*\prime}, z_h^{*\prime})|^2)^{1/2} \le C$.

5. An Exchange Economy with Finite p-Convexity

In this section, the existence of a general equilibrium of an exchange economy is investigated, which only includes consumers, brokers or dealers. For simplicity we omit productive firms. In this model, the trading technology of each broker is not necessarily convex. The model includes some cases of fixed costs, e.g., a model with set-up cost, as special cases. By introducing the more restrictive concept of finite p-convexity we are able to prove the existence of an exact equilibrium, even when there are some fixed costs in transacting.

A fixed cost is represented by an initial fixed amount of input before there is any output; after that, we can allow outputs and inputs to increase. For example, setting up an office with a computer etc., will require inputs independent of how much work is done in the office. Clearly, this gives a non-convex production set.

We will retain all the assumptions in Sec. 4 except (A.12), (A.13) and (A.14). It is not difficult to show the boundedness of sets $\hat{X}_i, \hat{Z}_h, \hat{\Psi}_i$ and $\hat{\Phi}_h$. Moreover, we will introduce another assumption called finite p-convexity. Finally, the existence of general equilibrium is proved. To this end, we give the following definition of finite p-convexity.

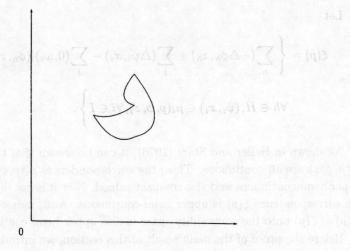

Fig. 1. Finite p-convex set.

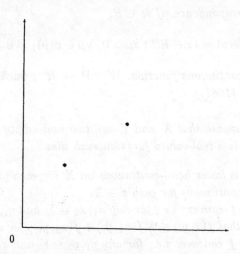

Fig. 2. Finite p-convex set.

Definition 5.1. Let X be a subset of R^n and $\triangle^{(n-1)}$ be the simplex of R^n. Then X is called finitely p-convex if for any x_1, $x_2 \in X$ and $p_1, \ldots, p_m \in \text{int}(\triangle^{(n-1)})$ there is $\bar{x} \in X$ such that $p_i \bar{x} \leq p_i(\frac{x_1+x_2}{2})$ for all $i = 1, 2, \ldots, m$ (see Figs. 1 and 2).

Let $(\phi, z) = (\phi_1, z_1, \ldots, \phi_h, z_h)$ and define the feasible set $\tilde{\Gamma}_i(p, \psi, z)$ (given price p and broker's plan (ϕ, z)) of consumer i analogously to $\tilde{\Gamma}_i^3(\tau)$; and the feasible set $\tilde{\Gamma}_h(\gamma_h)$ of broker h analogously to $\tilde{\Gamma}_h^2(\gamma_h)$ in Sec. 3. Define the demand function $\mu_i(p, \phi, z)$ of consumer i as $\mu_i(\tau)$ and the demand function $\tau_h(\gamma_h)$ of broker h as $\tau_h(\gamma_h)$ in Sec. 3.

Let

$$\xi(p) = \left\{ \sum_h (-\triangle\phi_h, z_h) + \sum_i (\triangle\psi_i, x_i) - \sum_I (0, \omega_i) | (\phi_h, z_h) \in \tau_h(\gamma_h), \right.$$

$$\left. \forall h \in H, (\psi_i, x_i) \in \mu_i(p, \phi, z), \forall i \in I \right\}.$$

As shown in Heller and Starr (1976), it can be shown that the sets $\tilde{\Gamma}_h(\gamma_h)$ and $\tilde{\Gamma}_i(p, \phi, z)$ are all continuous. Thus, the correspondences $\mu_i(p, \phi, z)$ and $\tau_h(\gamma_h)$ are upper hemi-continuous and also compact valued. Now it is not difficult to show that the correspondence $\xi(p)$ is upper hemi-continuous. And, moreover, the projection $\xi_0(p)$ of $\xi(p)$ onto the commodity space is also upper hemi-continuous.

Before the proof of the main result of this section, we introduce two lemmas.

Lemma 5.1. *Let $P \subseteq R^l$ be a compact set and let $\phi : P \to R^m$ be an upper hemi-continuous correspondence. If $\forall p \in P$,*

$$\Phi(p) = \{z \in R^m : z\mu > 0, \ \forall\mu \in \phi(p)\} \neq \emptyset,$$

then there exists a continuous function, $W : P \to R^m$, such that $W(p) \in \Phi(p)$, $\forall p \in P$ (cf. McCabe (1981)).

Lemma 5.2. *Suppose that X and Y are two non-empty compact spaces and that $f : X \times Y \to R$ is a real-valued function such that*

(i) *$x \to f(x, y)$ is lower hemi-continuous on X for each $y \in Y$; $y \to f(x, y)$ is upper hemi-continuous for each $x \in X$.*
(ii) *X is finitely f-convex: i.e., for any $x_1, x_2 \in X$ and $y_1, \ldots, y_n \in Y$ there is $\bar{x} \in X$ such that $f(\bar{x}, y_i) \leq \frac{1}{2}[f(x_1, y_i) + f(x_2, y_i)]$ for all $i = 1, \ldots, n$;*
(iii) *Y is finitely f-concave: i.e., for any $y_1, y_2 \in Y$ and $x_1, \ldots, x_m \in X$ there exist $\bar{y} \in Y$ such that $f(x_j, \bar{y}) \geq \frac{1}{2}[f(x_j, y_1) + f(x_j, y_2)]$ for all $j = 1, \ldots, m$.*

Then

$$\min_X \max_Y f(x, y) = \max_Y \min_X f(x, y)$$

(cf. Granas and Fon-Che Liu (1987)).

We now turn to the main result of this section.

Theorem 5.1. *Suppose that $\xi_0(p)$ is finitely p-convex and all assumptions omitting producers in Sec. 2 hold. Then there exists a general equilibrium $e^* = ((\psi_i^*, x_i^*)_{i \in I}, (\phi_h^*, z_h^*)_{h \in H}, p^*)$ in the non-convex exchange economy, i.e., e^* satisfies the following condition:*

(i) (ψ_i^*, x_i^*) solves problem $(***)$ for each $i \in I$;
(ii) (ϕ_h^*, z_h^*) solves problem $(**)$ for each $h \in H$;
(iii) e^* satisfies market clearance, i.e.,

$$\sum_h z_h^* + \sum_i x_i^* - \sum_i \omega_i = 0;$$

and

$$\sum_h \triangle \psi_h^* = \sum_i \triangle \phi_i^*.$$

Proof. We first truncate by a natural number n the set Z_h (defined in Sec. 3) and prove the existence of general equilibrium in the truncated economy $\mathcal{E}^n$. And then by taking limits, the existence of equilibrium can be obtained as in Geanakoplos and Polemarchakis (1986).

Furthermore, the cubes C^1 and C^2 are also chosen large enough to include the truncated feasible sets of all brokers.

Note that the consumption sets of all consumers are $R_+^{|\mathbf{E}| \times M}$. Hence, by the definition of $\xi(p)$, to prove the existence of general equilibrium it suffices to show that there exists $p_0 \in \triangle^{(|\mathbf{E}| \times M - 1)}$ such that $\xi_0(p_0) \cap (-R_+^{|\mathbf{E}| \times M}) \neq \emptyset$.

It is equivalent to that there exist $z_0 \in \pi \xi_0(p_0)$ such that $\max_{p \in \triangle^{(|\mathbf{E}| \times M - 1)}} p z^0 \leq 0$, where $\pi = (\pi(e))_{e \in \mathbf{E}}$ as defined in Sec. 2 and

$$\pi \xi_0(p_0) = \{(\pi(e) z_1(e), \dots, \pi(e) z_M(e))_{e \in \mathbf{E}} | z = (z_1(e), \dots, z_M(e))_{e \in \mathbf{E}} \in \xi_0(p_0)\}.$$

And it is easy to show that $\pi \xi_0(p)$ is upper hemi-continuous and finitely p-convex.

We will prove the conclusion of this theorem by a contradiction. To this end, let, for each $k \geq |\mathbf{E}| \times M$,

$$\triangle_k^{(|\mathbf{E}| \times M - 1)} = \left\{ p = (p_1(e), \dots, p_M(e)) \in \triangle^{(|\mathbf{E}| \times M - 1)} | p_i(e) \geq \frac{1}{k}, i = 1, \dots, |\mathbf{E}| \times M \right\}.$$

For each $p \in \triangle_k^{(|\mathbf{E}| \times M - 1)}$ and each $z \in \pi \xi_0(p)$, suppose that there exists a $p' \in \triangle_k^{(|\mathbf{E}| \times M - 1)}$ such that $p'z > 0$. Hence, $\max_{\triangle_k^{(|\mathbf{E}| \times M - 1)}} p'z > 0$. By the continuity of the function $\max_{\triangle_k^{(|\mathbf{E}| \times M - 1)}} p'z$, we have

$$\min_{\pi \xi_0(p)} \max_{\triangle_k^{(|\mathbf{E}| \times M - 1)}} p'z > 0.$$

In Lemma 5.2, by taking $X = \pi \xi_0(p), Y = \triangle_k^{(|\mathbf{E}| \times M - 1)}$ and $f(z, p') = p'z$, it is easy to verify that all conditions in this lemma are satisfied. And, particularly, the condition finite f-convexity of X corresponds to the finite p-convexity of $\pi \xi_0(p)$. Therefore,

$$\max_{\triangle_k^{(|\mathbf{E}| \times M - 1)}} \min_{\pi \xi_0(p)} p'z > 0,$$

which implies that there exist $p'_k \in \triangle_k^{(|\mathbf{E}|\times M-1)}$, such that $\min_{\pi\xi_0(p)} p'_k z > 0$ and moreover, $p'_k z > 0, \forall z \in \pi\xi_0(p)$. This is equivalent to that

$$\Phi(p) = \left\{ p' \in \triangle_k^{(|\mathbf{E}|\times M-1)} | p'z > 0, \forall z \in \pi\xi_0(p) \right\} \neq \emptyset$$

for each $p \in \triangle_k^{(|\mathbf{E}|\times M-1)}$.

Thus, by Lemma 5.1, there exists a continuous function $W(p) : \triangle_k^{(|\mathbf{E}|\times M-1)} \longrightarrow \triangle_k^{(|\mathbf{E}|\times M-1)}$ such that $W(p) \in \Phi(p), \forall p \in \triangle_k^{(|\mathbf{E}|\times M-1)}$. Then, by the Brouwer fixed point theorem, there is a $p_k^0 \in \triangle_k^{(|\mathbf{E}|\times M-1)}$ such that $p_k^0 = W(p_k^0)$. This means that $p_k^0 z > 0, \forall z \in \pi\xi_0(p_k^0)$, which contradicts the Walras' Law which has been established in Sec. 3. Therefore, for each $p_k \in \triangle_k^{(|\mathbf{E}|\times M-1)}$, there exists $z_k \in \pi\xi_0(p_k)$ such that $\max_{\triangle_k^{(|\mathbf{E}|\times M-1)}} pz_k \leq 0$.

Since $\{\pi\xi_0(p_k)|p_k \in \triangle_k^{(|\mathbf{E}|\times M-1)}, k = 1,2,\ldots)\}$ are compact subsets of a compact set, there is a convergent subsequence of (p_k, z_k) with limit (p^0, z^0).

Note that the set $\xi_0(p)$ is empty for each $p \in \partial\triangle^{(|\mathbf{E}|\times M-1)}$ since $\pi\xi$ is upper hemi-continuous (as proved in Lemmas 3.3 and 3.5) and the utility function of consumer 1 is strictly increasing. It is not difficult to show that $p^0 \in \mathrm{int}\triangle^{(|\mathbf{E}|\times M-1)}$, $z^0 \in \pi\xi_0(p^0)$ and $\max_{\triangle^{(|\mathbf{E}|\times M-1)}} pz^0 \leq 0$, proving the existence of equilibrium of the truncated economy.

In exactly the same method as that of Geanakoplos and Polemarchakis (1986), it can be shown that there exists a equilibrium in the original economy. $\square$

We will finish this section by an example. We will show that the commodity excess demand correspondence of a model with set-up costs satisfies finite p-convexity.

An example. Consider a one-period model in which there are two commodities, N securities and only one broker. We still retain the assumption of convexity for consumers. But the broker needs a set-up cost before trading. The trading technology of the broker is described as follows:

$$T = \left\{ (\theta^B, \theta^S, x_1, x_2) \in R_+^{N+1} : F(\theta^B, \theta^S, x_1, x_2) \leq 0, x_2 \geq k > 0 \right\} \cup \{(0,0,0,0)\},$$

where k is positive constant, that is, the broker needs k units of commodity 2 to set up his or her trading. Here the F is a convex function and strictly increasing with respect to each component. Including $(0,0,0,0)$ in the trading technology T means no trading.

Given a strictly positive commodity price, it suffices to show the commodity excess demand correspondence of broker is finitely p-convex since the commodity excess demand correspondences of all consumers are convex. Clearly, the commodity excess demand correspondence of the broker is a convex subset in T' (where T' is the projection of set $T - \{(0,0,0,0)\}$ onto the commodity space) or the union E of a convex subset of T' and $\{(0,0)\}$. The set E is finitely p-convex since, in the

Definition 5.1, for any $x_1, x_2 \in E$ and $p_1, \ldots, p_m \in \text{int}\,(\triangle^1)$, $p_i(0,0) \le p_i(\frac{x_1+x_2}{2})$ for all $i = 1, \ldots, m$.

Consequently, there exists an equilibrium in this model.

6. Conclusion

This paper has attempted to attain three objectives:

1. Prove the existence of equilibrium of an asset economy with transaction costs. The model is sufficiently general to cover most cases (finite states, time horizon) in the literature.
2. The method of proof proves some new results (see Ortu (1995)) extending arbitrage pricing dual results to cover transaction costs and different buying and selling prices.
3. In addition, two proofs are provided of existence of an equilibrium with non-convex transaction technologies. These proofs are important for addressing economies with fixed costs in transacting.

Two final comments: in Milne–Neave (1996), it is shown that the basic model can be adapted easily to accomodate a number of variations common in the literature. For example, by considering $I = J = \emptyset$, and brokers are considered as ordinary consumers with a "transaction technology" representing short-sales constraints on trading, the proofs can be interpreted as proving the existence of an equilibrium with trading constraints. A special case of this formulation is an exchange economy with incomplete asset markets. We have discussed an approximate equilibrium with non-convexities. It is possible to modify our model to allow a continuum of agents and obtain an exact equilibrium via the use of the Liapunov Convexity Theorem.

Finally observe that we can incorporate the recent model of Prechac (1996) by assuming that brokers have a linear technology. He assumes that this is a simple markup, but to avoid underpricing, assumes that banking or clearance house has a monopoly with an exogenous, fixed markup.

Appendix

Lemma A1. If the assumption (A.7) holds, then the set Z_h is a closed convex set.

Proof. The convexity of Z_h is obvious. It remains to show its closedness. Suppose $z_h^k \in Z_h$ and $z_h^k \to z_h$. For each k, there exists $\phi_h^{(k)}$ such that $(\triangle\phi_h^{(k)}, z_h^k) \in T_h(e)$, and, in particular, by (A.8), $(\triangle\phi_h^{(k)}, z_h') \in T_h(e)$, where $z_h' = (\max_k z_{h,1}^k, \ldots, \max_k z_{h,M}^k)$. If $\{\triangle\phi_h^{(k)}\}$ is unbounded, we may suppose $\triangle(\phi^{(k)})_{h,1}^B \to \infty$ without loss of generality.

But, by assumption (A.7),

$$\lim_{k\to\infty} |z|_h^k = \infty,$$

which provides a contradiction and proves the boundedness of $\{\triangle\phi_h^{(k)}\}$. Hence, we can choose a subsequence $\{\triangle\phi_h^{(k_n)}\}$ from $\{\triangle\phi_h^{(k)}\}$ such that

$$\lim_{n\to\infty} \triangle\phi_h^{(k_n)} = \triangle\phi_h \, ;$$

this implies, by closedness of $T_h(e)$, the closedness of Z_h.   □

References

[1] K. J. Arrow and G. Debreu, "Existence of an Equilibrium for a Competitive Economy", *Econometrica* **22** (1954) 265–290.

[2] B. Bensaid, J. Lesne, H. Pages and J. Scheinkman, "Derivative Asset Pricing with Transaction Costs", *Mathematical Finance* **2** (1992) 63–86.

[3] P. P. Boyle and T. Vorst, "Option Replication in Discrete Time with Transaction Costs", *J. Finance* **47** (1992) 272–293.

[4] G. Constantinides T. Zariphopoulou, "Universal Bounds on Option Prices with Proportional Transaction Costs", mimeo, University of Chicago, 1995.

[5] M. Davis and A. Norman, "Portfolio Selection with Transaction Costs", *Mathematics of Operations Research* **15** (1990) 676–713.

[6] P. DeMarzo, "An Extension of the Modigliani-Miller Theorem to Stochastic Economies with Incomplete Market", *Journal of Economic Theory* **45** (1988) 353–369.

[7] D. Duffie, "Stochastic Equilibria with Incomplete Financial Markets", *Journal of Economic Theory* **41** (1987) 405–416.

[8] D. Duffie, Dynamic Asset Pricing Theory, Princeton University Press, 1996.

[9] D. Duffie and T. Sun, "Transaction Costs and Portfolio Choice in a Discrete-Continuous Time Setting", *Journal of Economic Dynamics and Control* **14** (1990) 35–51.

[10] C. Edirisinghe, V. Naik and R. Uppal, "Optimal Replication of Options with Transaction Costs", *Journal of Quantative and Financial Analysis* **28** (1993) 117–138.

[11] W. Fleming, S. Grossman, J. L. Vila and T. Zariphopoulou, "Optimal Portfolio Rebalance with Transaction Costs", mimeo, 1989.

[12] J. Geanakoplos and H. Polemarchakis, "Existence, Regularity and Constrained Suboptimality of Competitive Allocations When the Asset Market Is Incomplete", in W. Heller, R. Starr and D. Starett (eds.), Uncertainty, Information and Communication, Vol. III, Cambridge University Press, 1986.

[13] A. Granas and Fon-Che Liu, "Some Minimax Theorems", in "Nonlinear and Convex Analysis", Lecture Notes in Pure and Appl. Math Series, 107, Springer-Verlag, 1987.

[14] W. Heller and R. Starr, "Equilibrium with Non-Convex Transaction Costs: Monetary and Non-Monetary Economics", *Review of Economic Studies* **43** (1976) 195–215.

[15] L. Hurwicz and S. Reiter, "On the Boundedness of the Feasible Set without Convexity Assumption", *International Economic Review* **14** (1973) 580–586.

[16] E. Jouini and H. Kallal, "Martingales and Arbitrage in Securities Markets with Transaction Costs", *Journal of Economic Theory* **66** (1995) 178–197.

[17] S. Kandell and S. Ross, "Some Intertemporal Models of Portfolio Selection with Transaction Costs", working paper 107, GRSP, University of Chicago.

[18] M. Magill and G. Constantinides, "Portfolio Selection with Transaction Costs", *Journal of Economic Theory* **13** (1976) 245–263.

[19] P. McCabe, "On Two Market Equilibrium Theorems", *Journal of Mathematical Economics* **8** (1981) 167–171.

[20] F. Milne and E. H. Neave, "Financial Asset Pricing and Transaction Costs", mimeo, Queen's University, 1996.

[21] F. Ortu, "Arbitrage and Martingales in Finite Dimensional Securities Markets with Transaction Costs: A Linear Programming Approach", mimeo, Universita di Trieste, 1995.

[22] J. Ostroy and R. Starr, "The Transaction Role of Money", Chapter 1 in B. Friedman and F. Hahn (eds.), *Handbook of Monetary Economies* Vol. 1, North-Holland, 1990.

[23] C. Prechac, "Existence of Equilibrium in Incomplete Markets with Intermediation Costs", *Journal of Mathematical Economics* **25** (1996) 373–380.

[24] M. Taksar, M. Klass and D. Assaf, "A Diffusion Model for Optimal Portfolio Selection in the Presence of Brokerage Fees, *Mathematics of Operations Research* **13** (1988) 277–294.

PORTFOLIO GENERATING FUNCTIONS

ROBERT FERNHOLZ

INTECH, One Palmer Square, Princeton, NJ 08542, USA

A general method is presented for constructing dynamic equity portfolios through the use of mathematical generating functions. The return on these functionally generated portfolios is related to the return on the market portfolio by a stochastic differential equation. Under appropriate conditions, this equation can be used to establish a dominance relationship between a functionally generated portfolio and the market portfolio.

Keywords: Portfolio generating function, diversity.

Classification codes: G11, C62.

1. Introduction

Functionally generated equity portfolios were used in Fernholz (1998b) to study market diversity and in Fernholz (1997) to establish conditions under which arbitrage will exist in equity markets. In this paper we present a general discussion of these portfolios and the mathematical functions which generate them. We show that the return on such a portfolio is related to the return on the market portfolio by a stochastic differential equation. Under appropriate conditions, this equation can be used to establish a dominance relationship between a functionally generated portfolio and the market portfolio.

Functionally generated portfolios are not merely mathematical curiosities. Besides having been applied in studying theoretical questions such as market diversity and arbitrage, these portfolios have been used for actual equity investments. In fact, an institutional investment product based on a functionally generated portfolio of the stocks in the S&P 500 Index was introduced in 1996 (see Fernholz, Garvy, and Hannon (1998)). Moreover, Fernholz (1998a) has shown that a phenomenon similar to that which governs the behavior of functionally generated portfolios may explain the size effect, the historical tendency of smaller stocks to have higher returns than larger stocks.

The rest of this paper is organized as follows: In Sec. 2 there is a review of those elements of stochastic portfolio theory which we shall need later. Section 3 presents a formal development of the theory of *portfolio generating functions* (Definition 3.1). In Sec. 4, functionally generated portfolios are used to determine conditions under which portfolio dominance will exist. In particular, we show that a class of functions called *measures of diversity* (Definition 4.3) generate portfolios which will dominate the market portfolio under antitrust type restrictions. Section 5 is a summary.

We shall use a model of stock price processes represented by continuous semi-martingales which is fairly standard in continuous-time financial theory (see, e.g., Karatzas and Shreve (1991) or Duffie (1992)). We shall make certain simplifying assumptions, among them:

1. Companies do not merge or break up, and the total number of shares of a company remains constant. The list of companies in the market is fixed.
2. There are no dividend payments.
3. There are no transaction costs, taxes, or problems with the indivisibility of shares.

2. Review of Stochastic Portfolio Theory

In this section we shall review the basic definitions and results needed in the later sections. Much of the material in this section can also be found in Fernholz (1998b), but for completeness it is presented here also. We shall generally follow the definitions and notation used in Karatzas and Shreve (1991).

Let

$$W = \{W(t) = (W_1(t), \ldots, W_n(t)), \mathcal{F}_t, t \in [0, \infty)\}$$

be a standard n-dimensional Brownian motion defined on a probability space $\{\Omega, \mathcal{F}, P\}$ where $\{\mathcal{F}_t\}$ is the augmentation under P of the natural filtration $\{\mathcal{F}_t^W = \sigma(W(s); 0 \le s \le t)\}$. We say that a process $\{X(t), \mathcal{F}_t, t \in [0, \infty)\}$ is *adapted* if $X(t)$ is $\mathcal{F}_t$-measurable for $t \in [0, \infty)$. If X and Y are processes defined on $\{\Omega, \mathcal{F}, P\}$, we shall use the notation $X = Y$ if

$$P\{X(t) = Y(t), \quad t \in [0, \infty)\} = 1.$$

For continuous, square-integrable martingales $\{M(t), \mathcal{F}_t, t \in [0, \infty)\}$ and $\{N(t), \mathcal{F}_t, t \in [0, \infty)\}$, we can define the *cross-variation process* $\langle M, N \rangle$. The cross-variation process is adapted, continuous, and of bounded variation, and the operation $\langle \cdot, \cdot \rangle$ is bilinear on the real vector space of continuous, square-integrable martingales. If $M = N$, we shall use the notation $\langle M \rangle = \langle M, M \rangle$; $\langle M \rangle$ is called the *quadratic variation process* of M, and has continuous, nondecreasing sample paths. The Brownian motion process defined above is a continuous, square-integrable martingale, and it is characterized by its cross-variation processes

$$\langle W_i, W_j \rangle_t = \delta_{ij} t, \quad t \in [0, \infty),$$

where $\delta_{ij} = 1$ if $i = j$, and 0 otherwise.

A *continuous semimartingale* $X = \{X(t), \mathcal{F}_t, t \in [0, \infty)\}$ is a measurable, adapted process which has the decomposition,

$$X(t) = X(0) + M_X(t) + V_X(t), \quad t \in [0, \infty), \quad \text{a.s.}, \tag{2.1}$$

where $\{M_X(t), \mathcal{F}_t, t \in [0,\infty)\}$ is a continuous, square-integrable martingale and $\{V_X(t), \mathcal{F}_t, t \in [0,\infty)\}$ is a continuous, adapted process which is locally of bounded variation. It can be shown that this decomposition is a.s. unique (see Karatzas and Shreve (1991)), so we can define the cross-variation process for continuous semimartingales X and Y by

$$\langle X, Y \rangle = \langle M_X, M_Y \rangle,$$

where M_X and M_Y are the martingale parts of X and Y, respectively.

Definition 2.1. Let X_0 be a positive number. A *stock* $X = \{X(t), \mathcal{F}_t, t \in [0,\infty)\}$ is a process of the form

$$X(t) = X_0 \exp\left(\int_0^t \gamma(s)\, ds + \int_0^t \sum_n^{\nu=1} \xi_\nu(s)\, dW_\nu(s) \right), \quad t \in [0,\infty), \qquad (2.2)$$

where $\gamma = \{\gamma(t), \mathcal{F}_t, t \in [0,\infty)\}$ is measurable, adapted, and satisfies $\int_0^t |\gamma(s)| ds < \infty$, for all $t \in [0,\infty)$, a.s., and for $\nu = 1,\ldots,n$, $\xi_\nu = \{\xi_\nu(t), \mathcal{F}_t, t \in [0,\infty)\}$ is measurable, adapted, and satisfies $\int_0^t \xi_\nu^2(s) ds < \infty$ for all $t \in [0,\infty)$, a.s., and such that there exists a number $\varepsilon > 0$ for which $\xi_1^2(t) + \cdots + \xi_n^2(t) > \varepsilon$, $t \in [0,\infty)$, a.s.

It follows directly from Definition 2.1 that X is adapted, that $X(t) > 0$ for all $t \in [0,\infty)$, a.s., and that X has initial value $X(0) = X_0$. We shall set the initial value X_0 to be the total capitalization of the company represented by X at time $t = 0$, and we shall assume that this total capitalization is positive. This is equivalent to assuming that there is a single share of stock outstanding, and $X(t)$ represents its price at time t. We assume that stock shares are infinitely divisible, so there is no loss of generality in assuming a single share outstanding. The process γ is called the *growth rate* (*process*) of X and, for each ν, the process ξ_ν represents the sensitivity of X to the νth source of uncertainty W_ν.

Suppose that we have a family of stocks X_i, $i = 1,\ldots,n$,

$$X_i(t) = X_0^i \exp\left(\int_0^t \gamma_i(s)\, ds + \int_0^t \sum_{\nu=1}^n \xi_{i\nu}(s)\, dW_\nu(s) \right), \quad t \in [0,\infty). \qquad (2.3)$$

Consider the matrix valued process ξ defined by $\xi(t) = (\xi_{i\nu}(t))_{1 \le i,\nu \le n}$ and define the *covariance process* σ where $\sigma(t) = \xi(t)\xi^T(t)$. The cross-variation processes for $\log X_i$ and $\log X_j$ are related to σ by

$$\langle \log X_i, \log X_j \rangle_t = \int_0^t \sigma_{ij}(s)\, ds, \quad t \in [0,\infty), \quad \text{a.s.} \qquad (2.4)$$

Since the processes $\xi_{i\nu}$ are assumed to be square integrable in Definition 2.1, it follows that for all i and j,

$$\int_0^t \sigma_{ij}(s)\, ds < \infty, \quad t \in [0,\infty), \quad \text{a.s.}$$

Definition 2.2. A *market* is a family $\mathcal{M} = \{X_i, \ldots, X_n\}$ of stocks, defined as in (2.3), for which there is a number $\varepsilon > 0$ such that

$$x\sigma(t)x^T \geq \varepsilon \|x\|^2, \quad x \in \mathbb{R}^n, t \in [0, \infty), \quad \text{a.s.} \tag{2.5}$$

The *strong nondegeneracy* condition (2.5) is fairly common and can be found, for example, in Karatzas and Shreve (1991) and Karatzas and Kou (1996), and as *uniform ellipticity* in Duffie (1992).

Definition 2.3. Let $\mathcal{M}$ be a market of n stocks. A *portfolio* in $\mathcal{M}$ is a measurable, adapted process $\pi = \{\pi(t) = (\pi_1(t), \ldots, \pi_n(t)), \mathcal{F}_t, t \in [0, \infty)\}$ such that $\pi(t)$ is bounded on $[0, \infty) \times \Omega$ and

$$\pi_1(t) + \cdots + \pi_n(t) = 1, \quad t \in [0, \infty), \quad \text{a.s.}$$

The processes π_i represent the respective *proportions*, or *weights*, of each stock in the portfolio. A negative value for $\pi_i(t)$ indicates a short sale. Suppose $Z_\pi(t)$ represents the value of an investment in π at time t. Then the amount invested in the ith stock X_i will be

$$\pi_i(t)Z_\pi(t),$$

so if the price of X_i changes by $dX_i(t)$, the induced change in the portfolio value will be

$$\pi_i(t)Z_\pi(t)\frac{dX_i(t)}{X_i(t)}.$$

Hence the total change in the portfolio value at time t will be

$$dZ_\pi(t) = \sum_{i=1}^n \pi_i(t)Z_\pi(t)\frac{dX_i(t)}{X_i(t)},$$

or, equivalently,

$$\frac{dZ_\pi(t)}{Z_\pi(t)} = \sum_{i=1}^n \pi_i(t)\frac{dX_i(t)}{X_i(t)}. \tag{2.6}$$

Since we are interested in the behavior of portfolios, we are interested in solutions to (2.6). The following proposition and corollary are proved in Fernholz (1998b).

Proposition 2.1. *Let π be a portfolio and let*

$$\gamma_\pi(t) = \sum_{i=1}^n \pi_i(t)\gamma_i(t) + \gamma_\pi^*(t), \tag{2.7}$$

where

$$\gamma_\pi^*(t) = \frac{1}{2}\left(\sum_{i=1}^n \pi_i(t)\sigma_{ii}(t) - \sum_{i,j=1}^n \pi_i(t)\pi_j(t)\sigma_{ij}(t)\right). \tag{2.8}$$

Then, for any positive initial value Z_0^π, the process Z_π defined by

$$Z_\pi(t) = Z_0^\pi \exp\left(\int_0^t \gamma_\pi(s)\,ds + \int_0^t \sum_{i,\nu=1}^n \pi_i(s)\xi_{i\nu}(s)\,dW_\nu(s)\right), \quad t \in [0,\infty), \quad (2.9)$$

is a strong solution of (2.6).

Corollary 2.1. *Let π be a portfolio and Z_π be its value process. Then for $t \in [0,\infty)$,*

$$d\log Z_\pi(t) = \sum_{i=1}^n \pi_i(t)\,d\log X_i(t) + \gamma_\pi^*(t)\,dt. \quad (2.10)$$

The process γ_π in (2.7) is called the *portfolio growth rate (process)* of the portfolio π, and γ_π^* in (2.8) is called the *excess growth rate (process)*. It was proved in Fernholz (1998b) that for portfolios with non-negative weights, the excess growth rate is non-negative, and is positive unless the portfolio consists of a single stock.

The total capitalization of the market can be represented by a portfolio. Let us assume from now on that the market is $\mathcal{M} = \{X_1, \ldots, X_n\}$, with n stocks.

Definition 2.4 The portfolio

$$\mu = \{\mu(t) = (\mu_1(t), \ldots, \mu_n(t)), \mathcal{F}_t, t \in [0,\infty)\},$$

where

$$\mu_i(t) = \frac{X_i(t)}{X_1(t) + \cdots + X_n(t)}, \quad (2.11)$$

for $i = 1, \ldots, n$, is called the *market portfolio (process)*.

It is clear that the μ_i defined by (2.11) satisfy the requirements of Definition 2.3. If we let

$$Z(t) = X_1(t) + \cdots + X_n(t), \quad (2.12)$$

then $Z(t)$ satisfies (2.6) with weights $\mu_i(t)$ given by (2.11). Hence, the value of the market portfolio represents the combined capitalization of all the stocks in the market. We shall let μ exclusively represent the market portfolio and Z its value process.

For any stock X_i and portfolio π we can consider the quotient process X_i/Z_π defined by

$$\log\left(X_i(t)/Z_\pi(t)\right) = \log X_i(t) - \log Z_\pi(t). \quad (2.13)$$

This process is a continuous semimartingale with

$$\langle \log(X_i/Z_\pi), \log(X_j/Z_\pi) \rangle_t = \langle \log X_i, \log X_j \rangle_t - \langle \log X_i, \log Z_\pi \rangle_t$$

$$- \langle \log X_j, \log Z_p \rangle_t + \langle \log Z_\pi \rangle_t. \quad (2.14)$$

If we define the process $\sigma_{i\pi}$ by

$$\sigma_{i\pi}(t) = \sum_{j=1}^{n} \pi_j(t)\sigma_{ij}(t),$$

for $i = 1, \ldots, n$, then

$$\langle \log X_i, \log Z_\pi \rangle_t = \int_0^t \sigma_{i\pi}(s)\, ds.$$

Define the *relative covariance (process)* τ^π to be the matrix valued process

$$\tau^\pi(t) = (\tau^\pi_{ij}(t))_{1 \le i, j \le n},$$

where

$$\tau^\pi_{ij}(t) = \sigma_{ij}(t) - \sigma_{i\pi}(t) - \sigma_{j\pi}(t) + \sigma_{\pi\pi}(t), \tag{2.15}$$

for $i, j = 1, \ldots, n$, where $\sigma_{\pi\pi}(t) = \pi(t)\sigma(t)\pi^T(t)$. Then for all i and j,

$$\langle \log(X_i/Z_\pi), \log(X_j/Z_\pi) \rangle_t = \int_0^t \tau^\pi_{ij}(s)\, ds. \tag{2.16}$$

In the case that $i = j$, we know that $\langle \log(X_i/Z_\pi) \rangle_t$ is non-decreasing, so

$$\tau^\pi_{ii}(t) \ge 0, \quad t \in [0, \infty), \quad \text{a.s.}$$

We shall use $\tau = \tau^\mu$ to represent the relative covariance process of the market portfolio μ, and τ_{ij} to represent its ijth component.

Let $\eta = \{\eta(t) = (\eta_1(t), \ldots, \eta_n(t)), \mathcal{F}_t, t \in [0, \infty)\}$ be a portfolio. Then, by Corollary 2.1,

$$d \log Z_\eta(t) = \sum_{i=1}^{n} \eta_i(t)\, d \log X_i(t) + \gamma^*_\eta(t)\, dt,$$

so

$$d \log (Z_\eta(t)/Z_\pi(t)) = \sum_{i=1}^{n} \eta_i(t)\, d \log (X_i(t)/Z_\pi(t)) + \gamma^*_\eta(t)\, dt. \tag{2.17}$$

The *relative variance (process)* of η and π is defined by

$$\tau^\pi_{\eta\eta}(t) = (\eta(t) - \pi(t))\sigma(t)(\eta(t) - \pi(t))^T$$

$$= \eta(t)\tau^\pi(t)\eta^T(t). \tag{2.18}$$

Lemma 2.1. *The rank of $\tau^\pi(t)$ is $n - 1$, for $t \in [0, \infty)$, a.s. The null space of $\tau^\pi(t)$ is spanned by $\pi(t)$.*

Proof. From (2.18) and condition (2.5), it follows that

$$\eta(t)\tau^\pi(t)\eta^T(t) = 0$$

if and only if $\eta(t) = \pi(t)$, for all $t \in [0, \infty)$, a.s. $\qquad \square$

The following lemma expresses the excess growth in terms of the relative covariance process.

Lemma 2.2. Let π and η be portfolios. Then for $t \in [0, \infty)$,

$$\gamma_\eta^*(t) = \frac{1}{2} \left(\sum_{i=1}^{n} \eta_i(t) \tau_{ii}^\pi(t) - \tau_{\eta\eta}^\pi(t) \right).$$

Proof. The proof is a direct calculation using (2.15) and (2.18). □

Lemma 2.3. *Let π be a portfolio. Then there exists an $\varepsilon > 0$ such that for $i = 1, \ldots, n$,*

$$\tau_{ii}^\pi(t) \geq \varepsilon \left(1 - \pi_i(t)\right)^2, \quad t \in [0, \infty), \quad \text{a.s.} \tag{2.19}$$

Proof. For any i and $t \in [0, \infty)$, let $x(t) = (\pi_1(t), \ldots, \pi_i(t) - 1, \ldots, \pi_n(t))$. Then,

$$\tau_{ii}^\pi(t) = \sigma_{ii}(t) - 2\sigma_{i\pi}(t) + \sigma_{\pi\pi}(t)$$

$$= x(t)\sigma(t)x^T(t)$$

$$\geq \varepsilon \|x(t)\|^2, \quad t \in [0, \infty), \quad \text{a.s.,}$$

where ε is chosen as in (2.5). Since,

$$\|x(t)\|^2 \geq (1 - \pi_i(t))^2,$$

the lemma follows. □

Lemma 2.4. *Let π be a portfolio with non-negative weights, and let $\pi_{\max}(t) = \max_{1 \leq i \leq n} \pi_i(t)$. Then there exists an $\varepsilon > 0$ such that*

$$\gamma_\pi^*(t) \geq \varepsilon (1 - \pi_{\max}(t))^2.$$

Proof. If we let $\eta = \pi$ in Lemma 2.2, then Lemma 2.1 implies that

$$\gamma_\pi^*(t) = \frac{1}{2} \sum_{i=1}^{n} \pi_i(t) \tau_{ii}^\pi(t)$$

$$\geq \frac{\varepsilon}{2} (1 - \pi_{\max}(t))^2,$$

where ε is chosen as in Lemma 2.3, since the $\pi_i(t)$ are non-negative. □

3. Portfolio Generating Functions

In this section we shall show that certain real-valued functions of the market weights can be used to generate portfolios, and we shall study the properties these functions and the portfolios they generate. Functionally generated portfolios are of interest because under certain market conditions it can be shown that a dominance relationship exists between such a portfolio and the market portfolio. This relationship will be discussed in the next section; this section will be devoted to the basic properties of generating functions and the portfolios they generate.

We shall consider real-valued functions defined on the open simplex

$$\Delta^n = \{x \in \mathbb{R}^n : x_1 + \cdots + x_n = 1, \quad 0 < x_i < 1, \quad i = 1, \ldots, n\}.$$

It will be convenient to use the standard coordinate system in $\mathbb{R}^n$, even though it is not a coordinate system on Δ^n. For this reason we shall consider functions that are defined in an open neighborhood $U \subset \mathbb{R}^n$ of Δ^n. A real-valued function defined on a subset of $\mathbb{R}^n$ is C^2 if it is twice continuously differentiable in all variables. We shall use the notation D_i for the partial derivative with respect to the ith variable, and D_{ij} for the second partial derivative with respect to the ith and jth variables.

Definition 3.1. Let U be an open neighborhood of Δ^n and $\mathbf{S}$ be a positive C^2 function defined in U. Then $\mathbf{S}$ is the *generating function* of the portfolio π if there exists a measurable, adapted process $\Theta = \{\Theta(t), \mathcal{F}_t, t \in [0, \infty)\}$ such that

$$d\log(Z_\pi(t)/Z(t)) = d\log \mathbf{S}(\mu(t)) + \Theta(t)dt, \quad t \in [0, \infty), \quad \text{a.s.} \qquad (3.1)$$

Θ is called the *drift process* corresponding to $\mathbf{S}$.

We shall also say that the function $\mathbf{S}$ *generates* π, and that π is *functionally generated*. Definition 3.1 can be extended to include time dependent generating functions, but we shall not consider such functions here.

Proposition 3.1. *Suppose that* $\mathbf{S}_1$ *and* $\mathbf{S}_2$ *generate* π_1 *and* π_2 *with drift processes* Θ_1 *and* Θ_2, *respectively. Then*

$$d\log(Z_{\pi_1}(t)/Z_{\pi_2}(t)) = d\log(\mathbf{S}_1(\mu(t))/\mathbf{S}_2(\mu(t)) + (\Theta_1(t) - \Theta_2(t))dt,$$

$$t \in [0, \infty), \quad \text{a.s.}$$

Proof. The proof follows directly from Definition 3.1. $\qquad\qquad \square$

What follows is the main theorem on portfolio generating functions.

Theorem 3.1. *Let* **S** *be a positive* C^2 *function defined on a neighborhood* U *of* Δ^n *such that for* $i = 1, \ldots, n$, $x_i D_i \log \mathbf{S}(x)$ *is bounded on* Δ^n. *Then* **S** *generates the portfolio* π *with weights*

$$\pi_i(t) = \left(D_i \log \mathbf{S}(\mu(t)) + 1 - \sum_{j=1}^{n} \mu_j(t) \, D_j \log \mathbf{S}(\mu(t)) \right) \mu_i(t), \quad t \in [0, \infty), \quad \text{a.s.},$$

(3.2)

for $i = 1, \ldots, n$, *and drift process*

$$\Theta(t) = \frac{-1}{2\,\mathbf{S}(\mu(t))} \sum_{i,j=1}^{n} D_{ij}\mathbf{S}(\mu(t)) \mu_i(t) \mu_j(t) \tau_{ij}(t), \quad t \in [0, \infty), \quad \text{a.s.} \quad (3.3)$$

Proof. The weight process μ_i is a quotient process with $\mu_i(t) = X_i(t)/Z(t)$ for all t. By (2.16) it follows that

$$d\langle \log \mu_i, \log \mu_j \rangle_t = \tau_{ij}(t)\, dt, \quad t \in [0, \infty), \quad \text{a.s.},$$

so by Itô's Lemma,

$$d\mu_i(t) = \mu_i(t)\, d\log \mu_i(t) + \frac{1}{2}\mu_i(t)\tau_{ii}(t)\, dt, \quad t \in [0, \infty), \quad \text{a.s.}, \quad (3.4)$$

and

$$d\langle \mu_i, \mu_j \rangle_t = \mu_i(t)\mu_j(t)\tau_{ij}(t)\, dt, \quad t \in [0, \infty), \quad \text{a.s.} \quad (3.5)$$

Itô's Lemma, along with (3.5), implies that a.s. for all $t \in [0, \infty)$,

$$d\log \mathbf{S}(\mu(t)) = \sum_{i=1}^{n} D_i \log \mathbf{S}(\mu(t))\, d\mu_i(t)$$

$$+ \frac{1}{2} \sum_{i,j=1}^{n} D_{ij} \log \mathbf{S}(\mu(t))\mu_i(t)\mu_j(t)\tau_{ij}(t)\, dt.$$

Now,

$$D_{ij} \log \mathbf{S}(\mu(t)) = \frac{D_{ij}\mathbf{S}(\mu(t))}{\mathbf{S}(\mu(t))} - D_i \log \mathbf{S}(\mu(t)) D_j \log \mathbf{S}(\mu(t)),$$

so, a.s., for all $t \in [0, \infty)$,

$$d\log \mathbf{S}(\mu(t)) = \sum_{i=1}^{n} D_i \log \mathbf{S}(\mu(t))\, d\mu_i(t)$$

$$+ \frac{1}{2\mathbf{S}(\mu(t))} \sum_{i,j=1}^{n} D_{ij}\mathbf{S}(\mu(t))\mu_i(t)\mu_j(t)\tau_{ij}(t)\, dt$$

$$- \frac{1}{2} \sum_{i,j=1}^{n} D_i \log \mathbf{S}(\mu(t)) D_j \log \mathbf{S}(\mu(t))\mu_i(t)\mu_j(t)\tau_{ij}(t)\, dt. \quad (3.6)$$

In order for (3.1) to hold, the martingale parts of $\log \mathbf{S}(\mu(t))$ and $\log(Z_\pi(t)/Z(t))$ must be equal. Corollary 2.1 implies that for the portfolio π, a.s. for all $t \in [0, \infty)$,

$$d\log(Z_\pi(t)/Z(t)) = \sum_{i=1}^{n} \pi_i(t)\, d\log(X_i(t)/Z(t)) + \gamma_\pi^*(t)\, dt$$

$$= \sum_{i=1}^{n} \pi_i(t)\, d\log\mu_i(t) + \gamma_\pi^*(t)\, dt$$

$$= \sum_{i=1}^{n} \frac{\pi_i(t)}{\mu_i(t)}\, d\mu_i(t) - \frac{1}{2}\sum_{i,j=1}^{n} \pi_i(t)\pi_j(t)\tau_{ij}(t)\, dt \qquad (3.7)$$

by Lemma 2.2. Suppose that

$$\pi_i(t) = (D_i \log \mathbf{S}(\mu(t)) + \varphi(t))\mu_i(t), \qquad (3.8)$$

where $\varphi(t)$ is chosen such that $\sum_{i=1}^{n} \pi_i(t) = 1$. Then, a.s. for all $t \in [0, \infty)$,

$$\sum_{i=1}^{n} \frac{\pi_i(t)}{\mu_i(t)}\, d\mu_i(t) = \sum_{i=1}^{n} D_i \log \mathbf{S}(\mu(t))\, d\mu_i(t) + \varphi(t)\sum_{i=1}^{n} d\mu_i(t)$$

$$= \sum_{i=1}^{n} D_i \log \mathbf{S}(\mu(t))\, d\mu_i(t),$$

since $\sum_{i=1}^{n} d\mu_i(t) = 0$. Hence, the martingale parts of $\log\mathbf{S}(\mu(t))$ and $\log(Z_\pi(t)/Z(t))$ are equal. If

$$\varphi(t) = 1 - \sum_{j=1}^{n} \mu_j(t)D_j \log \mathbf{S}(\mu(t)),$$

then $\sum_{i=1}^{n} \pi_i(t) = 1$, and (3.2) is proved.

If $\pi_i(t)$ satisfies (3.8), then a.s. for all $t \in [0, \infty)$,

$$\sum_{i,j=1}^{n} \pi_i(t)\pi_j(t)\tau_{ij}(t) = \sum_{i,j=1}^{n} D_i \log \mathbf{S}(\mu(t))D_j \log \mathbf{S}(\mu(t))\mu_i(t)\mu_j(t)\tau_{ij}(t)$$

$$+ 2\varphi(t)\sum_{i,j=1}^{n} D_i \log \mathbf{S}(\mu(t))\mu_i(t)\mu_j(t)\tau_{ij}(t)$$

$$+ \varphi^2(t)\sum_{i,j=1}^{n} \mu_i(t)\mu_j(t)\tau_{ij}(t)$$

$$= \sum_{i,j=1}^{n} D_i \log \mathbf{S}(\mu(t))D_j \log \mathbf{S}(\mu(t))\mu_i(t)\mu_j(t)\tau_{ij}(t),$$

since $\mu(t)$ is in the null space of $\tau(t)$ by Lemma 2.1. Hence, a.s. for all $t \in [0, \infty)$,

$$d\log(Z_\pi(t)/Z(t)) = \sum_{i=1}^{n} D_i \log \mathbf{S}(\mu(t)) \, d\mu_i(t)$$

$$-\frac{1}{2} \sum_{i,j=1}^{n} D_i \log \mathbf{S}(\mu(t)) D_j \log \mathbf{S}(\mu(t)) \mu_i(t)\mu_j(t)\tau_{ij}(t) \, dt \, .$$

This equation and (3.6) imply that a.s. for all $t \in [0, \infty)$,

$$d\log\left(Z_\pi(t)/Z(t)\right) = d\log \mathbf{S}(\mu(t))$$

$$-\frac{1}{2\,\mathbf{S}(\mu(t))} \sum_{i,j=1}^{n} D_{ij}\mathbf{S}(\mu(t))\mu_i(t)\mu_j(t)\tau_{ij}(t) \, dt \, ,$$

so (3.3) is proved. The process Θ defined by (3.3) is clearly measurable and adapted. $\qquad\square$

Example 3.1. Here are a few examples of generating functions and the portfolios they generate.

1. $\mathbf{S}(x) = 1$ generates the market portfolio μ with $\Theta(t) = 0$.
2. $\mathbf{S}(x) = c_1 x_1 + \cdots + c_n x_n$, where $c_1, \ldots, c_n$ are constants, generates the buy-and-hold portfolio which holds c_i of the capitalization of the the ith stock. Here $\Theta(t) = 0$. This type of portfolio is commonly held by investors, at least in a piecewise manner.
3. $\mathbf{S}(x) = (x_1 \cdots x_n)^{1/n}$ generates the equal weighted portfolio with $\Theta(t) = \gamma_\pi^*(t)$. The Value Line Index is such a portfolio.
4. $\mathbf{S}(x) = x_1^{p_1} \cdots x_n^{p_n}$, where $p_1, \ldots, p_n$ are constant and $p_1 + \cdots + p_n = 1$, generates the constant weight portfolio with weights $\pi_i(t) = p_i$ and $\Theta(t) = \gamma_\pi^*(t)$.

Corollary 3.1. *Let* $\mathbf{S}_1$ *and* $\mathbf{S}_2$ *generate portfolios* π_1 *and* π_2, *respectively. Then for constants* p_1 *and* p_2 *such that* $p_1 + p_2 = 1$, *the function*

$$\mathbf{S} = \mathbf{S}_1^{p_1} \mathbf{S}_2^{p_2}$$

generates a portfolio Y *with weights*

$$\pi_i = p_1 \pi_{1i} + p_2 \pi_{2i} \, ,$$

where π_{1i} *and* π_{2i} *are the weights of* π_1 *and* π_2, *respectively.*

Proof. The proof follows directly from (3.2) of Theorem 3.1. $\qquad\square$

Not all portfolios are functionally generated; let us now characterize those that are. Recall that a differential is *exact* if it is of the form $\sum_i D_i G(x)\, dx_i$ for some differentiable function G (see Spivac (1965)).

Proposition 3.2. *Let π be a portfolio. Then π is functionally generated if and only if there exist continuously differentiable real-valued functions $F, f_1, \ldots, f_n$ defined in a neighborhood of Δ^n such that $\pi_i(t) = f_i(\mu(t))$ for all $t \in [0, \infty)$, a.s., for $i = 1, \ldots, n$, and*

$$\sum_{i=1}^{n} \left(\frac{f_i(x)}{x_i} + F(x) \right) dx_i \qquad (3.9)$$

is an exact differential.

Proof. Suppose that $\pi_i(t) = f_i(\mu(t))$ for $i = 1, \ldots, n$, and the differential in (3.9) is exact. Then there is a function G such that

$$D_i G(x) = \frac{f_i(x)}{x_i} + F(x),$$

for $i = 1, \ldots, n$. Hence, Itô's Lemma implies that, a.s., for all $t \in [0, \infty)$,

$$dG(\mu(t)) = \sum_{i=1}^{n} \left(\frac{f_i(\mu(t))}{\mu_i(t)} + F(\mu(t)) \right) d\mu_i(t) + \frac{1}{2} \sum_{i,j=1}^{n} D_{ij} G(\mu(t)) \mu_i(t) \mu_j(t) \tau_{ij}(t)\, dt$$

$$= \sum_{i=1}^{n} \pi_i(t) \frac{d\mu_i(t)}{\mu_i(t)} + \frac{1}{2} \sum_{i,j=1}^{n} D_{ij} G(\mu(t)) \mu_i(t) \mu_j(t) \tau_{ij}(t)\, dt, \qquad (3.10)$$

since $\sum_i d\mu_i(t) = 0$. By (3.10) and (3.7), it follows that a.s., for all $t \in [0, \infty)$,

$$dG(\mu(t)) = d \log(Z_\pi(t)/Z(t))$$

$$+ \frac{1}{2} \sum_{i,j=1}^{n} (\pi_i(t)\pi_j(t)\tau_{ij}(t) + D_{ij} G(\mu(t)) \mu_i(t) \mu_j(t) \tau_{ij}(t))\, dt.$$

Hence, the function $\mathbf{S} = e^G$ generates π.

Now suppose that π has a generating function $\mathbf{S}$ defined in a neighborhood U of Δ^n. For $x \in U$, define

$$f_i(x) = \left(D_i \log \mathbf{S}(x) + 1 - \sum_{j=1}^{n} x_j D_j \log \mathbf{S}(x) \right) x_i,$$

for $i = 1, \ldots, n$, and let

$$F(x) = -1 + \sum_{j=1}^{n} x_j D_j \log \mathbf{S}(x).$$

By Theorem 3.1 the weights π_i satisfy $\pi_i(t) = f_i(\mu(t))$, for $i = 1,\ldots,n$, for all $t \in [0,\infty)$, a.s. Then for $x \in U$, the differential

$$\sum_{i=1}^{n} \left(\frac{f_i(x)}{x_i} + F(x) \right) dx_i = \sum_{i=1}^{n} D_i \log \mathbf{S}(x)\, dx_i$$

$$= d \log \mathbf{S}(x)$$

is exact, so the proposition is proved.   □

Example 3.2. Here is an example of a portfolio π whose weights depend differentiably on the market portfolio weights, but is not functionally generated. For $x \in \mathbb{R}^n$, let

$$f_1(x) = x_1,$$

$$f_2(x) = x_2 + \cdots + x_n,$$

$$f_i(x) = 0, \quad \text{for } i = 3,\ldots,n,$$

and let π be the portfolio defined by $\pi_i(t) = f_i(\mu(t))$, for $i = 1,\ldots,n$. It can be shown (see Spivac (1965)) that if the differential in (3.9) is exact, then for all i and j,

$$D_j \left(\frac{f_i(x)}{x_i} + F(x) \right) = D_i \left(\frac{f_j(x)}{x_j} + F(x) \right).$$

No function F will satisfy these equations for all i and j, so π has no generating function.

It would seem reasonable that only the values of a generating function on Δ^n should affect the portfolio it generates. To prove this, we first need a couple of lemmas.

Lemma 3.1. *Let $f_1,\ldots,f_n$ be continuous real-valued functions on Δ^n and g be a continuous real-valued function on $\Delta^n \times [0,\infty)$. If*

$$\sum_{i=1}^{n} f_i(\mu(t))\, d\mu_i(t) = g(\mu(t),t)\, dt, \quad t \in [0,\infty), \quad a.s.,$$

then for $x \in \Delta^n$, $f_i(x) = f_j(x)$ for all i and j, and $g = 0$.

Proof. For $t \in [0,\infty)$, let

$$Y(t) = \int_0^t \sum_{i=1}^{n} f_i(\mu(s))\, d\mu_i(s).$$

Then Y is a semimartingale with

$$d\langle Y \rangle_t = \sum_{i,j=1}^{n} f_i(\mu(t)) f_j(\mu(t)) \mu_i(t) \mu_j(t) \tau_{ij}(t)\, dt\,, \quad t \in [0,\infty)\,, \quad \text{a.s.,}$$

by (3.5) (see Karatzas and Shreve (1991)). By hypothesis

$$Y(t) = \int_0^t g(\mu(s), s)\, ds\,,$$

so the martingale part of Y in the decomposition (2.1) is null. Therefore, $\langle Y \rangle_t = 0$, so

$$\sum_{i,j=1}^{n} f_i(\mu(t)) f_j(\mu(t)) \mu_i(t) \mu_j(t) \tau_{ij}(t) = 0\,, \quad t \in [0,\infty)\,, \quad \text{a.s.,}$$

and $(f_1(\mu(t))\mu_1(t), \ldots, f_n(\mu(t))\mu_n(t))$ is in the null space of $\tau(t)$. By Lemma 2.1 the null space of $\tau(t)$ is spanned by $(\mu_1(t), \ldots, \mu_n(t))$, so

$$f_1(\mu(t)) = f_2(\mu(t)) = \cdots = f_n(\mu(t))\,.$$

Hence, a.s., for all $t \in [0,\infty)$,

$$\sum_{i=1}^{n} f_i(\mu(t))\, d\mu_i(t) = f_1(\mu(t)) \sum_{i=1}^{n} d\mu_i(t)$$

$$= 0\,,$$

since $\sum_{i=1}^{n} d\mu_i(t) = 0$, and therefore $g(\mu(t), t) = 0$. □

Lemma 3.2. *Let f be a continuously differentiable real-valued function defined in a neighborhood U of Δ^n. Then f is constant on Δ^n if and only if for all $x \in \Delta^n$, $D_i f(x) = D_j f(x)$ for all i and j.*

Proof. Parameterize Δ^n by positive real numbers $t_1, \ldots, t_{n-1}$ with $t_1 + \cdots + t_{n-1} < 1$ such that $x_i = t_i$ for $i = 1, \ldots, n-1$ and $x_n = 1 - t_1 - \cdots - t_{n-1}$. Then for $i = 1, \ldots, n-1$,

$$\frac{\partial f(x)}{\partial t_i} = D_i f(x) - D_n f(x)$$

for all $x \in \Delta^n$. If f is constant on Δ^n then all its partial derivatives with respect to the parameters t_i will be zero, which implies that $D_i f(x) = D_n f(x)$ for all i. Likewise if $D_i f(x) = D_j f(x)$ for all i and j, then the partial derivatives with respect to all the t_i will be zero, so f will be constant on Δ^n. □

Proposition 3.3. *Let S_1 and S_2 be positive C^2 functions defined in an open neighborhood of Δ^n. Then S_1 and S_2 generate the same portfolio if and only if S_1/S_2 is constant on Δ^n.*

Proof. Suppose S_1 and S_2 are defined on the open neighborhood U of Δ^n, and that S_1/S_2 is constant on Δ^n. Define the function f for $x \in U$ by

$$f(x) = \log S_1(x) - \log S_2(x).$$

Then f is constant on Δ^n, so Lemma 3.2 implies that $D_i f(x) = D_j f(x)$ for all i and j, for all $x \in \Delta^n$. Therefore,

$$D_i \log S_1(\mu(t)) - D_j \log S_1(\mu(t)) = D_i \log S_2(\mu(t)) - D_j \log S_2(\mu(t)),$$

for all i and j. Hence, the difference $D_i \log S_1(\mu(t)) - D_i \log S_2(\mu(t))$ is the same for all i, and (3.2) of Theorem 3.1 implies that the weights generated by S_1 and S_2 will be the same. It follows that S_1 and S_2 generate the same portfolio.

Now suppose that S_1 and S_2 generate the same portfolio. By Itô's Lemma, a.s., for all $t \in [0, \infty)$,

$$d\log\left(S_1(\mu(t))/S_2(\mu(t))\right) = \sum_{i=1}^{n} D_i \log\left(S_1(\mu(t))/S_2(\mu(t))\right) d\mu_i(t)$$

$$+ \frac{1}{2} \sum_{i,j=1}^{n} D_{ij} \log\left(S_1(\mu(t))/S_2(\mu(t))\right) \mu_i(t)\mu_j(t)\tau_{ij}(t)\, dt.$$

Since the values of the portfolios generated by S_1 and S_2 will be equal, (3.1) implies that

$$d\log\left(S_1(\mu(t))/S_2(\mu(t))\right) = (\Theta_1(t) - \Theta_2(t))\, dt, \quad t \in [0, \infty), \quad \text{a.s.},$$

so, a.s., for all $t \in [0, \infty)$,

$$\sum_{i=1}^{n} D_i \log\left(S_1(\mu(t))/S_2(\mu(t))\right) d\mu_i(t)$$

$$= \left(\Theta_1(t) - \Theta_2(t) - \frac{1}{2} \sum_{i,j=1}^{n} D_{ij} \log(S_1(\mu(t))/S_2(\mu(t)))\mu_i(t)\mu_j(t)\tau_{ij}(t)\right) dt.$$

Then Lemma 3.1 implies that for all $x \in \Delta^n$,

$$D_i \log(S_1(x)/S_2(x)) = D_j \log(S_1(x)/S_2(x)),$$

for all i and j. It follows by Lemma 3.2 that $\log\left(S_1(x)/S_2(x)\right)$ is constant for $x \in \Delta^n$. $\square$

This proposition implies that we could have defined generating functions on Δ^n rather than in a neighborhood. The difficulty is that on Δ^n there is no linear coordinate system which treats all the n stocks in the same manner.

For use in the next section, we need to establish conditions under which a generating function will have a positive drift process.

Proposition 3.4. *Let* **S** *be a generating function such that for all* $x \in \Delta^n$, *the matrix* $(D_{ij}\mathbf{S}(x))$ *has at most one positive eigenvalue, and if there is a positive eigenvalue it corresponds to an eigenvector perpendicular to* Δ^n. *Let* π *be the portfolio generated by* **S**. *Then* $\pi_i(t) \geq 0$, *for* $i = 1, \ldots, n$, *and* $\Theta(t) \geq 0$ *for all* $t \in [0, \infty)$, *a.s. If for all* $x \in \Delta^n$, $\mathrm{rank}(D_{ij}\mathbf{S}(x)) > 1$, *then* $\Theta(t) > 0$ *for all* $t \in [0, \infty)$, *a.s.*

Proof. Suppose that **S** is a generating function which satisfies the hypothesis of the proposition. For any $t \in [0, \infty)$, define $x(u) \in \Delta^n$ by

$$x(u) = uv_k + (1 - u)\mu(t)$$

for $0 \leq u < 1$, where $v_k = (0, \ldots, 1, \ldots, 0)$ with 1 in the kth position and 0 elsewhere. Let

$$f(u) = \mathbf{S}(x(u)),$$

so

$$f'(u) = D_k\mathbf{S}(x(u)) - \sum_{i=1}^{n} \mu_i(t)D_i\mathbf{S}(x(u)), \tag{3.11}$$

and

$$f''(u) = (v_k - \mu(t))\,(D_{ij}\mathbf{S}(x(u)))\,(v_k - \mu(t))^T$$
$$\leq 0,$$

since $v_k - \mu(t)$ is parallel to Δ^n and hence is composed of eigenvectors of $(D_{ij}\mathbf{S}(x(u)))$ which have non-positive eigenvalues. The usual convexity arguments imply that

$$f(u) \leq f(0) + uf'(0), \qquad 0 \leq u < 1,$$

so,

$$0 < f(0) + uf'(0), \qquad 0 \leq u < 1.$$

It follows from this and (3.11) that

$$0 \leq \mathbf{S}(\mu(t)) + D_k\mathbf{S}(\mu(t)) - \sum_{i=1}^{n} \mu_i(t)D_i\mathbf{S}(\mu(t)).$$

Hence, by (3.2) of Theorem 3.1, $\pi_k(t) \geq 0$, $k = 1, \ldots, n$, for all $t \in [0, \infty)$, a.s.

For any $t \in [0, \infty)$, let $\lambda_1, \ldots, \lambda_n$ be the eigenvalues of $(D_{ij}\mathbf{S}(\mu(t)))$. Let $e_k = (e_{k1}, \ldots, e_{kn})$ be a normalized eigenvector corresponding to the eigenvalue λ_k, for $k = 1, \ldots, n$. For any $x, y \in \mathbb{R}^n$,

$$x\left(D_{ij}\mathbf{S}(\mu(t))\right)y^T = \sum_{k=1}^{n}\lambda_k x e_k^T e_k y^T$$

$$= \sum_{k=1}^{n}\lambda_k \sum_{i,j=1}^{n} x_i y_j e_{ki} e_{kj}.$$

It follows that

$$\sum_{i,j=1}^{n} D_{ij}\mathbf{S}(\mu(t))\mu_i(t)\mu_j(t)\tau_{ij}(t) = \sum_{k=1}^{n}\lambda_k \sum_{i,j=1}^{n}\mu_i(t)\mu_j(t)e_{ki}e_{kj}\tau_{ij}(t). \qquad (3.12)$$

If one of the λ_i is positive, we can assume without loss of generality that it is λ_1 with $e_1 = \pm(n^{-1/2}, \ldots, n^{-1/2})$. If $(\tau_{ij}(t))$ is positive semi-definite with null space generated by $\mu(t)$, as in Lemma 2.1, then

$$\sum_{i,j=1}^{n}\mu_i(t)\mu_j(t)e_{ki}e_{kj}\tau_{ij}(t) \geq 0 \qquad (3.13)$$

for $k = 1, \ldots, n$, and

$$\sum_{i,j=1}^{n}\mu_i(t)\mu_j(t)e_{1i}e_{1j}\tau_{ij}(t) = 0$$

if $e_1 = \pm(n^{-1/2}, \ldots, n^{-1/2})$. Hence, by Lemma 2.1 and (3.12),

$$\sum_{i,j=1}^{n} D_{ij}\mathbf{S}(\mu(t))\mu_i(t)\mu_j(t)\tau_{ij}(t) \leq 0, \quad t \in [0,\infty) \quad \text{a.s.}$$

By (3.3) of Theorem 3.1, $\Theta(t) \geq 0$, for all $t \in [0,\infty)$, a.s.

If for all $x \in \Delta^n$, $\operatorname{rank}(D_{ij}\mathbf{S}(x)) > 1$, then at least one of the eigenvalues $\lambda_2, \ldots, \lambda_n$ is negative. Since the eigenvectors e_k are pairwise orthogonal, only e_1 can be perpendicular to Δ^n, so (3.13) will be positive for at least one $k \geq 2$. Therefore, Lemma 2.1 and (3.12) imply that

$$\sum_{i,j=1}^{n} D_{ij}\mathbf{S}(\mu(t))\mu_i(t)\mu_j(t)\tau_{ij}(t) < 0, \quad t \in [0,\infty)\text{a.s.}$$

Hence, $\Theta(t) > 0$, for all $t \in [0,\infty)$, a.s.   □

4. Portfolio Dominance and Measures of Diversity

In this section we shall use the stochastic differential equation (3.1) to establish dominance relationships between certain functionally generated portfolios and the market portfolio. We shall show that if there are appropriate bounds on a generating function and the corresponding drift process, then such a relationship will hold. In particular, we shall show that functions which are *measures of diversity* will generate portfolios that dominate the market portfolio if appropriate bounds exist on the concentration of market capital. Our purpose here is not to achieve maximum generality, but rather to explore the use of generating functions for constructing dominant portfolios. Accordingly, we shall entertain a bias toward simplicity over generality.

Definition 4.1. Let η and ξ be portfolios. Then η *strictly dominates* ξ if there is a number $t > 0$ such that

$$P\{Z_\eta(t)/Z_\eta(0) > Z_\xi(t)/Z_\xi(0)\} = 1. \tag{4.1}$$

This definition is stronger than the usual definition of "dominates" in which (4.1) is replaced by $P\{Z_\eta(t)/Z_\eta(0) \geq Z_\xi(t)/Z_\xi(0)\} = 1$ and $P\{Z_\eta(t)/Z_\eta(0) > Z_\xi(t)/Z_\xi(0)\} > 0$.

Lemma 4.1. *Let* $\mathbf{S}$ *be the generating function of the portfolio* π *with drift process* Θ, *and suppose that* c_1 *and* c_2 *are constants. If for all* $t > 0$, $\mathbf{S}(\mu(t)) > c_1 > 0$, *a.s.,* *and* $\Theta(t) > c_2 > 0$, *a.s., then* π *strictly dominates* μ. *If for all* $t > 0$, $\mathbf{S}(\mu(t)) < c_1$, *a.s., and* $\Theta(t) < c_2 < 0$, *a.s., then* μ *strictly dominates* π.

Proof. Suppose the set of conditions with $c_1 > 0$ and $c_2 > 0$ hold. It follows from (3.1) that a.s., for $T > 0$,

$$\log(Z_\pi(T)/Z_\pi(0)) - \log(Z(T)/Z(0)) > \log c_1 - \log \mathbf{S}(\mu(0)) + c_2 T.$$

Therefore, if

$$T > (\log \mathbf{S}(\mu(0)) - \log c_1)/c_2,$$

then,

$$\log(Z_\pi(T)/Z_\pi(0)) > \log(Z(T)/Z(0))$$

almost surely. The proof for the second set of conditions is similar. $\qquad\square$

Dominance relationships between pairs of functionally generated portfolios can similarly be established by applying Proposition 3.1 if appropriate bounds exist.

Example 4.1. Consider the function

$$\mathbf{S}(x) = x_1^2.$$

Theorem 3.1 implies that $\mathbf{S}$ generates a portfolio π with weights

$$\pi_1(t) = 2 - \mu_1(t),$$

and

$$\pi_i(t) = -\mu_i(t),$$

for $i = 2, \ldots, n$. The drift process is

$$\Theta(t) = -\tau_{11}(t).$$

In this case $\mathbf{S}(\mu(t)) < 1$, so we can apply Lemma 4.1 if we can establish a negative upper bound on Θ. Lemma 2.3 shows that for all $t > 0$, $\tau_{11}(t) > \varepsilon \left(1 - \mu_1(t)\right)^2$, a.s., where ε is that of (2.5). Hence, if $\mu_1(t)$ can be bounded away from 1, the market portfolio will strictly dominate π. This brings us to the concept of *market diversity*.

Definition 4.2. The market $\mathcal{M}$ is *diverse* if there exists a number $\delta > 0$ such that for $i = 1, \ldots, n$,

$$\mu_i(t) \leq 1 - \delta, \quad t \in [0, \infty), a.s. \tag{4.2}$$

This definition was given in Fernholz (1998b). The condition it imposes, (4.2), is observable and broadly consistent with the actual behavior of equity markets in industrial economies, especially if there are antitrust laws. It was shown in Fernholz (1998b) that a portfolio generated by the entropy function (see Example 4.2 below) will strictly dominate the market portfolio in a diverse market. The entropy function is the archetypal measure of diversity; here we wish to give a general definition of these measures. Recall that a real-valued function F defined on a subset of $\mathbb{R}^n$ is *symmetric* if it is invariant under permutations of the variables $x_i, i = 1, \ldots, n$, and *concave* if for $0 < p < 1$ and $x, y \in \mathbb{R}^n$, $F(px + (1 - p)y) > pF(x) + (1 - p)F(y)$.

Definition 4.3. A positive C^2 function defined on an open neighborhood of Δ^n is a *measure of diversity* if it is symmetric and concave.

In this definition, symmetry ensures that all stocks are treated in the same manner, and concavity implies that transferring capital from a larger company to a smaller one will increase the measure. The results of the previous section imply that measures of diversity can be used to generate portfolios.

Proposition 4.1. *Suppose that* $\mathbf{S}$ *is a measure of diversity which generates a portfolio* π *with drift process* Θ. *Then* $\Theta(t) \geq 0$ *for all* $t \in [0, \infty)$, *a.s., and* $\mu_i(t) \geq \mu_j(t)$ *implies that* $\pi_j(t)/\mu_j(t) \geq \pi_i(t)/\mu_i(t)$ *for all* $t \in [0, \infty)$, *a.s.*

Proof. If $\mathbf{S}$ is a measure of diversity, then by definition it is concave and C^2. It is well known that for such a function, the matrix $(D_{ij}\mathbf{S}(x))$ is negative semi-definite. A negative semi-definite matrix has no positive eigenvalues, so Proposition 3.4 implies that $\Theta(t) \geq 0$ for all $t \in [0, \infty)$, a.s.

Now suppose that $x = (x_1, \ldots, x_n) \in \Delta^n$ with $x_i \leq x_j$ for some $i < j$. Define

$$x(u) = (x_1, \ldots, (1 - u)x_i + ux_j, \ldots, ux_i + (1 - u)x_j, \ldots, x_n),$$

so $x(0) = x$ and $x(1)$ is x with the ith and jth coordinates reversed. Define $f(u) = S(x(u))$, so f is C^2 and concave, and since S is symmetric, $f(0) = f(1)$. Now,

$$f'(u) = (x_j - x_i)(D_i S(x(u)) - D_j S(x(u))) ,$$

and the concavity of f implies that $f'(0) \geq 0$. Since $x_i \leq x_j$, it follows that $D_i S(x) \geq D_j S(x)$. Then (3.2) of Theorem 3.1 implies that for $\mu_i(t) \geq \mu_j(t)$, $\pi_j(t)/\mu_j(t) \geq \pi_i(t)/\mu_i(t)$. $\qquad\square$

This proposition shows that the weight ratios $\pi_i(t)/\mu_i(t)$ decrease with increasing market weight. Hence, if a stock's market weight increases, i.e., the stock goes up relative to the market, then the portfolio π will sell some (fractional) shares of that stock.

Although Proposition 4.1 shows that a measure of diversity will have a non-negative drift process, this is not sufficient to ensure that the portfolio it generates will dominate the market portfolio. In order to apply Lemma 4.1, the drift process must have a positive lower bound. Let us now consider some examples of measures of diversity, and determine in which cases such a bound exists.

Example 4.2. The *entropy* function,

$$S(x) = -\sum_{i=1}^{n} x_i \log x_i ,$$

was studied in Fernholz (1998b). The weights of the portfolio it generates are

$$\pi_i(t) = -\frac{\mu_i(t) \log \mu_i(t)}{S(\mu(t))} ,$$

and the drift process is

$$\Theta(t) = \frac{\gamma^*(t)}{S(\mu(t))} .$$

Lemma 2.4 implies that in a diverse market $\gamma^*(t)$, and hence, $\Theta(t)$, has a positive lower bound. It is not difficult to show that the same is true for $S(\mu(t))$, so Lemma 4.1 can be applied.

Example 4.3. The geometric mean in Example 3.1 is a measure of diversity which generates a portfolio with all weights equal to n^{-1} and drift process $\Theta(t) = \gamma_\pi^*(t)$. The value of the geometric mean of the $\mu_i(t)$ will be bounded away from zero only if each individual $\mu_i(t)$ is so bounded. This condition is quite restrictive, and does not necessarily hold even in a diverse market.

Example 4.4. For $0 < p < 1$, let

$$D_p(x) = \left(\sum_{i=1}^{n} x_i^p\right)^{1/p} .$$

This is a measure of diversity, and has in fact been used to construct an institutional equity investment product (see Fernholz, Garvy, and Hannon (1998)). The portfolio generated by $\mathbf{D}_p$ has weights

$$\pi_i(t) = \frac{\mu_i^p(t)}{(\mathbf{D}_p(\mu(t)))^p} \,.$$

The drift process, which is positive, is

$$\Theta(t) = (1 - p)\gamma_\pi^*(t).$$

Here $\mathbf{D}_p(\mu(t)) > 1$. Moreover, Proposition 4.1 implies that $\pi_{\max}(t) \le \mu_{\max}(t)$, so if the market is diverse, Lemma 2.4 implies that $\gamma_\pi^*(t)$ will have a positive lower bound. Therefore, Lemma 4.1 can be applied.

Example 4.5. Let

$$\mathbf{S}(x) = 1 - \frac{1}{2} \sum_{i=1}^n x_i^2 \,.$$

This measure of diversity was used Fernholz (1997). Here,

$$\pi_i(t) = \left(\frac{2 - \mu_i(t)}{\mathbf{S}(\mu(t))} - 1 \right) \mu_i(t) \,,$$

for $i = 1, \ldots, n$, with drift process

$$\Theta(t) = \frac{1}{2\,\mathbf{S}(\mu(t))} \sum_{i=1}^n \mu_i^2(t) \tau_{ii}(t) \,.$$

$\mathbf{S}(\mu(t)) > 1/2$ and in a diverse market, $\Theta(t)$ has a positive lower bound by Lemma 2.3.

Example 4.6. The *Gini coefficient* is frequently used to measure diversity. It is usually defined as

$$G(x) = \frac{1}{2} \sum_{i=1}^n |x_i - n^{-1}| \,.$$

If we modify it to

$$\mathbf{S}(x) = 1 - \frac{1}{2} \sum_{i=1}^n |x_i - n^{-1}| \,,$$

this comes closer to Definition 4.3, but fails to be C^2. Nevertheless a version of Theorem 3.1 is probably valid with portfolio weights

$$\pi_i(t) = \left(\frac{\operatorname{sign}(n^{-1} - \mu_i(t))}{2\,\mathbf{S}(\mu(t))} + 1 - \sum_{j=1}^n \frac{\mu_j(t)\operatorname{sign}(n^{-1} - \mu_j(t))}{2\,\mathbf{S}(\mu(t))} \right) \mu_i(t) \,,$$

and a non-negative drift process Θ which depends on a local time measuring the time $\mu_i(t)$ spends near n^{-1}, for $i = 1, \ldots, n$. See Karatzas and Shreve (1991), Chapter 6, for details about Brownian local time.

Rather than attempt to analyze the Gini coefficient here, we shall settle for a quadratic version of it. Let

$$\mathbf{S}(x) = 1 - \frac{1}{2} \sum_{i=1}^{n} (x_i - n^{-1})^2 .$$

This is a measure of diversity with portfolio weights

$$\pi_i(t) = \left(\frac{n^{-1} - \mu_i(t)}{\mathbf{S}(\mu(t))} + 1 - \sum_{j=1}^{n} \frac{\mu_j(t)\left(n^{-1} - \mu_j(t)\right)}{\mathbf{S}(\mu(t))} \right) \mu_i(t) ,$$

and positive drift process

$$\Theta(t) = \frac{1}{2\,\mathbf{S}(\mu(t))} \sum_{i=1}^{n} \mu_i^2(t)\tau_{ii}(t) .$$

Here $\mathbf{S}(\mu(t)) > 1/2$, and as in the previous example, in a diverse market $\Theta(t)$ will have a positive lower bound.

Example 4.7. The function

$$\mathbf{S}(x) = 1 - \frac{1}{4} \sum_{i=1}^{n} (x_i - n^{-1})^4$$

is quite similar to that of Example 4.6, and is indeed a measure of diversity. Here $3/4 < \mathbf{S}(\mu(t)) < 1$, and the drift process is

$$\Theta(t) = \frac{3}{2\,\mathbf{S}(\mu(t))} \sum_{i=1}^{n} \left(\mu_i(t) - n^{-1}\right)^2 \mu_i^2(t)\tau_{ii}(t) .$$

Now suppose that the relative covariance processes τ_{ii} are uniformly bounded, say by $M > 0$. Then

$$\Theta(t) < 2M \max_{1 \le i \le n} \left(\mu_i(t) - n^{-1}\right)^2 .$$

Since there is a positive probability that all the $\mu_i(t)$ will remain arbitrarily close to n^{-1} for an arbitrarily long time, the contribution from $\Theta(t)\,dt$ in (3.1) can be minimal. If then one of the $\mu_i(t)$ quickly increases to almost $1 - \delta$, $\mathbf{S}(\mu(t))$ will decrease and π will have lower return than the market over the period. Hence, even in a diverse market, π does not dominate μ.

Example 4.8. The *Rényi entropy* is a generalization of the entropy function we considered above (see Rényi (1960)). For $p \neq 1$ it is defined by

$$\mathbf{S}_p(x) = \frac{1}{1-p} \log \sum_{i=1}^{n} x_i^p \,.$$

As $p \to 1$, $\mathbf{S}_p$ tends to the usual entropy function. It can be shown that for $p < 1$, $\mathbf{S}_p$ is a measure of diversity, but for $p > 1$, $\mathbf{S}_p$ is not concave. For $p > 1$, the weights of the portfolio $\mathbf{S}_p$ generates may be negative and the corresponding drift process may take on negative values.

Example 4.9. Suppose that $x_{(i)}$, $i = 1, \ldots, n$, are the order statistics, $x_{(i)} \leq x_{(i+1)}$. The function
$$\mathbf{S}(x) = x_{(1)} + \cdots + x_{(k)} \,,$$

for $k < n$, is symmetric and concave, but not C^2. Nevertheless, it probably can be proved that $\mathbf{S}$ generates a portfolio π such that for $i = 1, \ldots, n$,

$$\pi_i(t) = \begin{cases} \mu_i(t)/\mathbf{S}(\mu(t)) & \text{if } \mu_i(t^-) \leq \mu_{(k)}(t^-), \\ 0 & \text{otherwise,} \end{cases}$$

where $\mu_i(t^-) \leq \mu_{(k)}(t^-)$ if there is an $\varepsilon > 0$ such that $\mu_i(s) \leq \mu_{(k)}(s)$ for $0 < t - s < \varepsilon$. Then,

$$d \log(Z_\pi(t)/Z(t)) = d \log \mathbf{S}(\mu(t)) + d\Lambda_k(t) \,,$$

where Λ_k is a local time measuring the time $\mu_{(k)}(t)$ spends near $\mu_{(k+1)}(t)$. This construction is related to the *size effect* discussed in Fernholz (1998a).

5. Conclusions

We have shown that positive C^2 functions of the market weights generate equity portfolios, and that the return on these portfolios is related to the market return by a stochastic differential equation. Under appropriate conditions, this equation can be used to establish a dominance relationship between a functionally generated portfolio and the market portfolio. In a diverse market, certain measures of market diversity generate portfolios that dominate the market portfolio.

References

D. Duffie, *Dynamic Asset Pricing Theory*, Princeton, NJ: Princeton University Press, 1992.

R. Fernholz, Arbitrage in equity markets. Technical report, INTECH, Princeton, NJ, 1997.

R. Fernholz, Crossovers, dividends and the size effect, *Financial Analysts Journal* **54**(3) (1998a, May/June) 73–78.

R. Fernholz, On the diversity of equity markets, *Journal of Mathematical Economics*, to appear (1998b).

R. Fernholz, R. Garvy and J. Hannon, Diversity weighted indexing, *Journal of Portfolio Management* **24**(2) (1998, Winter) 74–82.

I. Karatzas and S. G. Kou, On the pricing of contingent claims under constraints, *The Annals of Applied Probability* **6** (1996) 321–369.

I. Karatzas and S. Shreve, *Brownian Motion and Stochastic Calculus*, New York: Springer-Verlag, 1991.

A. Rényi, On measures of entropy and information, in *Proceedings of the 4th Berkeley Symposium on Mathematics, Statistics, and Probability* (1960), pp. 547–561.

M. Spivac, *Calculus on Manifolds*, New York: Benjamin, 1965.